SOIL
PHYSICS
with
HYDRUS

Modeling and Applications

SOIL PHYSICS
with
HYDRUS
Modeling and Applications

David E. Radcliffe and Jiří Šimůnek

CRC Press
Taylor & Francis Group
Boca Raton   London   New York

CRC Press is an imprint of the
Taylor & Francis Group, an **informa** business

CRC Press
Taylor & Francis Group
6000 Broken Sound Parkway NW, Suite 300
Boca Raton, FL 33487-2742

© 2010 by Taylor and Francis Group, LLC
CRC Press is an imprint of Taylor & Francis Group, an Informa business

No claim to original U.S. Government works

Printed in the United States of America on acid-free paper
10 9 8 7 6 5 4 3 2 1

International Standard Book Number: 978-1-4200-7380-5 (Hardback)

Visit the Taylor & Francis Web site at
http://www.taylorandfrancis.com

and the CRC Press Web site at
http://www.crcpress.com

Contents

Preface

In 1985, Gaylon Campbell published a slim volume entitled *Soil Physics with Basic* (Campbell, 1985). This textbook was one of the first and best publications to show the potential for numerical computer models to solve applied problems in the field of soil physics. Since that time, numerical models and computers have come a long way. Numerical models have become much more efficient and can now be applied to two- and three-dimensional problems of saturated and unsaturated water flow, heat flow, and solute transport. User-friendly interfaces have been developed that make the setup of the model much easier and more intuitive. The interfaces are especially helpful in displaying the model output of two- and three-dimensional simulations in a way that is immediately intelligible to students, including the use of computer animation. Computer speed has increased to the point that very difficult problems can be solved within a matter of hours, including powerful inverse methods that could only be applied to analytical solutions in the past.

One of the most advanced and popular numerical computer models for the field of soil physics is the HYDRUS series: HYDRUS-1D and HYDRUS (2D/3D). In our conversations with soil physicists teaching undergraduate and graduate courses in soil physics and vadose zone hydrology across the United States, Europe, Australia, and Asia we have found that many are using HYDRUS models in some portion of their course. We developed this textbook to use the HYDRUS models to teach the principles and applications of soil physics. The HYDRUS models include the Rosseta-based pedo-transfer functions, which we use to illustrate soil water characteristic curve relationships and unsaturated hydraulic conductivity functions. HYDRUS-1D is used for hands-on problems that demonstrate infiltration, evaporation, and percolation of water through soils of different textures and layered soils. It is also used to show heat flow and solute transport in these systems, including the effect of physical and chemical nonequilibrium conditions. Examples of two-dimensional flow in fields, hillslopes, boreholes, and capillary fringes are included using HYDRUS (2D/3D). Use of two other software packages that complement the HYDRUS series is shown. One of these is the Code for Quantifying the Hydraulic Functions of Unsaturated Soils (RETC; van Genuchten et al., 1991), used to fit soil water retention curves. The other is the Studio of Analytical Models for Solving the Convection-Dispersion Equation (STANMOD; Toride et al., 1999), which uses analytical solutions to solve solute transport problems. The HYDRUS models, RETC, and STANMOD (as well as many examples, tutorials, references, and applications for further illustration on how to use the codes) are available on the HYDRUS website (http://www.hydrus3D.com). All of the software is free except for HYDRUS (2D/3D). Software simulations described in the text (identified as HYDRUS simulation 5.1, Parameter Optimization, etc.) are available on the accompanying website (http://www.pc-progress.com/en/Default.aspx?h3d-book). These are:

- RETC Simulation 2.1 – Cecil Ap Horizon
- HYDRUS Simulation 3.1 – Rosetta-Lite Module
- HYDRUS Simulation 4.1 – Heat Flow Without Convection
- HYDRUS Simulation 5.1 – Ponded Infiltration
- HYDRUS Simulation 5.2 – Flux Infiltration
- HYDRUS Simulation 5.3 – Infiltration into Layered System
- HYDRUS Simulation 5.4 – 2D Infiltration into Layered System
- HYDRUS Simulation 5.5 – Borehole Infiltration
- HYDRUS Simulation 5.6 – Subsurface Drip Irrigation
- HYDRUS Simulation 5.7 – Redistribution
- HYDRUS Simulation 5.8 – Evaporation
- HYDRUS Simulation 5.9 – Transpiration
- HYDRUS Simulation 5.9 – Data
- HYDRUS Simulation 5.10 – Scaling
- HYDRUS Simulation 5.11 – Capillary Barrier
- HYDRUS Simulation 5.12 – Hillslope A
- HYDRUS Simulation 5.13 – Hillslope B
- HYDRUS Simulation 5.14 – Parameter Estimation
- HYDRUS Simulation 5.14 – Data
- STANMOD Simulation 6.1 – Effect of Beta
- STANMOD Simulation 6.2 – Parameter Optimization
- HYDRUS Simulation 6.1 – Nonlinear Adsorption and Transport
- HYDRUS Simulation 6.2 – Transport of Nitrogen Species
- HYDRUS Simulation 6.3 – Parameter Optimization of Nitrogen Species Model
- HYDRUS Simulation 6.4 – Capillary Fringe

There are many excellent packages such as Mathcad (Parametric Technology Corporation) and MATLAB® (The Mathworks, Inc.) that can be used in conjunction with a soil physics textbook. However, as discussed by Wraith and Or (1998), the learning curve for using these packages is steep and the packages can be expensive. On the other hand, most students are familiar with Microsoft Excel and have it on their computers. Our experience has been that students are very receptive to learning "new tricks" with Excel. With a little work, Excel can be used to accomplish many of the tasks that the math software packages would be used for in a soil physics course (including matrix computations) and we use it throughout this textbook.

Our textbook should be suitable for teaching an undergraduate or a lower level graduate course in soil physics or vadose zone hydrology. It assumes that students are familiar with differential and integral calculus. However, a course could be taught to students with only differential calculus since derivations involving integral calculus appear at the end of each chapter and could be optional for students. The textbook should also be suitable for an individual who wants to teach herself/himself about the field of soil physics and how to use the HYDRUS models.

Comparing our textbook to the other excellent soil physics textbooks available, our book is not as quantitative as the books by Warrick (2003) and Tindall and

Kunkel (1999). It is comparable to the books by Hillel (2004) and Jury and Horton (2004). Little detail on measurement methods is included as these methods change rapidly and there are excellent references (e.g., Dane and Topp, 2002). Gas flow is not treated in this book (except for the volatilization of solutes). This is because the HYDRUS models do not simulate gas flow, other than carbon dioxide transport in the specialized UNSATCHEM module. In our opinion, the primary contribution of this textbook is in the extensive use of the HYDRUS and related models to teach soil physics.

MATLAB® is a registered trademark of The Math Works, Inc. For product information, please contact:

The Math Works, Inc.
3 Apple Hill Drive
Natick, MA
Tel: 508-647-7000
Fax: 508-647-7001
E-mail: info@mathworks.com
Web: http://www.mathworks.com

1 Soil Solid Phase

1.1 INTRODUCTION

The near-surface region of Earth, extending from plant canopies through the soil matrix to underlying bedrock, has been referred to as the "critical zone" (NRC 2001; Brantley et al. 2007). It is a focal point of human activities in terms of both its support of life processes and its response to impacts of land and water management. The quality of human life and the global environment are highly dependent on processes in soils; rapidly expanding systems of food, fiber, water, and energy production and waste disposal are now based on the management of Earth's near-surface. The emerging paradigm is that of the critical zone as the controlling medium for surface and subsurface water dynamics; transformation, storage, and release of bioactive elements; maintenance of biological productivity and diversity; and regulation of gas fluxes between the biosphere and atmosphere (Brantley et al. 2007).

The chapter begins with a description of the three soil phases, followed by a discussion of the primary particles that make up the solid phase of soil. Then, the effect of mineralogy on the properties of primary particles is discussed, especially clay particles. Primary particles combine together to make secondary particles, creating soil structure, which is described in the last section. Much of this chapter should be a review for students who have had an introductory soil science course or read an introductory soil science textbook, such as Brady and Weil (2008).

A word on units: the units for variables in equations can usually be reduced to three *fundamental units* of length [L], mass [M], and time [T]. In the metric system, these are meters, kilograms, and seconds (mks), or centimeters, grams, and seconds (cgs). For each variable in an equation, the fundamental units in square brackets are given. To the three fundamental units of length, mass, and time, it is sometimes helpful to add temperature [K] and charge [Q]. Unit conversion tables are provided in the appendix for metric and English units. Also, throughout the book we will use a dot (.) to represent a decimal point and a comma (,) to indicate multiples of 10^3, e.g., 2,441.3 (contrary to the convention in Europe where a comma is used to represent a decimal point).

1.2 SOIL PHASES

Bulk soil consists of three dispersed phases: solid, gas, and liquid. The total volume of soil can be quantified in terms of the volume of each phase:

$$v = v_g + v_l + v_s, \tag{1.1}$$

where v is the total volume [L^3], v_g is the volume of gas or air [L^3], v_l is the volume of liquid or water [L^3], and v_s is the volume of solids [L^3]. The volume of pores is $v_g + v_l$,

1

which is usually about 50% of the total volume. Dividing both sides of Equation 1.1 by v, the equation is

$$1 = a + \theta + \frac{v_s}{v},$$
(1.2)

where a is the *volumetric air content* [L^3L^{-3}]:

$$a = \frac{v_g}{v},$$
(1.3)

and θ is the *volumetric water content* [L^3L^{-3}]:

$$\theta = \frac{v_l}{v}.$$
(1.4)

Porosity (φ) [L^3L^{-3}] is the sum of the volumetric air and water contents:

$$\varphi = a + \theta.$$
(1.5)

Bulk soil can also be quantified in terms of the mass of each phase:

$$m = m_g + m_l + m_s,$$
(1.6)

where m is the total mass [M], and m_g, m_l, and m_s are the respective masses of gas, liquid, and solid [M]. The mass of gas is negligible compared to the other two phases. The solid mass divided by the volume of solid (v_s) is the *particle density* (ρ_s) [ML^{-3}]:

$$\rho_s = \frac{m_s}{v_s}.$$
(1.7)

The particle density of most soil particles is relatively constant, approximately 2.65 g cm^{-3}, the density of quartz. Soil samples with high iron content have slightly higher particle densities; samples with high organic matter content have lower particle densities. The procedure for measuring particle density is described by Flint and Flint (2002), but the measurement is not often made due to the consistency in the value. The solid mass divided by the total volume of soil is the dry *bulk density* (ρ_b) [ML^{-3}]:

$$\rho_b = \frac{m_s}{v}.$$
(1.8)

Unlike ρ_s, bulk density varies widely, depending on the porosity, texture, and structure of a soil. In swelling soils, where the bulk volume changes substantially with water content, a wet bulk density is often used, in which case the mass of liquid is included in the numerator of Equation 1.8. Bulk densities typically range from 1.2 to 1.8 g cm^{-3}. It may be counterintuitive that clay soils have the lowest bulk densities, despite their heavy, sticky nature. The highest bulk densities occur not with pure sands, but with soils that are about 80%–90% sand. At this point, the sand particles are in contact with each other and can't be packed more densely. A small amount of finer material, 10%–20% clay or silt, can fill in the voids between the sand particles and produce a maximum density. Very low bulk densities, even less than 1 g cm^{-3}, can occur near the surface of forest soils or volcanic soils. Methods for measuring bulk density are described by Grossman and Reinsch (2002) and include the core and clod methods.

The relationship between porosity, particle density, and dry bulk density is

$$\varphi = 1 - \frac{\rho_b}{\rho_s}. \tag{1.9}$$

To the extent that ρ_s is constant, Equation 1.9 shows that porosity is determined by bulk density.

Example 1.1

What fraction of a soil with a bulk density of 1.50 g cm^{-3} consists of pore space? Since ρ_s is not given, assume a value of 2.65 g cm^{-3} and use Equation 1.9:

$$\varphi = 1 - \frac{1.50 \frac{g}{cm^3}}{2.65 \frac{g}{cm^3}} = 0.43.$$

1.3 SOIL TEXTURE

Soil particles can be separated into different size classes. According to the U.S. Department of Agriculture (USDA) classification system, particles with a diameter smaller than 0.002 mm (2 μm) are *clay* size, 0.002–0.05 mm are *silt* size, and 0.05–2 mm are *sand* size (Figure 1.1). The sand-size fraction can be subdivided into fractions labeled coarse, medium, fine, etc. Particles larger than 2 mm are gravel size and usually not considered part of the soil. When larger particles are common, such as in streambeds, other classes are included, such as cobbles and boulders (Knighton 1984). Classification systems differ on where they make the divisions between silt and sand, but most systems use 0.002 mm as the upper limit for clay (Figure 1.1). This is because it is the approximate upper limit for the size of a particle that will stay in suspension indefinitely, diffract a beam of light, and exhibit Brownian motion. This is one definition of a *colloid*, and the clay fraction is often called the *colloidal fraction*.

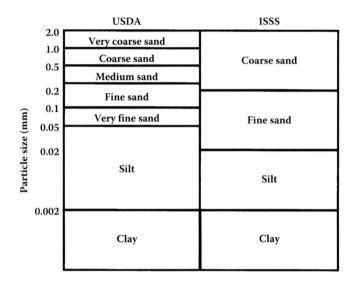

FIGURE 1.1   The U.S. Department of Agriculture (USDA) and International Soil Science Society (ISSS) classification systems for soil particle-size fractions. (From Jury, W.A., and Horton, R., *Soil Physics*, John Wiley, Hoboken, NJ, Inc., 2004. With permission.)

The sand, silt, and clay fractions of a soil are used with a textural triangle to determine the *textural class* or *texture* of a soil (Figure 1.2). Soils that consist predominately of one size fraction are classified as sands, silts, or clays (the apexes of the triangle). Soils with roughly equal percentages of different-sized particles are classified as *loams* (the center of the triangle).

Texture is one of the most important factors affecting the physical characteristics of a soil. For example, golf greens are made almost entirely of sand because the sand drains water well and provides a uniform bounce for golf balls. Silty soils form where there are large loess deposits (e.g., in Mississippi and China) and these soils erode very easily. Clay soils are to be avoided if driving a tank or heavy agricultural machinery. Clay- and silt-sized particles are easily transported in streams, but sand- and gravel-sized particles settle to the streambed and are only transported in the largest storms.

Instead of three distinct sizes of particles, soils usually contain particles that vary across a continuous range of sizes. This is known as the *particle-size distribution*, and it can be shown as a cumulative distribution function (Figure 1.3). Sands usually have a narrow particle-size distribution; loams and clays usually have a wide particle-size distribution. The median particle size has a diameter such that 50% of the particles have a larger diameter and 50% have a smaller diameter. This is referred to as the d_{50} in disciplines studying stream and lake sediments (Knighton 1984).

Particle-size distribution (whether it is used to determine the textural class, Figure 1.2, or the cumulative distribution function, Figure 1.3, of a soil) is determined through a combination of *sieving* and *sedimentation*. The first step is to sieve out the gravel. Then, the organic matter is removed by adding hydrogen peroxide (H_2O_2)

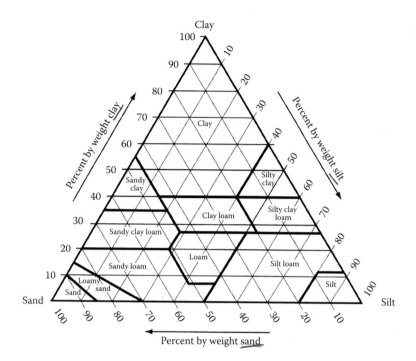

FIGURE 1.2 Textural triangle showing the percentages of clay (<0.002 mm), silt (0.002–0.05 mm), and sand (0.05–2.0 mm) in the conventional soil textural classes. (From Hillel, D., *Introduction to Environmental Soil Physics*, Elsevier Academic Press, Amsterdam, 2004. With permission.)

and letting it boil off as CO_2. Next, any secondary particles must be broken down into primary particles by immersing the soil sample in a solution that contains a dispersing agent. Then, the sand fraction is collected by passing the solution through a 0.05 mm (300 mesh) sieve (or a series of sieves of different sizes to subdivide the sand fraction into coarse, medium, fine, etc.). The solution that passes through the sieve (containing silt- and clay-sized particles) is vigorously stirred, placed in a large sedimentation cylinder, and allowed to sit. Since the particle size determines how quickly a particle settles, the largest silt-sized particles immediately begin to settle out.

Two methods are used to determine the percentage of clay in the solution after the silt has settled below the depth of sampling. In the *pipette method*, a sample of known volume is collected from the solution using a pipette, the sample is placed in an oven, and water is evaporated so that the mass of clay can be measured (Gee and Or 2002). In the *hydrometer method*, a calibrated bulb that sinks to deeper depths in less dense solutions is placed in the cylinder to determine the density (Gee and Or 2002). Once the percentage of sand and clay fractions are known, the silt fraction is calculated by difference.

It is important to know the rate at which the silt particles will settle out. In using the pipette, the sample must be taken from a shallow enough depth in the cylinder that all

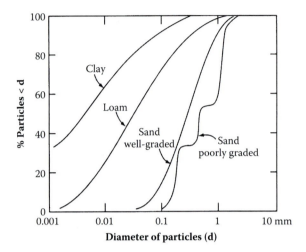

FIGURE 1.3 Particle-size distribution of a clay, loam, well-graded sand, and poorly graded sand. (From Hillel, D., *Introduction to Environmental Soil Physics*, Elsevier Academic Press, Amsterdam, 2004. With permission.)

the silt-sized particles will have settled below this depth. Similarly, the hydrometer must not sink to a depth that contains silt particles. It can take more than 8 h for all the silt particles to settle to the bottom of a standard 1000 mL sedimentation cylinder, depending on the temperature of the solution. If the rate of settling is known, then a sample can be taken from a depth of 10 cm below the surface, for example, after the appropriate amount of time. Similarly, if the hydrometer usually sinks to a depth of 20 cm, then the earliest time to take a reading can be determined.

Stokes's law was developed in 1851 and is used to determine the rate of settling and the time to sample. A particle falling through a fluid will accelerate until all of the forces on the particle are equal, at which point it will reach a constant or terminal velocity (acceleration is zero):

$$V = \frac{(\rho_s - \rho_w)d^2 g}{18\eta},$$ (1.10)

where V is the terminal settling velocity [LT^{-1}], ρ_s is the density of the particle [ML^{-3}], ρ_w is the density of water [ML^{-3}], d is the diameter of the particle [L], g is the acceleration due to gravity [LT^{-2}], and η is the viscosity of water [$ML^{-1}T^{-1}$]. At this velocity, a particle will settle through a height h [L] in a given time t [T]. Substituting h/t for V and solving for t, the equation is

$$t = \frac{18\eta h}{(\rho_s - \rho_w)d^2 g}.$$ (1.11)

Note that Stokes's law assumes a spherical shape for the particle. Therefore, in discussing particle size, one refers to an *effective diameter* or the diameter of a spherical particle that would fall at the same rate as the actual particle, which in the case of a clay-sized particle is often not spherical. As it is written, Stokes's law applies to particles smaller than about 0.06 mm. The law can be modified for larger particles by accounting for turbulent flow (Guy 1969).

Stokes's law has many applications beyond determining particle-size distribution. It is used to determine the rate at which suspended sediment particles of different sizes settle on the Continental Shelf in river plumes (Perianez 2005). It is also used to estimate the time it takes for particulate organic matter to settle in oceans, which is an important part of the global cycling of carbon and affects climate change (Gardner 1997).

Example 1.2

Calculate how long it will take for all the silt particles to fall below a depth of 10.0 cm if the temperature of the solution is 25°C.

The smallest silt particle will fall at the slowest velocity, so the calculation should be made with the lower limit for silt, 0.002 mm (Figure 1.1). At 25°C, the viscosity of water is 0.0089 g cm^{-1} s^{-1} and the density of water is 0.997 g cm^{-3} (see Chapter 2, Table 2.1). Acceleration due to gravity is 980.6 cm s^{-2}. Using Equation 1.11:

$$t = \frac{18\left(0.0089\,\frac{g}{cms}\right)(10.0\,cm)}{\left[\left(2.65\,\frac{g}{cm^3}\right)-\left(0.997\,\frac{g}{cm^3}\right)\right](0.0002\,cm)^2\left(980.6\,\frac{cm}{s^2}\right)} = 2.47\cdot10^4\,s = 6.86\,h.$$

A word on significant digits: how many significant digits should one show in an answer? The answer can be no more accurate than the input data. Therefore, the number of significant digits in the answer must be the same as the input data with the least number of significant digits. For example:

$$y = \sin(kx)$$

where $k=0.097$ m^{-1} (two significant digits) and $x=4.73$ m (three significant digits). The answer should have two significant digits, $y=\sin(kx)=0.0080$.

For intermediate calculations, keep one more significant digit than will appear in your final answer and then round off at the end. See http://www.physics. uoguelph.ca/tutorials/sig_fig/SIG_dig.htm.

1.4 SOIL MINERALOGY

The colloid- or clay-size fraction differs from the silt and sand fraction in a very important way. Sand- and silt-sized particles are, for the most part, made of quartz, often approximately spherical in shape. Clay-sized particles consist of complex aluminosilicate minerals, such as kaolinite, vermiculite, and smectite. This size fraction also contains aluminum oxides, iron oxides, and fine organic matter such as humus. Aluminosilicate minerals are platelike rather than spherical in shape. Due to their complex rigid crystalline structure, cations of similar size but different charge may substitute for each other, causing a net negative charge in a process known as *isomorphic substitution*. This charge imbalance is satisfied by soil solution cations that form a loose adsorbed layer on clay surfaces. This is the all-important property of clay minerals, known as the cation exchange capacity (CEC), usually expressed in units of centimoles of positive charge per kilogram of soil, $cmol_c$ kg^{-1}.

Clay minerals such as kaolinite, iron oxides, and aluminum oxides have little isomorphic substitution and, consequently, a low CEC. However, they do develop a net positive charge that is dependent on the soil solution pH and total ion concentration. This charge imbalance is satisfied by adsorbed anions, producing an anion exchange capacity (AEC).

Another important source of charge in the colloid fraction is organic matter. Since clay mineralogy can't be changed and pH can only be manipulated within a narrow range because of plant tolerances, adding organic matter is a very important way to increase CEC. The CECs and AECs of various minerals in the clay-size soil fractions are shown in Table 1.1.

The clay-size fraction differs in another important way from the sand and silt fractions. Because of its fine size and irregular shape, a given mass of clay has a much greater surface area than the same mass of sand or silt. Surface area per unit of mass is expressed as *specific surface*, s [L^2M^{-1}]:

$$s = \frac{a_s}{m_s},\qquad(1.12)$$

where a_s is the solid surface area [L^2]. The ranges of typical values for some clay minerals and organic matter are shown in Table 1.2.

TABLE 1.1
Approximate charge characteristics of soil colloids

Colloid type	CEC ($cmol_c$ kg^{-1})	AEC ($cmol_c$ kg^{-1})
Organic matter	200	0
Smectite	100	0
Vermiculite	150	0
Fine-grained micas	30	0
Chlorite	30	0
Kaolinite	8	2
Aluminum oxide	4	5
Iron oxide	4	5
Allophane	30	15

Source: Adapted from Brady, N.C., and Weil, R.R., *The Nature and Property of Soils,* Pearson Prentice Hall, Upper Saddle River, NJ, 2008.

TABLE 1.2
Specific surface ranges for selected colloids

Colloid	s (m^2 g^{-1})
Kaolinite	15–20
Illite	80–100
Bentonite	115–260
Montmorillonite	280–500
Crystalline iron oxides	116–184
Amorphous iron oxides	305–412
Organic matter	560–800

Source: Adapted from Skopp, J.M., *Handbook of Soil Science*, CRC Press, Boca Raton, FL, 2000.

Specific surface can be measured by determining the volume of nitrogen gas that adsorbs to the surface of a soil sample and calculating from that the volume that would be required to form a layer one molecule thick, using the equations of Brunauer et al. (1938). This and other methods are described by Pennell (2002).

It is important to realize that the values in Table 1.2 are for the clay-size fraction only. Specific surfaces of bulk soils are lower because the sand and silt fractions have negligible specific surface and serve to dilute the clay and organic matter values. Typical values for soils of different soil textural classes are shown in Table 1.3.

Specific surface is a critical property of soils. Important physical and chemical processes occur at surfaces and the more surface area that is exposed, the more reactive the soil. CEC is an example. Since it is the charge per mass of soil, it is the product of specific surface and charge density σ (charge per unit surface area) [QL^{-2}]:

$$CEC = s\sigma. \tag{1.13}$$

Example 1.3

Calculate the specific surface area of a soil in m^2 g^{-1} that consists of 80% sand, 10% silt, and 10% clay. The clay fraction is predominately illite.

TABLE 1.3
Specific surfaces for bulk soils

Soil textural class	s (m^2 g^{-1})
Sands	<10
Sandy loams and silt loams	5–20
Clay loams	15–40
Clays	>25

Source: Adapted from Skopp, J.M., *Handbook of Soil Science*, CRC Press, Boca Raton, FL, 2000.

Assume an average value of s for illite from Table 1.2 of 90 m^2 g^{-1} and zero for the silt and sand fractions. Using Equation 1.13:

$$s = \left(90 \frac{m^2}{\text{gram of clay}}\right)\left(\frac{10 \text{ g of clay}}{100 \text{ g of soil}}\right) = 9.0 \frac{m^2}{\text{gram of soil}}.$$

Other important processes that depend on specific surface are adsorption of contaminants, retention of water, adsorption of gases, and plasticity. Tables 1.1 through 1.3 show how important physical and chemical properties vary within the clay-size fraction. It is apparent that simply using texture or particle-size distribution to differentiate among soils is a rather crude approximation. Specific surface would probably be a better way to differentiate among soils, but it is more difficult to measure (Hillel 2004).

Mineralogy and its interaction with water also affect soil color (Brady and Weil 2008). Soils and horizons with high organic matter content tend to be darker. Soils with high water contents can have reduced oxygen levels that promote the accumulation of organic matter. Soils that contain iron oxides vary in color, depending on water content and oxygen levels. In well-drained soils, oxidized iron compounds impart bright red and brown colors. In poorly drained soils, reduced iron is removed, often revealing gray colors of silica minerals.

1.5 SOIL STRUCTURE

Few soils exist as a collection of primary particles, the subject of the discussion so far. Primary particles join together to form secondary particles or *soil structure*. Soil structure is one of the most important physical properties of soil and is often discussed. However, it is poorly defined and very difficult to quantify. For example, *good tilth* is a term used by farmers to indicate that the soil is easily tilled and penetrated by roots, but no agreed-upon methods exist to measure tilth.

Peds, *aggregates*, and *clods* are "secondary structural units made up of primary particles that are distinguished from adjacent structures on the basis of failure zones" (Kay and Angers 2001) (Figure 1.4). Secondary structural units in the topsoil are usually referred to as aggregates if they form naturally or clods if they are formed artificially (by plowing when a soil is too wet, for example). Below the topsoil, secondary structural unit s are usually referred to as peds. How distinct a ped or aggregate is depends on bonding within the structural unit and bonding in the failure zone.

One classification system for structure is that used by the U.S. Soil Survey (Soil Survey Division Staff, 1993). Some soils have no apparent structure and these are referred to as *structureless*. The soil in this case is either *single-grained* (no secondary units) or *massive* (all one unit). In structured soils, the aggregates or peds are classified in terms of their *shape*, *size*, and *grade*. There are five basic shapes:

- *Platy*: the units are flat, platelike, and generally oriented horizontally (Figure 1.5).
- *Columnar*: the units appear as vertical columns with the tops rounded and distinct.

FIGURE 1.4 A soil aggregate broken open along a failure zone.

- *Prismatic*: the units are similar to columnar peds, but the tops are indistinct and normally flat (Figure 1.6).
- *Blocky*: the units are blocklike or polyhedral and described as angular blocky if the faces intersect at relatively sharp angles or subangular blocky if the intersections are rounded (Figure 1.7).
- *Granular*: the units are approximately spherical (Figure 1.8).

Five classes are used for size and are shown in Table 1.4. The size limits differ by the shape category and refer to the smallest dimension of the ped. Grade describes the distinctness of individual units and how the cohesion within the units compares to the adhesion between units. Three classes are used: weak, moderate, and strong.

FIGURE 1.5 Strong, thin, platy structure. (Courtesy of Soil Survey Division Staff, *Soil Survey Manual*, Soil Conservation Service, U.S. Department of Agriculture Handbook 18, 1993.)

FIGURE 1.6   Strong, medium, prismatic structure. (Courtesy of Soil Survey Division Staff, *Soil Survey Manual*, Soil Conservation Service, U.S. Department of Agriculture Handbook 18, 1993.)

Structure affects porosity (the total pore space) and the pore-size distribution (how the pore space is divided among pores of different sizes). For example, a soil with granular structure would have few large pores and a narrow pore-size distribution. The failure zones that form aggregates can become large pores, often referred to as one type of *macropore*. Macropores are large, continuous voids in soil and include structural, shrink-swell, and tillage fractures; old root channels; and soil fauna burrows. Suggested lower limits for macropore diameters and widths are in the 0.03–3.00 mm range (Luxmoore 1981; Beven and Germann 1982; White 1985). Macropores are important because they can increase infiltration and may result in

FIGURE 1.7   Strong, medium, coarse angular blocky structure. (Courtesy of Soil Survey Division Staff, *Soil Survey Manual*, Soil Conservation Service, U.S. Department of Agriculture Handbook 18, 1993.)

FIGURE 1.8 Strong, fine, medium granular structure. (Courtesy of Soil Survey Division Staff, *Soil Survey Manual*, Soil Conservation Service, U.S. Department of Agriculture Handbook 18, 1993.)

bypass flow where water and solutes move rapidly through the profile and do not interact with the soil matrix (Quisenberry and Phillips 1976).

A soil profile usually consists of a series of horizons that can differ in texture, structure, and color (Figure 1.9). The profile description for a pedon of the Cecil series is shown in Table 1.5.

1.6 SUMMARY

Soils are a dispersed, three-phase system of solid material, water, and air. In this chapter, the focus has been on the solid phase. It consists of primary particles of

TABLE 1.4

Size classes of soil structure

Size class	Platy[a] (mm)	Prismatic or columnar (mm)	Blocky (mm)	Granular (mm)
Very fine	<1	<10	<5	<1
Fine	1–2	10–20	5–10	1–2
Medium	2–5	20–50	10–20	2–5
Coarse	5–10	50–100	20–50	5–10
Very coarse	>10	>100	>50	>10

Source: Soil Survey Division Staff, 1993.

[a] In describing plates, "thin" is used instead of "fine" and "thick" instead of "coarse."

FIGURE 1.9 Nashville silt loam soil profile. (Courtesy of Soil Survey Division Staff, *Soil Survey Manual*, Soil Conservation Service, U.S. Department of Agriculture Handbook 18, 1993.)

different sizes and mineralogy. The clay-size fraction and organic fraction are especially important because of their charge and high specific surface. Primary particles join together to form secondary particles, which result in different soil structure. Soil structure is difficult to quantify, but it has a very important effect on many soil physical processes. Hence, the most important general characteristics of the soil phase are texture, structure, and mineralogy.

In the next chapter, we begin our study of the soil water phase.

1.7 DERIVATIONS

In this section, Stokes's law (Equation 1.10) is derived. Three forces act on a soil particle placed in a sedimentation cylinder. The force of gravity F_g [MLT^{-2}] acts in the downward direction:

$$F_g = m_s g = \rho_s v_s g = \rho_s \frac{4\pi r^3}{3} g,$$

where m_s is the mass of the particle [M], g is the acceleration due to gravity [LT^{-2}], ρ_s is the particle density [ML^{-3}], and v_s is volume of the particle [L^3]. Since Stokes's law assumes the particle is a sphere, the formula for the volume of a sphere, $4\pi r^3/3$, has been substituted for v_s, where r is the radius [L]. Acting against the force of gravity is the force of buoyancy, F_b [MLT^{-2}]. Archimedes's principle states that this force is equal to the weight of the water displaced by the particle:

$$F_b = m_l g = \rho_l v_s g = \rho_l \frac{4\pi r^3}{3} g,$$

where m_l is the mass of the displaced water [M] and ρ_l is the density of water. Also acting against gravity is the viscous drag force, F_d [MLT^{-2}], exerted on the particle by the surrounding liquid as it settles:

$$F_d = 6\pi r \eta V,$$

where η is the viscosity of water [$ML^{-1}T^{-1}$] and V is the settling velocity of the particle (the variable of interest).

TABLE 1.5
Cecil soil series description

Horizon	Depth (cm)	Textural class	Structure	Color
Ap	0–21	Loamy sand	Weak, medium granular	Brown
BA	21–26	Clay loam	Weak, medium, subangular blocky	Red
Bt1	26–102	Clay	Strong, medium, subangular blocky	Red with few strong brown mottles
Bt2	102–131	Clay	Strong, medium, subangular blocky	Red
BC	131–160	Clay loam	Weak, medium, subangular blocky	Red
C	251+	Sandy clay loam	Massive	Red with light red and weak red mottles

Source: Plot 4 adapted from Bruce, R.R., et al., *Physical Characteristics of Soils in the Southern Region: Cecil*, Georgia Agricultural Experiment Stations, Athens, GA, 1983.

The particle stops accelerating when the opposing forces are equal:

$$F_g = F_b + F_d$$

$$\rho_s \frac{4\pi r^3}{3} g = \rho_l \frac{4\pi r^3}{3} g + 6\pi r \eta V$$

$$(\rho_s - \rho_l) \frac{4\pi r^3}{3} g = 6\pi r \eta V$$

$$V = \frac{(\rho_s - \rho_l) 4 r^2 g}{18\eta}.$$

After substituting particle diameter $(d=2r)$, Stokes's law (Equation 1.10) is obtained:

$$V = \frac{(\rho_s - \rho_w) d^2 g}{18\eta}.$$

1.8   PROBLEMS

1. Calculate the time in minutes at which all silt particles will have settled below a depth of 10 cm in a column of well-stirred, dilute soil and water. Temperature$=20°C$. Show that the units cancel.
2. If a soil consists of 13% clay and 22% silt, estimate the specific surface in square meters per gram assuming that the entire clay fraction consists of montmorillonite and the silt and sand fractions have negligible specific surface.
3. What is the specific surface in square meters per gram of a sphere with a diameter of 0.002 mm and a particle density of 2.65 g cm^{-3}? Volume of a sphere$=4\pi r^3/3$. Surface area of a sphere$=4\pi r^2$.
4. How long will it take the smallest sand particle to settle from the surface of a lake to the bottom at 10 m if the temperature is 25°C?
5. Derive Equation 1.9. Start with Equation 1.2 and substitute Equations 1.5, 1.7, and 1.8.

	Bulk density ($\frac{kg}{m3}$)	Porosity value (%)
Gravel	1870	30 - 40
Fine to medium mix sand	1850	30 - 35
Uniform sand	1650	30 - 40
Medium to coarse mix sand	1530	35 - 40
Silt	1280	40 - 50
Clay	1220	45 - 55

2 Soil Water Content and Potential

2.1 INTRODUCTION

Soil physics is the study of the physical properties and processes that occur in soil. These are mediated in large part by soil water. The two most important characteristics of water in soil are the amount of water (*water content*) and the energy status of the water (*water potential*). These two factors affect many soil processes, including soil temperature, oxygen and other gas movement, water and solute movement, runoff and infiltration, and plant uptake of water. This chapter covers the unique properties of water, soil water content, soil water potential, and the relationship between water content and potential.

2.2 ENERGY AND WORK

To discuss the unique properties of water, it is important to understand energy and work. There are two types of energy: *kinetic* and *potential*. The kinetic energy of an object is due to motion:

$$KE = \frac{1}{2}mv^2, \tag{2.1}$$

where KE is the kinetic energy [ML^2T^{-2}], m is the mass [M], and v is velocity [LT^{-1}] of the object. If KE is in joules (J), then m is in kilograms (kg) and v is in meters per second (m s^{-1}).

Potential energy is the energy of an object due to its position in a force field. For example, the potential energy that an object gains as it is raised to a given height in the earth's gravitational field is

$$PE = mgh, \tag{2.2}$$

where PE is the potential energy [ML^2T^{-2}], g is the acceleration due to gravity [LT^{-2}], and h is the height [L]. If PE is in Joules, then m is in kilograms, $g = 9.81$ m s^{-2}, and h is in meters.

Work (W) [ML^2T^{-2}] is done when an object is moved by a force through a given distance and is the product of the force (F) [MLT^{-2}] in Newtons (N) and the distance (d) [L] in meters:

$$W = Fd. \tag{2.3}$$

Work has the same units as energy, Newton-meters (N-m), which is defined as a Joule, in the meters-kilograms-seconds (mks) metric system. This is because work and energy are interchangeable. When an object is lifted to a higher elevation in the earth's gravitational field, work is done and that work becomes the increased gravitational potential energy of the object. If the object is released, the potential energy is converted to kinetic energy as the object falls.

The molecules of any substance are in constant motion, so they possess kinetic energy. Temperature is a measure of this motion and the only exception to the above statement occurs at a temperature of zero kelvin ($-273°C$) when all molecular motion stops.

2.3   PROPERTIES OF BULK WATER

Before discussing water in soils, it is useful to review several of the unique properties of bulk water and how water interacts with surfaces. Many of the bulk properties are a consequence of water's molecular structure. Water's unique properties make life on Earth possible. Hence the excitement in July 2008, when the *Mars Phoenix Lander* sampled Martian soil and conclusively showed the presence of water (Figure 2.1) (Chang 2008).

The water molecule is asymmetrical with an angle of 105° between the hydrogen atoms attached to the oxygen atom, instead of 180° (Figure 2.2). The water molecule is about 3 angstroms (Å) in diameter. The bonds between the H and O atoms are covalent and quite strong. The asymmetrical arrangement results in an *electrical dipole*, which creates an *electrical field* around water molecules. Within this field, any charged particle experiences a force toward or away from the water molecule, just as a body with mass experiences a force in the earth's gravitational field.

FIGURE 2.1   *Mars Phoenix Lander's* solar panel and robotic arm with a soil sample in the scoop. Measurements in July 2008 showed that the soil sample contained water. (From http://www.nasa.gov/mission_pages/phoenix/main/index.html.)

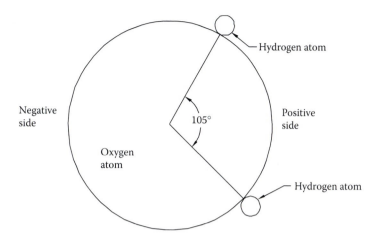

FIGURE 2.2 Asymmetric arrangement of water molecule.

Because of the electrical dipole, water molecules interact with other water molecules, dissolved ions in solution, gas interfaces, and mineral and organic surfaces. Water molecules interact with other water molecules by forming hydrogen bonds. These are simple electrostatic attractions between oppositely charged poles of water molecules. Hydrogen bonds are considerably weaker than the covalent bonds within water molecules. The attraction between water molecules is called *cohesion*. Hydrogen bonding is best seen in ice, where it causes an open rigid structure (Figure 2.3). As temperature increases, molecules acquire more thermal energy and many of the hydrogen

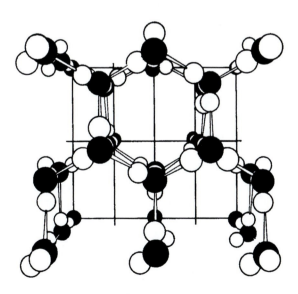

FIGURE 2.3 Molecular structure of ice. Oxygen atoms are shown in black and hydrogen atoms are show in white. (From Hillel, D., *Fundamentals of Soil Physics*, Academic Press, New York, 1980. With permission.)

bonds are broken. The ice structure collapses into a less ordered arrangement. As a consequence, the density of water increases as ice melts (Table 2.1).

As a liquid, water molecules bond together to form polymers or *flickering clusters* (Hillel 1980) (Figure 2.4). It is estimated that a cluster has on average 40 water molecules at 20°C. The clusters are constantly forming and breaking apart. It is only when water becomes a gas that water molecules exist as independent molecules.

Because of H-bonding, it takes a lot of energy (heat) to break enough bonds to destroy the open rigid structure in ice and change it to a liquid (333 J g^{-1} at 0°C). This energy requirement is known as the *heat of fusion* and water has an unusually high value compared to other fluids (Table 2.1). For the same reason, it takes a lot of heat to break the remaining bonds and change liquid water to gas or *water vapor* (2441 J g^{-1} at 25°C). This is known as the *heat of vaporization* and it too is relatively high for water (Table 2.1).

H-bonding also restricts molecular velocities within the liquid phase so that it requires a great deal of energy (heat) to raise the temperature of water. This is known as the *specific heat*: the energy required to raise 1 g of water 1°C (Table 2.1). At 25°C, it is 4.18 J g^{-1} °C^{-1}. These large values (specific heat and heat of fusion) have a very stabilizing effect on the climate of Earth as compared to other planets and the moon.

Each liquid water molecule has an "average" energy, but the energy at any given moment changes, so it is best to think of a probability distribution of energies. The average is the level with the most likely probability, but there is a small probability of acquiring a very high level of energy for an instant. Some molecules acquire enough energy to break the covalent bond with one of the hydrogen atoms, leaving it behind with the water molecule to which it was H-bonded. This produces a hydronium ion and a hydroxyl ion:

$$2H_2O \leftrightarrow H_3O^+ + OH^-. \tag{2.4}$$

These new charged water molecules have considerably different electrical fields compared to the field around a dipole water molecule. This field has very important effects on ions in solution and mineral and organic surfaces. The abbreviation H$^+$ is used for H$_3$O$^+$ and its concentration is characterized using pH, which is defined as:

$$pH = -\log_{10}(H^+), \tag{2.5}$$

where H$^+$ is the hydronium concentration in moles per liter (mol L^{-1}).

Another unique property of water is that it is a good solvent. Consider NaCl added to water. There is an attraction between Na$^+$ and Cl$^-$ ions, but there is also an attraction between these ions and surrounding water molecules due to their dipole. Because of the attraction to water, the attraction between Na$^+$ and Cl$^-$ ions is overcome and the compound *dissolves* or *ionizes* in water. The water molecules orient around each ion and form a hydrated layer (Figure 2.5). The bonding in this hydration sphere restricts molecular motion, lowers the energy level of water, and the lost energy is given off as *heat of hydration*.

Table 2.1

TABLE 2.1
Physical properties of water

Temperature T ($^\circ$C)	Density ρ_w (g cm^{-3})	Coefficient of dynamic viscosity η (g cm^{-1} s^{-1})	Heat of fusion (J g^{-1})	Heat of vaporization (J g^{-1})	Specific heat (J g^{-1} $^\circ$C^{-1})	Surface tension σ (J m^{-2})	Apparent relative permittivity or dielectric constant ε_{ra}
-10	0.99794	–	–	2524	4.27	–	–
-5	0.99918	–	–	2511	4.23	$7.64 \cdot 10^{-2}$	3.2
0	0.99987	0.01787	333	2500	4.22	$7.56 \cdot 10^{-2}$	88.00
4	1.00000	0.01567	–	2491	4.21	$7.50 \cdot 10^{-2}$	–
5	0.99999	0.01519	–	2488	4.20	$7.48 \cdot 10^{-2}$	86.04
10	0.99973	0.01307	–	2477	4.19	$7.42 \cdot 10^{-2}$	84.11
15	0.99913	0.01139	–	2465	4.19	$7.34 \cdot 10^{-2}$	82.22
20	0.99823	0.01002	–	2453	4.18	$7.27 \cdot 10^{-2}$	80.36
25	0.99708	0.00890	–	2441	4.18	$7.19 \cdot 10^{-2}$	78.54
30	0.99568	0.00798	–	2429	4.18	$7.11 \cdot 10^{-2}$	76.75
35	0.99406	0.00719	–	2418	4.18	$7.03 \cdot 10^{-2}$	75.00

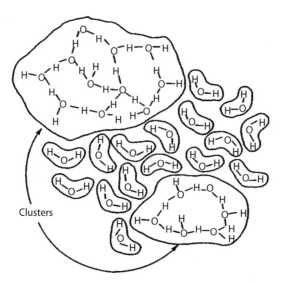

FIGURE 2.4 Molecular structure of liquid water. (From Hillel, D., *Fundamentals of Soil Physics*, Academic Press, New York, 1980. With permission.)

The presence of salts in water causes *boiling point elevation*. Hydrated water has more bonding so that a higher molecular velocity (temperature) is required for molecules to break away and become individual vapor molecules. Salts also cause *freezing point depression*. It doesn't take as much energy to go from ice to a solution of water and NaCl (and the restricted molecular velocities in the hydrated water) as

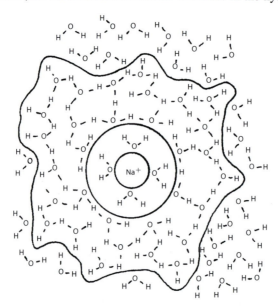

FIGURE 2.5 Hydrated sphere surrounding a sodium ion in water. (From Hillel, D., *Fundamentals of Soil Physics*, Academic Press, New York, 1980. With permission.)

it does to go from ice to pure water. Hence, adding salt to ice on highways causes ice to melt when the temperature is below 0°C. Also, adding salt to ice surrounding an ice-cream maker causes ice to melt faster, which takes heat from the system and freezes the ice cream.

The attraction of water to dissolved salts creates an increase in pressure if the solution is confined. For example, consider two containers of water separated by a *semipermeable membrane* (Figure 2.6). One container has a salt solution and the other is salt free. The membrane will allow water molecules to move across, but does not allow hydrated ions to pass through. The attraction of water molecules to the dissolved ions will cause water to move from the salt-free container to the container with salt. This will cause an increase in pressure in the container with salt and this pressure is called *osmotic pressure*. The pressure is related to the difference in the height of the water columns in the manometers connected to each container in Figure 2.6:

$$\pi = \rho g h, \tag{2.6}$$

where ρ is the density of the solution $[ML^{-3}]$ in kilograms per cubic meter, g is the acceleration due to gravity $[LT^{-2}]$ in meters per second squared, h is the difference in height $[L]$ in meters, and π is the osmotic pressure $[ML^{-1}T^{-2}]$ in Pascals (Pa). The osmotic pressure is also related to the concentration of the salt solution:

$$\pi = C_s RT, \tag{2.7}$$

where C_s is the ion concentration $[ML^{-3}]$ in moles per cubic meter, R is the universal gas constant $[L^2T^{-2}K^{-1}]$, 8.314 Pa m³ mol⁻¹ K⁻¹, and T is the temperature $[K]$ in the kelvin (K) scale. *Reverse osmosis* is a process for desalinating ocean water that uses two chambers separated by a semipermeable membrane. In this process, a pressure is applied to the salt water chamber that exceeds the osmotic pressure and drives water through the membrane to the salt-free chamber.

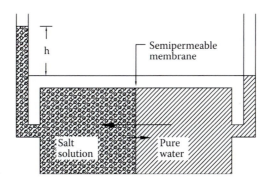

FIGURE 2.6 Two containers separated by a semipermeable membrane. On one side of the membrane is a salt solution. On the other side is pure water. Osmosis causes an increase in pressure in the container with the salt solution.

In Stokes law (Chapter 1, Equation 1.10), one of the variables was water viscosity. This is another property of bulk water that can be understood in terms of the unique molecular structure of water. When adjacent layers of water molecules are made to slide over each other (move in shear), there is a large drag force due to the cohesion between water molecules (this can be thought of as an internal friction in water). The drag force per unit area between layers or shear stress, τ [MLT^{-2}], is proportional to the rate of change in velocity of subjacent water layers, V [LT^{-1}], in the direction perpendicular to the movement of water, y [L]:

$$\tau = -\eta \frac{dV}{dy}, \tag{2.8}$$

where dV/dy is the velocity gradient and the proportionality constant is η, the *coefficient of dynamic viscosity* of water [$ML^{-1}T^{-1}$]. The negative sign in Equation 2.8 indicates that the drag force acts in the opposite direction to the movement of the water layers. Viscosity decreases as temperature increases (Table 2.1). The ratio of dynamic viscosity of a fluid to its density is called the *kinematic viscosity* [L^2T^{-1}] (Hillel 2004).

2.4   PROPERTIES OF WATER AT AIR AND SOLID INTERFACES

Water displays several unique properties at interfaces with gases and solids, which will be important when soil water is considered. *Surface tension* occurs at a gas (air) and water interface. If the forces experienced by water molecules in bulk water are contrasted with molecules at an air–water interface, a difference is seen (Figure 2.7). Molecules in the bulk water experience random pulls in all directions due to cohesion, whereas molecules at the surface experience more pull into the interior. There are gaseous water molecules (water vapor) above the interface that exert a pull on the water molecule at the interface, but there are relatively few of these molecules compared to those in liquid water. The result is that the surface acts like a membrane that can be stretched. Another way to think about it is that it requires more energy for water molecules to remain at the interface. Surface tension (σ) [MT^{-2}] is a measure of this energy per unit area. At 20°C, σ has a value of 7.27×10^{-2} J m^{-2}. Because of H-bonding, water has a relatively high surface tension compared to other fluids. As temperature increases, there is less H-bonding and σ decreases (Table 2.1).

Because of the charged surfaces and large specific surface of clays and organic matter (Chapter 1, Tables 1.1 and 1.2), the interaction of water molecules with clay and organic matter surfaces and adsorbed ions is of special interest. Due to the dipole, water molecules are attracted to any charged solid surface. This attraction is known as *adhesion*. The attraction to clay surfaces causes an *ice-like structure* in water molecules. The thickness of this structure ranges from 8 Å (2–3 water molecular layers) in kaolinite to 68 Å (23 molecular layers) in montmorillonite (Jury and Horton 2004).

Water is also attracted to adsorbed ions on the surfaces of clay particles (called *micelles* or *tactoids*). When a soil is wetted, water moves into these areas of high salt

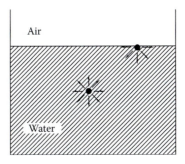

FIGURE 2.7 Water molecule at the air–water interface and water molecule in the interior of the water in a container. (After Hillel, D., *Fundamentals of Soil Physics*, Academic Press, New York, 1980.)

concentration and creates an osmotic pressure. This pressure acts against clay particles and can cause them to move apart if the pressure is sufficiently large to overcome the bonding between clay particles. The amount of pressure depends on the ion concentration, which is a function of cation exchange capacity, specific surface, and ion valence. The result is the swelling associated with certain types of 2:1 clays. Water is also attracted to ions in the interlayer spaces within clay micelles and this creates an osmotic pressure that can affect interlayer spacing. This principle is used in the identification of clay minerals in x-ray diffraction techniques.

When water is in contact with air and a solid (a three-phase system such as soil), there is a *contact angle* between the solid and the air–water interface (Figure 2.8). The angle (α) is measured from the solid surface through the liquid. The forces of adhesion pull water molecules in the direction of the unwetted solid surface and favor a small contact angle. The forces of cohesion resist this pull and favor a large contact angle. A water droplet on a clean *hydrophilic* surface has a small contact angle, whereas a droplet on a waxy *hydrophobic* surface has a large contact angle. Liquid mercury has very large cohesive forces and forms small beads on a solid surface.

The curved water surface or *meniscus* forms at the air–water interface in a narrow tube (Figure 2.9). The surface is curved because there is a difference in pressure between the air and the water. The pressure is greater on the concave side. The difference is a function of the radius of curvature (R) [L] and the surface tension (σ) [MT^{-2}]:

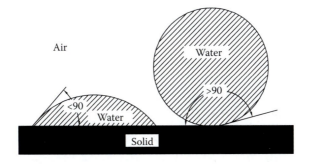

FIGURE 2.8 Two water droplets with different contact angles on a solid surface.

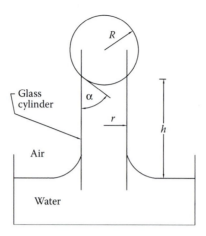

FIGURE 2.9   Capillary rise of water in a cylinder of radius r with a meniscus of curvature R and a contact angle of α.

$$\Delta P = \frac{2\sigma}{R},\qquad(2.9)$$

where ΔP is the difference in pressure [ML^{-1}T^{-2}]. This difference is always positive so it is calculated as the pressure on the concave side, less the pressure on the convex side. For water rising in a small glass cylinder, for example, it would be $\Delta P = P_{air} - P_{water}$.

The phenomenon of *capillary rise* is due to many of the surface effects discussed so far. Water rises in a narrow glass cylinder due to adhesion and cohesion (Figure 2.9). Water molecules are attracted to glass and tend to rise up the side, pulling other water molecules behind them. This creates an area of tension or low pressure in the water and a concave meniscus forms. The curved meniscus indicates that the pressure in the water (P_{water}) near the top of the cylinder is less than atmospheric pressure (P_{air}). Initially, water enters the cylinder opening because it is an area of low pressure when the meniscus is near the bottom of the cylinder. Water continues to enter and the meniscus rises until the weight of the water column in the cylinder exactly offsets the decrease in pressure at the bottom of the cylinder. At this point, water stops rising. An analysis of the forces at equilibrium (see Section 2.12.1) leads to the equation for the height of capillary rise (h):

$$h = \frac{2\sigma\cos(\alpha)}{\rho_w g r},\qquad(2.10)$$

where ρ_w is the density of water [ML^{-3}], g is the acceleration due to gravity [LT^{-2}], and r is the radius of the cylinder [L].

As σ increases, capillary rise increases (and it varies with temperature since σ is a function of temperature, as is ρ_w, Table 2.1). As r increases, capillary rise decreases. The effect of a contact angle is not so apparent due to the cosine function. This function has a value of 1 at 0°, 0 at 90°, and −1 at 180°. So capillary rise decreases as the contact angle increases, and it can be negative for liquids such as mercury with a contact angle >90°.

The capillary rise equation shows that water will rise from a free water surface into small soil pores. The smaller the pore, the greater the height of rise. The capillary rise equation will be used later in the chapter to discuss the effect of capillarity on soil water energy and to determine pore-size distributions.

Capillary rise is also important in many biological systems (Figure 2.10).

Example 2.1

Calculate the height of rise of water in a clean glass tube (contact angle of zero) with a radius of 0.50 mm. Assume a temperature of 20°C. What would the height of rise be if the contact angle was 60°?

From Table 2.1, $\sigma = 7.27 \times 10^{-2}$ J m^{-2} or kilogram per second squared, and $\rho_w = 998$ kg m^{-3}. Substituting into Equation 2.10 and recognizing that $\cos(0) = 1$:

$$h = \frac{2\left(7.27 \cdot 10^{-2}\, \frac{kg}{s^2}\right)1}{\left(998\, \frac{kg}{m^3}\right)\left(9.81\, \frac{m}{s^2}\right)(0.00050\, m)} = 2.97 \cdot 10^{-2}\, m = 30\, mm.$$

For a contact angle of 60°,

$$h = 30\, mm \cdot \cos(60) = 30\, mm \cdot 0.50 = 15\, mm,$$

the height of rise is 15 mm.

FIGURE 2.10 Phalaropes use capillarity with a tweezering motion of their beaks to "ratchet up" droplets of water with tiny bits of food. (From Rainey Shuler Photography, Santa Cruz, CA. With permission.)

One of the best ways to understand an equation is to make graphs of the dependent variable (h in the case of Equation 2.10) as a function of the independent variables (r, α, etc.). In the following example, Microsoft Excel is used to examine the capillary rise equation.

Example 2.2

Use Microsoft Excel to determine the height of rise of water under the same conditions as in Example 2.1 for cylinders with radii from 0.1 to 5.0 mm in steps of 0.1 mm. Show the effect of different contact angles.

First, a spreadsheet is set up with the values of the constants (σ, ρ_w, and g) (Figure 2.11). To the right, create a column of values of r (only the first 10 values are shown) and, below a cell with a contact angle value of zero, enter the formula for the height of capillary rise (see the Formula Bar at the top of the spreadsheet) and copy it into the cells below to create a second column with the value of h. To use the cosine function, convert the angle from degrees to radians in Excel (see the Formula Bar).

To see the effect of different contact angles, add another column using a new contact angle of 60°. Select the three columns including the headings and then click on the Insert tab on the toolbar. Select the Scatter plot with lines connecting points. Click on the graph and then select Layout under Chart Tools at the top of the toolbar to add and edit titles for the axes. Make a plot of h vs. r with separate curves for the two different contact angles that shows the effect of r and α on h (Figure 2.12). You can add axis titles using the Layout tab on the toolbar. Select Axis Title from the Labels section.

2.5   SOIL WATER CONTENT

Soil water content can be defined in four ways. One of these definitions was seen in Chapter 1 (Equation 1.4), *volumetric soil water content* (θ). It is a ratio of volumes and therefore unitless, but it is usually expressed in units of cubic centimeter per

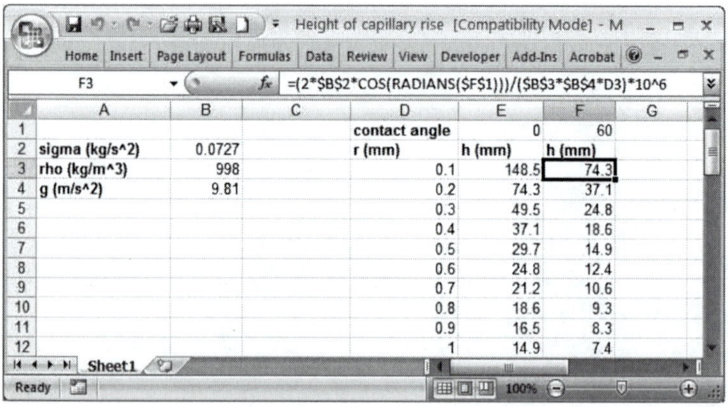

Formula Bar: F3   =(2*B2*COS(RADIANS(F1)))/(B3*B4*D3)*10^6

	A	B	C	D	E	F	G
1				contact angle	0	60	
2	sigma (kg/s^2)	0.0727		r (mm)	h (mm)	h (mm)	
3	rho (kg/m^3)	998		0.1	148.5	74.3	
4	g (m/s^2)	9.81		0.2	74.3	37.1	
5				0.3	49.5	24.8	
6				0.4	37.1	18.6	
7				0.5	29.7	14.9	
8				0.6	24.8	12.4	
9				0.7	21.2	10.6	
10				0.8	18.6	9.3	
11				0.9	16.5	8.3	
12				1	14.9	7.4	

FIGURE 2.11   Microsoft Excel spreadsheet for calculating the height of rise for a contact angle of 0° and 60°.

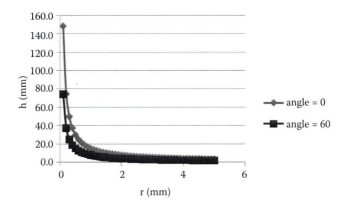

FIGURE 2.12 Plot of height of capillary rise as a function of cylinder radius for a contact angle of 0° and 60°.

cubic centimeter or cubic meter per cubic meter to emphasize that it is based on volumes.

Soil water content can also be defined on a mass basis and this is usually referred to as *gravimetric water content* (θ_g):

$$\theta_g = \frac{m_l}{m_s},\qquad(2.11)$$

where m_l is the mass of water (or liquid) and m_s is the mass of solid (or soil) [M]. It is also unitless, but it is usually expressed in units of gram per gram or kilogram per kilogram to emphasize that it is based on mass. The definition of m_s is the *oven-dry* mass, or the mass of a sample of soil after being placed in an oven at a temperature of 105°C for 24–48 h (Topp and Ferre 2002a). These conditions are specified because most of the water in soil evaporates under these conditions and the soil reaches a steady value of mass.

The relationship between volumetric and gravimetric water contents is

$$\theta = \theta_g \frac{\rho_b}{\rho_w}.\qquad(2.12)$$

In the cm-gram-second (cgs) unit system, $\rho_w \approx 1$ g cm^{-3} at normal temperatures and θ is just the product of θ_g and ρ_b. This is not true of the meters-kilograms-seconds (mks) unit system where $\rho_w \approx 1000$ kg m^{-3}.

The third method of expressing soil water content is *relative water content* or *effective saturation* (S_e):

$$S_e = \frac{\theta - \theta_r}{\theta_s - \theta_r},\qquad(2.13)$$

where θ_s is the saturated volumetric water content (equal to φ if there is no entrapped air and all pores are filled with water) and θ_r is the residual volumetric water content. The residual water content is somewhat arbitrarily defined as the water content where hydraulic conductivity (to be discussed in Chapter 3) approaches zero (van Genuchten 1980a). It is often an empirical constant found by fitting water characteristic curve data (discussed in Section 2.9.4). Equation 2.13 is often used with $\theta_r = 0$. Regardless, S_e is unitless [–] and varies between 0 and 1. The symbol Θ (uppercase θ) is often used instead of S_e.

The last definition for soil water content is *equivalent depth of water* (D_e) [L], which for a soil profile with a uniform water content is

$$D_e = \theta D, \tag{2.14}$$

where D is the depth of soil under consideration. D_e is the equivalent depth that soil water would have if the water was extracted and ponded over the soil surface (Hillel 1980). It is especially convenient for expressing the amount of water in a soil to a given depth (the top meter, for example). For soils with multiple (n) layers of different water contents, D_e is calculated as:

$$D_e = \sum_{i=1}^{n} \theta_i D_i, \tag{2.15}$$

where θ_i and D_i are the volumetric water contents and thicknesses, respectively, of each layer.

2.6 MEASURING SOIL WATER CONTENT

There are a number of methods for measuring soil water content, but there is *no best way*. Each method has its advantages and disadvantages. The various methods are described by Topp and Ferré (2002c). A good Web site on commercial equipment to measure water content (and potential) is the Soil Water Content Sensors and Measurement home page, www.sowacs.com.

2.6.1 GRAVIMETRIC METHODS

This method is described in detail by Topp and Ferré (2002a) and is the simplest method of measuring water content. It is usually done by taking small cores with a sampling tube, placing them in a container, determining the *wet weight*, drying the samples in an oven, and determining the *dry weight*. Then, the gravimetric water content is

$$\theta_g = \frac{\text{wet weight} - \text{dry weight}}{\text{dry weight} - \text{container weight}}. \tag{2.16}$$

This can be converted to a volumetric basis using bulk density and Equation 2.12. The method has been adapted for the use of microwave ovens (Topp and Ferré 2002b).

2.6.2 TIME DOMAIN REFLECTOMETRY

This method is described in detail by Ferré and Topp (2002). The bulk electrical and magnetic properties of a soil affect the propagation of an electromagnetic (EM) wave through the soil. The most important bulk soil properties affecting EM response are *electrical conductivity* (σ) and *dielectric permittivity* (ε). Electrical conductivity is a measure of how free electrons flow when exposed to an electric field. The dielectric permittivity is a measure of the displacement of constrained charges when exposed to an electric field. The high dielectric permittivity and electrical conductivity of water have a dominating effect on bulk soil properties. As such, measurements of EM wave propagation through soil can be used to infer soil water content.

The dielectric properties of water are due to the electrical dipole of the water molecule (Figure 2.2). Dielectric permittivity is a complex variable consisting of a real (ε') and an imaginary component (ε''). These components are usually divided by the dielectric permittivity of free space (ε_0) and referred to as real and imaginary *relative permittivity*, $\varepsilon'_r = \varepsilon'/\varepsilon'_0$ and $\varepsilon''_r = \varepsilon''/\varepsilon''_0$, respectively. Relative permittivity (sometimes called the *dielectric constant*) is not constant, but varies with the frequency of the EM signal. When an EM wave passes through soil, energy is lost through several mechanisms that involve soil water (Robinson et al. 2003). Water molecules become oriented as the EM wave passes through soil and the real component is a measure of the energy stored in these oriented water molecules. The imaginary component is a measure of energy loss due to conductance (called dielectric losses) and energy lost as heat by water molecules that cannot orient fast enough to the changing electrical field at high frequencies (called relaxation losses). For frequencies below 10 GHz, liquid water has a very large value of ε'_r (approximately 80) compared to the values for mineral (3–7) and organic (2–5) components of soils. Ice has a ε'_r of approximately 3 and air has a value of 1 (Table 2.1).

To measure water content with time domain reflectometry (TDR), a two- or three-rod waveguide is inserted in the soil and connected via coaxial cable to a TDR instrument. The instrument transmits a voltage step to the waveguide. Changes in impedance cause reflections of the input signal at times corresponding to the beginning and end of the waveguide. These reflections can be used to determine the travel time of the EM signal in soil to the end of the waveguide and back. The velocity (v) of the EM signal [LT^{-1}] is twice the length (L) of the waveguide [L], divided by the travel time (t) [T]:

$$v = \frac{2L}{t}. \tag{2.17}$$

The velocity of the EM signal is inversely related to the relative permittivity of the soil as measured by TDR, which is referred to as the *apparent relative permittivity* (ε_{ra}) [–] or apparent dielectric constant:

$$v = \frac{c}{\sqrt{\varepsilon_{ra}}}, \qquad (2.18)$$

where c is the velocity of light in free space (3×10^8 m s^{-1}). Combining these equations, an equation for ε_{ra} as a function of the waveguide length and measured EM travel time is obtained:

$$\varepsilon_{ra} = \left(\frac{ct}{2L} \right)^2. \qquad (2.19)$$

It is usually assumed that the relative permittivity measured by TDR is equal to the real component of relative permittivity ($\varepsilon_{ra} \approx \varepsilon_r'$).

Topp et al. (1980) showed that for a wide range of mineral soils, a single empirical calibration equation could be used to relate ε_{ra} to θ:

$$\theta = -5.3 \cdot 10^{-2} + 2.92 \cdot 10^{-2} \varepsilon_{ra} - 5.5 \cdot 10^{-4} \varepsilon_{ra}^2 + 4.3 \cdot 10^{-6} \varepsilon_{ra}^3. \qquad (2.20)$$

Subsequent studies have confirmed the wide applicability of this "universal calibration equation," but there are exceptions for low density soils, soils with high organic or clay content, and artificial soils such as glass beads. Dasberg and Hopmans (1992) showed that the ice-like structure of water near the surface of clays with high charge density and specific surface caused Equation 2.20 to underestimate θ.

Topp and Reynolds (1998) suggested an alternative to Equation 2.20:

$$\theta = 0.115\sqrt{\varepsilon_{ra}} - 0.176, \qquad (2.21)$$

which deviates from the original universal equation by < 0.01 m^3 m^{-3} over the range of $\theta = 0.05$ to 0.45 m^3 m^{-3}. The advantages of Equation 2.21 are that it is more closely related to theoretical mixture model equations (that relate θ to the apparent permittivity of different components of soil including water, mineral, and air), it allows a two-point calibration check for a given soil or medium, and it leads to less potential for error if one extrapolates to water contents beyond the range over which Equation 2.20 was developed.

Example 2.3

Two TDR waveguides are installed vertically from the soil surface, one waveguide is 30 cm in length and the other is 45 cm in length. A TDR instrument is used to determine EM signal travel times. The travel time for the 45 cm waveguide is 1.31×10^{-8} s and the travel time for the 30-cm waveguide is 0.76×10^{-8} s. Use Equation 2.21 to determine the water content in the 0–30 cm depth increment (θ_{0-30}) and the water content in the 0–45 cm depth increment (θ_{0-45}). What is the water content in the 30–45 cm depth increment (θ_{30-45})?

Calculate ε_{ra} for each waveguide using Equation 2.19. For the 30-cm waveguide:

$$\varepsilon_{ra} = \left(\frac{ct}{2L}\right)^2 = \left[\frac{\left(3\cdot10^8\,\frac{m}{s}\right)\left(0.76\cdot10^{-8}\,s\right)}{2(0.30\,m)}\right]^2 = 14.4.$$

For the 45-cm waveguide:

$$\varepsilon_{ra} = \left(\frac{ct}{2L}\right)^2 = \left[\frac{\left(3\cdot10^8\,\frac{m}{s}\right)\left(1.31\cdot10^{-8}\,s\right)}{2(0.45\,m)}\right]^2 = 19.1.$$

Calculate θ for each waveguide using Equation 2.21.

$$\theta_{0-30} = 0.115\sqrt{\varepsilon_{ra}} - 0.176 = 0.115\sqrt{14.4} - 0.176 = 0.26,$$
$$\theta_{0-45} = 0.115\sqrt{\varepsilon_{ra}} - 0.176 = 0.115\sqrt{19.1} - 0.176 = 0.33.$$

To determine the water content in the 30–45 cm increment, a relationship between θ_{0-30}, θ_{0-45}, and θ_{30-45} is needed. Equation 2.15 can be used for the equivalent depth of water in the soil profile:

$$D_e = \sum_{i=1}^{n} \theta_i D_i = \left(\theta_{0-30}\cdot30\,cm\right) + \left(\theta_{30-45}\cdot15\,cm\right) = \theta_{0-45}\cdot45\,cm.$$

Solving for θ_{30-45}:

$$\theta_{30-45} = 3\theta_{0-45} - 2\theta_{0-30} = \left(3\cdot0.327\right) - \left(2\cdot0.260\right) = 0.46.$$

2.6.3 Neutron Thermalization

Neutron moisture meters were very common in the past, but they are being used less now because other methods, such as TDR, have been developed that are not a radiation hazard. This method is described by Hignett and Evett (2002). A neutron meter is placed above a hollow access tube installed in soil. From within the meter, a neutron probe is lowered on a cable to the desired depth within the access tube. The probe contains a *fast neutron source* consisting of radioactive Americium (^{241}Am) and Berylium (Be). The ^{241}Am continuously emits α particles, which bombard the Be and cause it to emit high energy (*fast*) neutrons. The fast neutrons pass through the access tube into the soil where they strike nuclei of various elements in soil. Most of the nuclei of these elements (Al, Si, and O) have much greater mass than the neutrons, so there is little energy lost in collisions. However, when neutrons strike H nuclei (which have the same mass as a neutron) they are slowed substantially and after about 19 collisions with H, they reach thermal (room temperature) energy levels. The slowed neutrons form a cloud around the access tube and a small fraction return to the *slow-neutron detector*, also in the probe. The detector counts slow neutrons over a 30 s interval. As such, the count is directly related to the number of H nuclei in the sphere of measurement of the probe. Since most of the H in soil is due to water, the count is proportional to θ. There are other sources of hydrogen in soil, such as clay and organic matter, so a calibration curve has to be developed for each soil, which relates slow neutron count to θ.

2.6.4 CAPACITANCE DEVICES

Capacitance probes are described by Starr and Paltineanu (2002). The measurement is related to soil relative permittivity, but determines the *resonance frequency* of soil, rather than making a direct measurement of ε_{ra} as is done with TDR. The capacitance method is sometimes referred to as frequency domain reflectometry (FDR).

The approach is based on the principle that the resonance frequency F (in Hz) $[T^{-1}]$ of an alternating current circuit with a capacitor is a function of the relative permittivity of the material between and adjacent to the capacitor electrodes:

$$F = \frac{1}{2\pi\sqrt{LG\varepsilon_{ra}}}, \tag{2.22}$$

where L (in Henrys, H) is the circuit inductance $[ML^2Q^{-2}]$ and G is a geometric constant (in s^2 H^{-1}) $[T^2Q^2M^{-1}L^{-2}]$ based on the electrode configuration.

The two most common configurations of the capacitor electrodes are as pairs of parallel metal rods and as pairs of cylindrical metal rings. With the parallel rod configuration, the rods are pushed or buried in the soil so that soil is the medium between the capacitor electrodes. These types of sensors are well suited to surface or shallow depth measurements. The rings are contained within a probe and separated by an insulator. The probe is lowered into the soil via an access tube in a manner similar to that used with neutron probes. The soil is a medium for these types of capacitors only to the extent that the fringe field for the capacitor rings extends into the soil. This type of configuration is suited for deep measurements.

With both configurations, the electrodes are connected to an instrument that determines the resonance frequency of the circuit in a range from about 38 to 150 MHz. Since the geometric factor is not generally known for these configurations, Equation 2.22 cannot be used to determine ε_{ra} (and, in turn, θ using Equations 2.20 or 2.21) given a measurement of F. A calibration equation is used instead that relates F to θ directly.

A third type of capacitance sensor is not designed for use with access tubes and consists of a 5–20 cm long, thin piece of fiberglass with an embedded circuit containing a capacitor. This sensor is buried in the soil and measures the time it takes to charge the capacitor, which is a function of the dielectric constant (and water content) of the surrounding soil. These sensors require calibration, are insensitive to salinity and temperature, and are relatively inexpensive. However, they are not as accurate as the other capacitance devices, TDR, or neutron probes and they are sensitive to changes in bulk density (Parsons and Bandaranayake 2009).

2.7 SOIL WATER POTENTIAL

The problem with using soil water content as the sole measure of moisture status is that one cannot compare different soils. For example, it will be seen that a sandy soil might have a lower θ than a clayey soil, but a plant would have more water available in the sandy soil or water might move from the sandy soil to the clayey soil. The solution to this problem is to evaluate soil water status in terms of the energy level of water.

The concept of *soil water potential* is perhaps the most important contribution of soil physicists. It allows one to compare water in aquifers, soils, plants, and the atmosphere. The concept was developed by Buckingham (1907). Much of the difficulty in understanding the water potential approach comes from the different symbols used for the same quantity by different disciplines. It is a tribute to the usefulness of the approach that it is used by hydrologists, soil scientists, plant physiologists, and climatologists.

At the beginning of this chapter, it was shown that energy can take two forms: kinetic energy and potential energy. Kinetic energy is the energy due to velocity, including molecular velocity. For soil water, the kinetic component of energy is difficult to quantify and, of course, changes with temperature. For many problems in soil water, it is possible to ignore kinetic energy and only consider the potential energy: the energy that water has because of its position in a force field. Numerous forces act on water in soil in addition to gravity (solutes, capillarity, surface tension, etc.). These determine the potential energy of water at various positions (depths) in the soil. If the energy state of soil water is known, then so is the *direction of water flow*: from an area of high energy to an area of low energy. One also knows something about the *rate of water flow*. This depends on the *energy gradient*, or how much energy changes in a given distance, among other things (more on this in Chapter 3).

Only *relative energy states* must be known to predict water movement. Unlike kinetic energy, which is zero at a temperature of zero kelvin, there is no such thing as zero potential energy. On Earth, water is always in some force field. Soil water potential energy is compared to water in a *standard state*, which is arbitrarily defined as having a potential energy of zero:

Standard state water is pure (no solutes), free (no external forces other than gravity) water at a reference pressure P_0, reference temperature T_0, and reference elevation z_0. P_0 is usually atmospheric pressure.

Soil water potential is defined as the difference in potential energy per unit volume, mass, or weight of water compared to the standard state. Depending on the choice of volume, mass, or weight as a basis, the units and symbols used for soil water potential change.

For soil water potential on a volume basis, the units in the mks system are

$$\frac{\text{Energy}}{\text{Volume}} = \frac{\text{J}}{\text{m}^3} = \frac{\text{N} \cdot \text{m}}{\text{m}^3} = \frac{\text{N}}{\text{m}^2} = \text{Pa}, \tag{2.23}$$

where J is Joule, N is Newton, and Pa is Pascal.

The result is units of pressure to express energy per unit volume. We will follow the conventions of Jury and Horton (2004) and refer to soil water potential on a volume basis as *soil water potential* and use the symbol ψ_t for total soil water potential.

For soil water potential on a weight basis, the mks units are

$$\frac{\text{Energy}}{\text{Weight}} = \frac{\text{J}}{\text{N}} = \frac{\text{N} \cdot \text{m}}{\text{N}} = \text{m}. \tag{2.24}$$

In this case, the result is units of length to express energy per unit weight. We will refer to soil water potential on a weight basis as *soil water potential head* and use the symbol H for total soil water potential head. This is also referred to as the *hydraulic head*. The relationship between total soil water potential and total soil water potential head is

$$\psi_t = \rho_w g H, \tag{2.25}$$

where ρ_w is the density of water $[ML^{-3}]$ and g is the acceleration due to gravity $[LT^{-2}]$.

Soil water potential can also be expressed per unit mass of soil, and in this case the units are Joule per kilogram. This approach is seldom used and we will not discuss it further here. The three different systems for expressing total soil water potential and their units are shown in Table 2.2.

Example 2.4

Convert 100 kPa in total soil water potential to total soil water potential *head* in centimeters at a temperature of 25°C.

Solving Equation 2.25 for H:

$$H = \frac{\psi_t}{\rho_w g} = \frac{100 \text{ kPa}}{\left(997 \frac{\text{kg}}{\text{m}^3}\right)\left(9.81 \frac{\text{m}}{\text{s}^2}\right)} = \frac{100,000 \text{ Pa}}{9780.5 \frac{\text{N}}{\text{m}^3}} \cdot \frac{\frac{\text{N}}{\text{m}^2}}{\text{Pa}} = 10.22 \text{ m} = 1022 \text{ cm}.$$

TABLE 2.2

Alternative definitions of the total soil water potential and their units

Name	Definition	Symbol	Dimensions	SI units
Soil water potential	Energy per volume	ψ_t	$ML^{-1}T^{-2}$	$N\ m^{-2}$
Soil water potential head	Energy per weight	H	L	M
Chemical potential	Energy per mass	μ_t	L^2T^{-2}	$J\ kg^{-1}$

Source: Adapted from Jury, W.A., and Horton, R., *Soil Physics*, John Wiley, Hoboken, NJ, 2004.

Hence, a soil water potential of 100 kPa is the equivalent of 1,022 cm of soil water potential head. It is also 1000 hPa (hecta or 100 Pa), 0.1 MPa, 1 bar, and 0.987 atmospheres.

To simplify the evaluation of soil water potential, total potential is divided into the components caused by each force field acting on soil water. These are the forces caused by gravity, hydrostatic pressure, capillarity, solutes, soil air pressure, and swelling (Jury and Horton 2004). The component due to swelling soils is seldom used and we will not discuss it here.

2.7.1 GRAVITATIONAL COMPONENT

This component quantifies the effect of the gravitational force field on the energy of soil water:

gravitational potential is defined as the difference in energy per unit volume or weight between standard water and soil water due to gravity.

We will use the symbol ψ_z for gravitational potential and the letter z for gravitational potential head. This component of soil water potential is easily calculated by considering the elevation of the soil water sample. For gravitational potential, the equation is

$$\psi_z = \rho_w g \left(z_{soil} - z_0 \right),\tag{2.26}$$

where z_{soil} is the elevation of the soil water [L], and z_0 is the elevation of standard water (the reference elevation). If ρ_w [ML^{-3}] is in units of kilograms per cubic meter and g [LT^{-2}] is in units of meters per second squared then ψ_z [$ML^{-1}T^{-2}$] will be in Newtons per meter squared, or Pascals, and can be converted to kilopascals, megapascals, bars, atmospheres, etc. (see Appendix).

The equation for gravitational potential head is obtained by dividing Equation 2.26 by $\rho_w g$ and is simply:

$$z = z_{soil} - z_0.\tag{2.27}$$

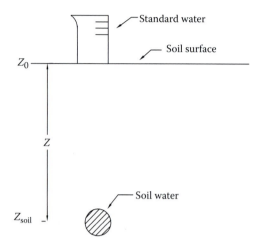

FIGURE 2.13 Elevation of standard water and soil water and their effect on gravitational potential head.

Quite often the soil surface is chosen as the reference elevation. One can think of standard water as a beaker of pure, free water placed at the soil surface in this case. This is shown in Figure 2.13. Since the energy due to gravity of the soil water is less than the energy due to gravity of the standard water when the soil water is at a lower elevation, ψ_z must be negative. If the soil surface is chosen as the reference elevation and that depth is designated a depth of zero ($z_0 = 0$), then z_{soil} must be negative in Equations 2.26 and 2.27.

Example 2.5

Calculate the gravitational potential and gravitational potential head of soil water at a depth of 50 cm below the surface. Use the soil surface as the reference elevation. Use a temperature of 25°C.

For gravitational potential, use Equation 2.26:

$$\psi_z = \rho_w g \left(z_{soil} - z_0 \right)$$

$$= \left(997 \, \frac{kg}{m^3} \right) \left(9.81 \frac{m}{s^2} \right) \left(-0.50 \, m - 0.00 \, m \right)$$

$$= \left(-4890.3 \, \frac{N}{m^2} \right) \left(\frac{1 \, Pa}{\frac{N}{m^2}} \right) \left(\frac{1 \, kPa}{10^3 \, Pa} \right)$$

$$= -4.9 \, kPa.$$

For gravitational potential head, use Equation 2.27:

$$z = z_{soil} - z_0 = -0.50 \, m - 0.00 \, m = -0.50 \, m.$$

2.7.2 HYDROSTATIC COMPONENT

This component quantifies the effect of pressure from overlying, free (hydrostatic) water on the energy of soil water. This pressure increases the energy of soil water relative to standard (free) water:

> *hydrostatic potential* is defined as the difference in energy per unit volume or weight between standard water and soil water due to the pressure exerted by overlying free water.

We will use the symbol ψ_p for hydrostatic potential. This component only occurs where the soil is saturated with water, as it would be below the *water table*.

In unconfined *aquifers* (soils or geologic material that can store and transmit water to wells), the water table is the top of the *groundwater system* and $\psi_p = 0$ at this elevation. If there is no vertical movement of water, it is the height to which water will fill a borehole, well, or pit that penetrates below the water table (Figure 2.14 and Figure 2.15). Under these conditions, it is also called the *potentiometric surface* or free water surface in hydrology textbooks (Fetter 1988). Above the water table is a zone of nearly saturated soil that is known as the *capillary fringe* (more about this in Chapter 3). Above the capillary fringe is the unsaturated or *vadose* zone where water contents are less than saturation (θ_s).

Saturated conditions can occur above the water table for short periods of time, such as might happen if infiltrating water encounters a less permeable layer. This is called a *perched water table* and ψ_p can be a component of water potential in this zone (Figure 2.15). In a *confined aquifer*, groundwater cannot rise into an overlying impermeable layer (an *aquiclude*). If boreholes or wells are drilled to penetrate the overlying layer, water will rise to the height of the free water or potentiometric surface. This may be above the soil surface and produce *artesian wells*. In these cases, the concept of a water table is not very useful, but it is still true that the height to the free water (potentiometric) surface represents the hydrostatic pressure acting on soil water at the point where the height is measured (base of a well, for example).

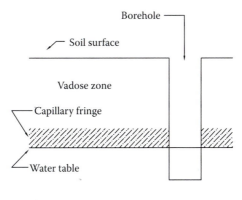

FIGURE 2.14 Vadose (unsaturated) zone, groundwater, water table, and the capillary fringe.

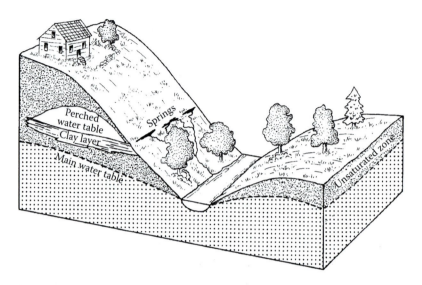

FIGURE 2.15 Unconfined aquifer with main water table (potentiometric surface) separating groundwater and the unsaturated zone in a humid region. A clay layer causes a perched water table. Wherever the water table intersects the soil surface, free water discharges as a spring or stream. (From Fetter, C.W., *Applied Hydrogeology*, Merrill, Columbus, OH, 1988. With permission.)

If there is no vertical movement of groundwater, then the hydrostatic component of soil water potential is easily calculated by considering the height of the column of water extending from the elevation of the soil water sample to the elevation of the water table. The hydrostatic potential (ψ_p) [$ML^{-1}T^{-2}$] is

$$\psi_p = \rho_w g \left(z_{\text{water table}} - z_{\text{soil}} \right), \qquad (2.28)$$

where $z_{\text{water table}}$ is the elevation of the water table (Figure 2.16). Water in the unsaturated zone has no effect on the overlying pressure since it is under tension (more about this later).

Since the energy due to the hydrostatic pressure of soil water is greater than that of standard water (free water) when the soil water is below the water table, ψ_p must be positive. If the soil surface is taken as a depth of zero, then the elevations in Equation 2.28 must be given as negative depths (z is negative in the downward direction). This will produce a positive ψ_p as long as z_{soil} is below $z_{\text{water table}}$. Hydrostatic potential can only be positive or zero; it cannot be negative in soil.

We will use the letter h for *hydrostatic potential head* or *pressure head* [L]. The equation for this component is obtained by dividing Equation 2.28 by $\rho_w g$ and is simply:

$$h = z_{\text{water table}} - z_{\text{soil}}. \qquad (2.29)$$

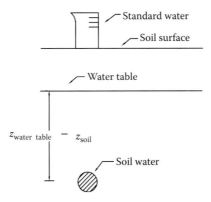

FIGURE 2.16 Hydrostatic component of soil water increases the potential energy of water below the water table compared to standard water.

If there is vertical movement of groundwater, then ψ_p or h can't be calculated using these equations, but must be measured using *piezometers* or *tensiometers*, for example (more on these later).

Example 2.6

Calculate the hydrostatic potential and hydrostatic potential head of soil water at a depth of 1 m. The water table is at a depth of 0.75 m and there is no vertical movement of water. Use a temperature of 25°C.

For hydrostatic potential, use Equation 2.28:

$$\psi_p = \rho_w g \left(z_{\text{water table}} - z_{\text{soil}} \right)$$

$$= \left(997 \frac{kg}{m^3} \right) \left(9.81 \frac{m}{s^2} \right) \left[-0.75\,m - (-1.00\,m) \right]$$

$$= 2445.1 \frac{kg \cdot m}{m^2 \cdot s^2}$$

$$= 2445.1 \frac{N}{m^2}$$

$$= 2445.1\,Pa$$

$$= 2.4\,kPa.$$

For pressure head, use Equation 2.29:

$$h = z_{\text{water table}} - z_{\text{soil}} = -0.75\,m - \left(-1.00\,m \right) = 0.25\,m$$

2.7.3 SOLUTE COMPONENT

This component quantifies the effect of solutes on the energy of soil water. As seen in Section 2.3, solutes cause increased hydrogen bonding around each dissolved ion and lower the energy of water compared to solute-free (standard) water:

> *solute potential* is defined as the difference in energy per unit volume or weight between standard water and soil water due to the presence of solutes.

We will use the symbol ψ_s for solute potential. Solute potential pressure can be calculated if the total dissolved ion concentration in soil water is known:

$$\psi_s = -RTC_s, \tag{2.30}$$

where C_s is the total dissolved ion concentration [ML^{-3}] in moles per cubic meter, R is the gas constant (8.314 Pa m^3 mol^{-1} K^{-1}), and T is the temperature [K] in kelvin (Figure 2.17).

We will use the symbol s for solute potential head and we obtain an equation for s by dividing Equation 2.30 by $\rho_w g$:

$$s = -\frac{RTC_s}{\rho_w g}. \tag{2.31}$$

Note the negative sign. Solute potential is either zero or negative because solutes lower the energy of soil water compared to standard water. C_s is the *total* ion concentration, so the moles of dissolved ions, not just the moles of a compound added to water, must be considered. See the following example.

Example 2.7

Calculate the solute potential of soil in kilopascals with a soil solution concentration of 0.010 mol L^{-1} CaCl$_2$ at a temperature of 25°C.

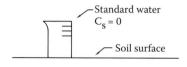

FIGURE 2.17 Solutes reduce the potential energy of soil water compared to standard water.

Convert the temperature to K:

$$T = 25°C = (273 + 25)\,K = 298\,K.$$

Calculate C_s:

$$CaCl_2 \rightarrow Ca + 2Cl$$

$$\left(0.010\frac{mol}{L}CaCl_2\right) \rightarrow \left(0.010\frac{mol}{L}Ca\right) + \left(0.020\frac{mol}{L}Cl\right)$$

$$C_s = \left(0.030\frac{mol}{L}\right)\left(\frac{1000L}{m^3}\right)$$

$$= 30\frac{mol}{m^3}.$$

Use Equation 2.30 to calculate solute potential:

$$\psi_s = -RTC_s = -\left(8.314\frac{Pa\cdot m^3}{mol\cdot K}\right)(298\,K)\left(30\frac{mol}{m^3}\right) = -74,327\,Pa = -74.4\,kPa.$$

Note in this example that a dilute concentration of solutes can lower the energy of water by about three quarters of an atmosphere (1 atm~100 kPa).

2.7.4 MATRIC COMPONENT

This component quantifies the effect of *capillarity* and *adsorption* on the energy of soil water under unsaturated conditions. The term *matric* comes from soil *matrix*, which is used to refer to the system of soil pores and solid particles, especially the finer pores. As seen at the beginning of this chapter, capillarity causes water to rise in a cylinder with a small radius. In soils, capillarity draws water into the fine pores of a dry soil, just as a dry sponge attracts water. Capillarity puts water under *tension* (a negative pressure), so it lowers the energy of soil water compared to standard (free) water. Also, water near the surface of charged particles, such as clay and organic matter, experiences a force that we will refer to here as adsorption and this also lowers the potential energy of soil water:

> *matric potential* is defined as the difference in energy per unit volume or weight between standard water and soil water due to capillarity and adsorption.

We will use the symbol ψ_m for matric potential and h for matric potential head or pressure head. Note that we refer to both hydrostatic potential head (saturated conditions) and matric potential head (unsaturated conditions) as pressure head and use the same symbol h. If h is positive, it is a hydrostatic potential head and if it is negative it is a matric potential head. In many flow problems, both saturated and unsaturated conditions occur and it is cumbersome to use different names and symbols for the pressure component of soil water potential.

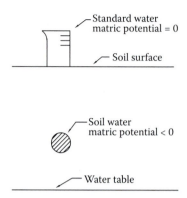

FIGURE 2.18   If the soil water content is less than saturation, then matric forces reduce the potential energy of soil water compared to standard water.

Matric potential is often the most important component of total soil water potential in an unsaturated soil (matric potential can be very negative) and changes dramatically with water content (becomes more negative as water content decreases). Matric potential only occurs in unsaturated soils (you cannot have capillarity in saturated soils where there are no menisci). It is always negative in unsaturated soils (lowers the potential energy of soil water compared to standard water) (Figure 2.18).

For short distances above a free water surface where a continuous column of water occurs and equilibrium has been reached (such as in the capillary fringe above a water table), the pressure head is simply the negative of the height of the soil water above the free water surface.

In most cases, however, one cannot assume a continuous column of water from the soil volume of interest to the water table and equilibrium (no water movement) conditions. Under these circumstances, there is no simple equation for calculating matric potential. Instead, it must be measured using a tensiometer (or one of the other devices discussed later in this chapter) or estimated from a relationship between θ and ψ_m or h (more on this in Section 2.9).

2.7.5   AIR PRESSURE COMPONENT

This component quantifies the effect of the air pressure in soil pores on the potential energy of soil water. Since air must be present, it only occurs in unsaturated soils:

air potential is defined as the difference in energy per unit volume or weight between standard water and soil water due to the effect of soil air pressure.

The air pressure of standard water is written as P_0 and it is usually taken to be atmospheric pressure ($P_0 = 1$ atm or nearly 100 kPa).

We will use the symbol ψ_a for soil air potential $[ML^{-1}T^{-2}]$. It is simply:

$$\psi_a = P_{\text{soil air}} - P_0, \qquad (2.32)$$

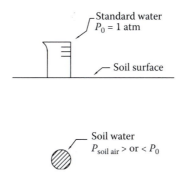

FIGURE 2.19 If soil air pressure is not equal to atmospheric pressure, then the potential energy of soil water will be affected.

where $P_{\text{soil air}}$ is the pressure of the air within soil pores in contact with the soil water. For soil air potential head, we will use the letter a [L]. It is simply Equation 2.32 divided by $\rho_w g$:

$$a = \frac{P_{\text{soil air}} - P_0}{\rho_w g}.$$

(2.33)

It may not be clear why $P_{\text{soil air}}$ might be different from atmospheric pressure, but this is often the case (Figure 2.19). The barometric pressure in the atmosphere above the soil surface is constantly changing as the weather changes ("highs" and "lows" associated with fronts that move through a region). Since the connecting pathways between soil air and the air above the soil may be tortuous or discontinuous, soil air pressure may lag atmospheric pressure as changes occur. This is best seen in large caves with a small entrance. There is usually a *cave wind* that expels or inhales air, depending on whether the pressure inside the cave is greater or less than atmospheric. If atmospheric pressure were constant, eventually the pressures would equilibrate and the wind would stop.

If soil air pressure is less than that above the soil surface (P_0), then the potential energy of soil water will be reduced compared to standard water. If soil air pressure is more than P_0, then the potential energy of soil water will be greater (Equations 2.32 and 2.33).

In practice, there is no easy way to measure $P_{\text{soil air}}$, so these equations are seldom used and the air potential component is usually ignored. However, it will be seen that this component affects measurements of soil water potential using several devices discussed in Section 2.8.

2.7.6 TOTAL SOIL WATER POTENTIAL

The energy level of soil water considering all the forces, or the total soil water potential, is the sum of all the components that are active. These will vary primarily between saturated and unsaturated conditions and are shown for total soil water potential and total soil water potential head in Table 2.3. Note that the gravitational

TABLE 2.3

Total water potential and its components under saturated and unsaturated conditions

Water content	Water potential	Water potential head
Saturated	$\psi_t = \psi_z + \psi_p + \psi_s$	$H = z + h + s$
Unsaturated	$\psi_t = \psi_z + \psi_m + \psi_s + \psi_a$	$H = z + h + s + a$

and solute components can occur under both saturated and unsaturated conditions. Hydrostatic potential (ψ_p or positive h) occurs only under saturated conditions. Matric potential (ψ_m or negative h) and air potential (ψ_a or a) occur only under unsaturated conditions.

Often the solute concentration in soil water is so low that its effect can be neglected. As mentioned earlier, there is no easy way to isolate the effect of soil air pressure, so this too is usually neglected. Under these circumstances, total soil water potential head or hydraulic head is

$$H = h + z. \tag{2.34}$$

Thus, hydrostatic pressure and gravity dominate the potential energy of water under saturated conditions (groundwater) and gravity and matric potential dominate under unsaturated conditions (vadose zone water). Gravitational potential head z may be positive or negative, depending on the elevation that is chosen as the reference (z_0 could be the soil surface or some other elevation). Pressure head h is positive under saturated conditions and negative under unsaturated conditions. Depending on the relative magnitudes of the components, hydraulic head H will be positive or negative. For saturated soils, h and z usually have similar magnitudes. For dry soils, h is so negative that it often dominates the effect of gravity (z).

For most of the other chapters in this book and especially the chapters on water flow, hydraulic head (Equation 2.34) will be used to represent total soil water potential. Thus, it will be common to express the energy status of soil in terms of centimeters or meters of water head. Soil water potential *head* is preferred for water flow because one is often dealing with fairly wet soils where small differences in energy potential can cause movement.

2.8 MEASURING SOIL WATER POTENTIAL COMPONENTS

There are a number of methods available for measuring soil water potential components. Each method has its advantages and disadvantages, so no single method is best.

2.8.1 TENSIOMETERS

Tensiometers are perhaps the most common method of measuring matric potential in the field and are described in detail by Young and Sisson (2002). A tensiometer consists of an airtight, water-filled, plastic tube with a porous ceramic or steel cup at the bottom (Figure 2.20). The pores in the cup are small enough to retain water under pressures (positive or negative) of 50–200 kPa, usually. Water moves in or out of the tube in response to the soil matric potential of unsaturated soils. A static equilibrium is reached (water stops moving in or out of the tube) when the potential energy of the water inside the tensiometer at the depth of the cup is the same as the potential energy of the soil water just outside the cup. This increases or reduces the air pressure in the head space at the top of the tube, which is measured with a vacuum gauge, manometer (water or mercury), or pressure transducer.

Strictly speaking, a tensiometer also measures the effect of air potential in an unsaturated soil. If there is a difference between soil air pressure in the soil at the cup and the pressure in the atmosphere above the soil surface, that will affect the reading on the vacuum gauge, or any of the other methods used to measure the pressure in the head space relative to atmospheric pressure. Tensiometers do not measure solute potential (ψ_s) because the openings in the porous cup are not small enough to exclude hydrated solutes (i.e., not a semipermeable membrane). Consequently, the concentration of solutes inside the tensiometer equilibrates with the concentration in the soil.

For vertically installed tensiometers, the pressure in the head space differs from the pressure at the bottom of the tube due to the weight of the column of water within the tensiometer. Therefore, in unsaturated soil, the relationship between the gauge reading (ψ_{gauge}) in units of pressure [$ML^{-1}T^{-2}$], soil matric potential, and soil air potential is

$$\psi_m + \psi_a = \psi_{gauge} + \rho_w g L, \tag{2.35}$$

where L is the length of the column of water. If the tensiometer is installed horizontally, then there is no effect on the column of water and $L = 0$. If the tensiometer is installed at an angle from vertical, then L is the difference in elevation between the cup and the gauge.

If the soil is unsaturated, ψ_m will be negative but the gauge reading may be positive, depending on ψ_a (which can be positive or negative depending on whether soil air pressure is greater than or less than atmospheric). Since there is no way to

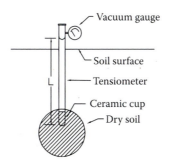

FIGURE 2.20   Tensiometer with a porous cup at the bottom and a vacuum gauge at the top. The length of the water column in the tensiometer is L.

differentiate between ψ_m and ψ_a using a tensiometer, the air pressure component is often ignored and it is assumed that:

$$\psi_m = \psi_{gauge} + \rho_w g L. \tag{2.36}$$

To get matric potential head or pressure head, Equation 2.36 is divided by $\rho_w g$:

$$h = \frac{\psi_{gauge}}{\rho_w g} + L. \tag{2.37}$$

Tensiometers have a vacuum limit of about −80 kPa. At more negative pressures, gases dissolved in water start to boil and the column of water is broken in the tensiometer.

Tensiometers can be used in saturated soils below the water table and in this case they measure hydrostatic potential. If the hydrostatic pressure in the soil water near the cup is greater than that of the water inside the tensiometer, then water will move from the soil into the cup and that will be reflected in the pressure in the head space. Since there is no air potential in saturated soils, a tensiometer measures only the hydrostatic potential in this case. Under these circumstances, Equation 2.36 applies, but the left side of the equation is ψ_p instead of ψ_m. Pressure head measured by tensiometers under saturated conditions is described by Equation 2.37.

Example 2.8

What is the total water potential head in centimeters of a soil at a depth of 30 cm, if a tensiometer shows a gauge reading of −450 hPa (hectapascals or 100 Pa)? The tensiometer is installed vertically and the water column has a length of 40 cm. Use a temperature of 25°C and the soil surface as the reference elevation.

Use Equation 2.37 to calculate pressure head:

See notes

$$h = \frac{\psi_{gauge}}{\rho_w g} + L$$

$$= \frac{-45,000\,Pa}{\left(997\,\frac{kg}{m^3}\right)\left(9.81\,\frac{m}{s^2}\right)} + 0.40\,m$$

$$= -4.6\,\frac{Pa}{\frac{kg \cdot m}{m^3 \cdot s^2}} + 0.40\,m$$

$$= -4.6\,\frac{Pa}{\frac{N}{m^3}} + 0.40\,m$$

$$= -4.60\,m + 0.40\,m$$

$$= -4.20\,m.$$

Use Equation 2.27 to calculate gravitational potential head:

$$z = z_{soil} - z_0 = -30.0 \text{ cm} = -0.30 \text{ m}.$$

Use Equation 2.34 to calculate hydraulic head or total potential head for an unsaturated soil, ignoring solute and air potential head (no information on these is given):

$$H = h + z = -4.20 \text{ m} - 0.30 \text{ m} = -4.50 \text{ m}.$$

Note that pressure head dominates total potential in this relatively dry soil.

2.8.2 PIEZOMETERS

Piezometers are used only below the water table and they measure hydrostatic potential (ψ_p) and positive pressure head (h) (Young 2002). The measurement can be easily converted to hydraulic head (H). The term piezometer comes from the Greek verb to press, *piezein*. Another word based on this derivation is *piezoelectric*, which describes the phenomenon of some materials that change electrical resistance in response to pressure (used in pressure transducers).

A piezometer is a hollow (usually plastic) tube placed in the soil, open to the atmosphere at the top and open to saturated soil at the bottom, often through a screened section. Water is under positive hydrostatic pressure below the water table and enters the piezometer through the screen. It rises in the tube until the hydrostatic pressure of the water inside the tube is the same as the hydrostatic pressure of the water in the soil just outside the screen. If there is no vertical movement of water, then the water will rise inside the piezometer until it coincides with the depth of the water table (Figure 2.21). If there *is* vertical movement of water, then the level of water in the piezometer will be above or below the water table depth (more on this at the end of this section).

The hydrostatic potential of the water at the base of the piezometer can be calculated from the water level in the piezometer:

$$\psi_p = \rho_w g h, \tag{2.38}$$

where h is height [L] of the water in the piezometer (always positive) above the bottom (Figure 2.21). Hydrostatic potential head or pressure head is simply h.

The total soil water potential in a saturated soil with negligible solutes would be the hydrostatic potential head plus gravitational head, or simply the hydraulic head ($H = h + z$), as seen before (Equation 2.34). One is free to choose the reference depth for gravitational potential (z_0), and although the soil surface is often chosen, suppose a depth below the piezometer is chosen in this case (Figure 2.21). In this case, H is just the elevation of water in the piezometer above the reference depth. That is, the elevation of water in a piezometer relative to the reference depth is the total soil water potential head. It is important to realize that this describes the potential energy of the

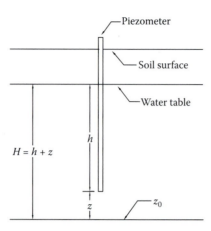

FIGURE 2.21 Total potential head of soil water at the base of a piezometer is the height of the water table above a reference depth z_0.

water at the base of the piezometer. The same thing happens if the reference depth is chosen so that it is above the bottom of the piezometer. The only difference is that total potential is negative, but for evaluating energy status, only a relative value is needed. So it does not matter.

Thus, one can determine the direction of water movement simply by measuring the elevation of water in multiple piezometers relative to a single reference depth (Figure 2.22). This is the elevation of the water table if there is no vertical movement, and the potentiometric surface in groundwater terminology. But it is important to remember that it is simply the total potential energy of the soil water. Note that the same reference elevation, z_0, must be used for all piezometers. For example, if piezometers are installed at multiple locations in a field and the soil surface is not completely flat, then the soil surface can't be used as z_0 at each piezometer. In this case, a reference depth below all piezometers (perhaps the bottom of the deepest

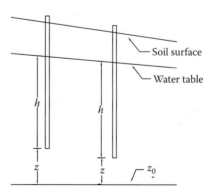

FIGURE 2.22 Multiple piezometers in a sloping field using the same reference elevation for gravitational head. Water would flow from the base of the piezometer on the left to the base of the piezometer on the right.

piezometer, simply 100 m below the lowest point in the field, or mean sea level) is often chosen so that all values of H will be positive.

With measurements from a set of piezometers in a field, contours of total potential head or the potentiometric surface can be drawn. Groundwater will move perpendicular to the contours from high to low total potential head. These types of measurements are used to determine the direction that pollutant plumes will travel and to decide where to place pumping wells that will capture the plume (Figure 2.23).

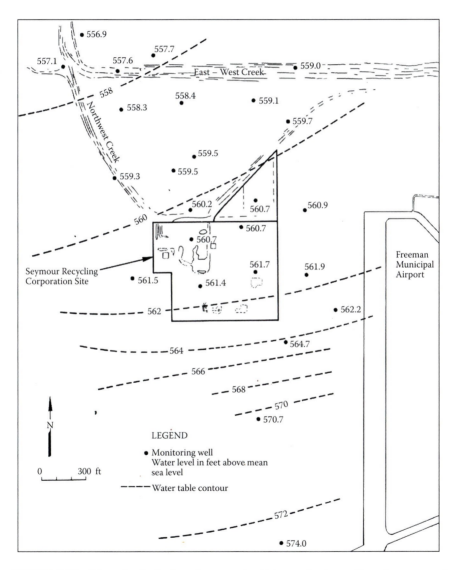

FIGURE 2.23 Water levels (in feet above sea level) in monitoring wells and contours of total potential (potentiometric surface of water table height) at a contaminated site. (From Fetter, C.W., *Applied Hydrogeology*, Merrill, Columbus, OH, 1988. With permission.)

Example 2.9

Two piezometers (A and B) are installed in a field, as shown in Figure 2.24. The depth to the water level in piezometer A is 145 cm. The depth to the water level in piezometer B is 122 cm. The soil surface at piezometer A is 31 cm higher than the surface at piezometer B. The length of both piezometers (from the soil surface to the base) is 300 cm. Is water flowing from A to B or vice versa?

Use the soil surface at piezometer A as the reference elevation for gravitation potential head z_0. At the base of piezometer A, the hydraulic head is

$$H = h + z = (300\,cm - 145\,cm) + (-300\,cm) = -145\,cm.$$

At the base of piezometer B, the hydraulic head is

$$H = h + z = (300\,cm - 122\,cm) + (-300\,cm - 31\,cm) = -153\,cm.$$

Since the hydraulic head is lower (more negative) at the base of piezometer B, water will flow from the base of piezometer A to the base of piezometer B.

Banks of piezometers (multiple installations at different depths) can be used to determine the direction of vertical groundwater flow, as well. Figure 2.25 shows a cross-section of a hill with dashed contour lines of equal total potential (Fetter 1988). The solid lines cross the contour lines at a right angle and show the path that water will follow. The water table is also shown and it follows the soil surface elevation in a *dampened* manner. Water moves vertically downward on the ridgetop, so this is an area of *groundwater recharge*. At midslope, the water moves roughly parallel to the soil surface. At the bottom of the hill, water again moves vertically but this time toward the surface, so this is an area of *groundwater discharge*. Figure 2.26 shows the water level in piezometers installed in the same landscape.

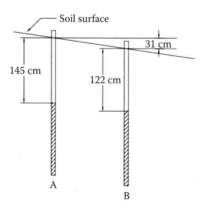

FIGURE 2.24 Two piezometers installed in a sloping field.

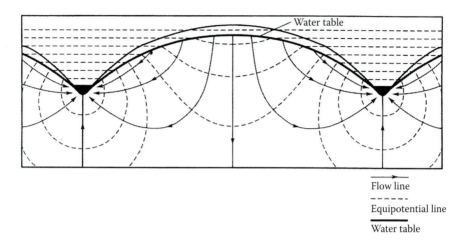

FIGURE 2.25 Lines of equal soil water potential (dashed lines) and the paths that groundwater will follow (solid lines) in an idealized landscape. (From Fetter, C.W., *Applied Hydrogeology*, Merrill, Columbus, OH, 1988. With permission.)

2.8.3 THERMOCOUPLE PSYCHROMETERS

In unsaturated soils, the potential energy of soil water is intimately related to the relative humidity of the soil air, and thermocouple psychrometers use this principle to measure soil matric potential (Andraski and Scanlon 2002). In the soil at an air–water interface, some water molecules are constantly gaining enough energy to escape to the vapor phase. Others lose energy and are captured in the liquid phase. The lower the energy level of the liquid water, the fewer water molecules that escape and the lower the relative humidity (*RH*) of the air in contact with the water. As seen in Section 2.7, capillary forces, adsorption, and solutes can affect the energy level of water, and in turn, the *RH*.

A thermocouple psychrometer for measuring soil water potential consists of a meter and soil sensor. The sensor is usually a small ceramic cup that is buried in the

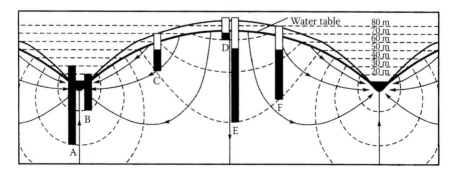

FIGURE 2.26 Elevation of water levels in piezometers installed in an idealized landscape. (From Fetter, C.W., *Applied Hydrogeology*, Merrill, Columbus, OH, 1988. With permission.)

soil and connected to the meter via a cable. The meter, cable, and sensor contain an electrical circuit known as a *thermocouple*. Thermocouple circuits are usually used to measure temperature. They are made up of dissimilar wires welded together at two or more junctions. One junction is considered the reference junction (which may be in the sensor or the meter) and remains at a constant temperature. The other junction is inside the ceramic cup cavity. If there is a difference in temperature between the two junctions, a voltage will be generated in the circuit, which can be measured. This is called the *Seebeck effect*. The opposite of the Seebeck effect is the *Peltier effect*. That is, if the circuit is closed and a voltage applied, current running through the sensor junction will cause it to be heated or cooled, depending on the direction of the current. A thermocouple psychrometer uses both the Seebeck and the Peltier effects in a two-step cycle.

In the first step, Peltier cooling is used to make a thin film of water condense on the junction by cooling the junction below the dewpoint of the soil air. In the second step, the sensor junction starts to warm back toward ambient soil temperature until a point is reached where the water film starts to evaporate. This causes the sensor junction to remain at a constant temperature (and the voltage to plateau) until all the water evaporates. The output of the psychrometer is this plateau voltage, which represents the difference in temperature between the reference junction and the sensor junction: the dryer the soil, the greater the output voltage. A calibration equation is used to convert the output voltage to water potential.

Thermocouple psychrometers measure matric potential *plus* solute potential. This is because the air–water interface at the cup acts as a semipermeable membrane and confines solutes to the liquid water. The *RH* and energy of the vapor molecules inside the cup will be reduced in proportion to the solute concentration in soil water, according to Equation 2.30. Therefore, the output from a thermocouple psychrometer is equal to the sum of soil matric and solute potentials:

$$\psi_m + \psi_s = \psi_{psychrometer}. \tag{2.39}$$

However, there is no way to separate the effect of the components using only a psychrometer. Thermocouple psychrometers are accurate in the potential range from approximately −8 MPa to −30 kPa (Andraski and Scanlon 2002).

2.8.4 HEAT DISSIPATION SENSORS

Heat dissipation sensors, described by Scanlon et al. (2002), consist of a heating element and a temperature sensing device (thermocouple) embedded in a porous ceramic sensor placed in soil. The heating element is turned on for a short interval and the change in temperature of the thermocouple is measured over time. The thermal conductivity of the porous ceramic material is a function of the water content of the ceramic. If the ceramic is wet, most of the heat dissipates into the soil through the water (water is a good conductor of heat as will be seen in Chapter 4). If the ceramic is dry, the thermocouple heats up. The sensor is connected to a data logger that controls the sensor. It goes through a cycle of heating the element and then measuring the temperature at some constant interval after heating.

When the sensor is embedded in soil, the matric potential is the same in the porous material and the soil, so the temperature of the thermocouple can be related to soil matric potential by calibration. At very negative matric potentials, the temperature will stay high. At less negative matric potentials, the temperature will drop. Each sensor must be calibrated. Heat dissipation sensors are accurate in the matric potential range from approximately −1 MPa to −10 kPa (Scanlon et al. 2002).

2.8.5   ELECTRICAL RESISTANCE SENSORS

Electrical resistance sensors are also described by Scanlon et al. (2002). These methods measure the electrical resistance of a porous block that is in contact with the soil. The most common sensor is a *gypsum block*. The block (made of gypsum) is part of a simple DC circuit; two leads are embedded in the block, but not in contact with each other. The resistance in the circuit depends on the electrical resistance of the block. When the block is in contact with wet soil, it absorbs water and the resistance between the leads is low. The block is connected via a cable to a meter that measures voltage (proportional to resistance) in the circuit and this is converted to matric potential through a calibration equation. The water content of soil and block may be different, but at equilibrium the matric potentials are the same. Electrical resistance depends on the solute concentration in the block as well as the water content. Gypsum is used because it buffers the solute concentration by dissolving slowly. As such, the sensor readings are not affected by changes in soil solute concentration, but the block dissolves over time. Some sensors have blocks made of nylon or fiberglass, but these sensors are affected by changes in soil solution concentrations. Granular matrix sensors have been developed recently, which consist of a gypsum wafer embedded in a granular matrix.

For accurate readings with this method, each block should be calibrated separately. Electrical resistance blocks work in the matric potential range from approximately −1 MPa to −10 kPa, but they are not very accurate (Scanlon et al. 2002). These sensors are probably best suited to irrigation management where a precise measure of soil water potential is not needed.

2.8.6   TENSION PLATES AND PRESSURE CHAMBERS

One of the most common methods for determining matric potential (ψ_m or h) is an indirect method. The relationship between θ and ψ_m or h is determined for a soil in the laboratory (Dane and Hopmans 2002). Then θ is measured in the field and converted to ψ_m or h using the known relationship.

This relationship is called a *soil water retention curve* and is written as $\theta(\psi_m)$ or $\theta(h)$. Many other names may be found in the literature, including soil water characteristic curve, capillary pressure-saturation relationship, and pF curve. The retention curve, historically, was often given in terms of picofarad (pF), which is defined as the negative logarithm of the absolute value of the pressure head in centimeters.

As alluded to earlier in this chapter, as water content decreases, the matric potential decreases sharply (becomes very negative). As a soil drains, the largest pores empty first and water is held in the smaller pores. The forces of capillarity are

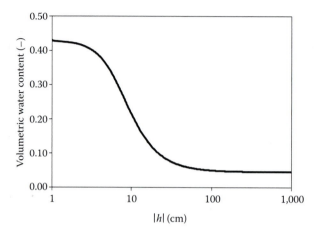

FIGURE 2.27 Typical water retention curve for a sand; volumetric water content as a function of the absolute value of pressure head in centimeters.

strongest in the small pores (see Equation 2.10). Also, as water content decreases, the water that is in an adsorbed state near charged surfaces becomes a larger percentage of the total water in soil and this contributes to the lower potential energy. The shape of the retention curve depends on the pore-size distribution and specific surface of the soil. As such, it is unique and has to be measured for each soil (a type of calibration between θ and h). A typical soil water retention curve for a sand is shown in Figure 2.27.

On the x-axis, the absolute value of pressure head (h) is plotted on a log scale, so the right side of the graph represents very dry soil and very low (negative) potential energy. Without a log scale for h, the sigmoid shape of the curve and the plateau in θ on the left side of the graph (logs stretch out a graph at low values) is not apparent. The argument of a log must always be positive, so that is the reason for plotting $|h|$ on the x-axis instead of h.

2.8.6.1   Tension Plates

The typical range for a measured retention curve is from $\psi_m = 0$ to -1500 kPa or $h = 0$ to $-15,330$ cm (the negative limit is explained in the next section). *Tension plates* can be used to get very accurate measurements in the wet range of the curve. They are not used very often, but this is a good place to start in trying to understand how a water retention curve is measured. A tension plate consists of a porous ceramic plate in contact with a *hanging water column* terminating in a reservoir open to the atmosphere (Figure 2.28). The hanging water column is created by connecting the plate and reservoir with plastic tubing. Intact soil cores contained in metal cylinders and saturated with water are placed on the plate with the reservoir water level even with the plate. The reservoir is then lowered to a distance L below the center of the core.

Water will flow from the sample to the reservoir until equilibrium in potential energy between the soil core and the reservoir is achieved. At this point, the negative

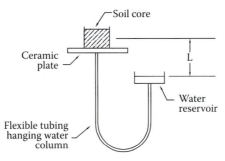

FIGURE 2.28 Tension plate apparatus for measuring soil water retention curve for matric potentials near zero.

matric potential in the core (caused by the lower θ as the sample loses water) offsets the higher gravitational potential of the sample relative to the reservoir. Suppose the reference height for gravitational potential (z_0) is set at the height of the reservoir. One can then think of the reservoir as standard water where the total water potential is zero, $\psi_{reservoir} = 0$. The total water potential in the soil core consists of a matric and gravitational component (solute concentrations inside the sample and in the water draining from the core will be the same, so there is no effect of solute potential) and at equilibrium it must be equal to $\psi_{reservoir}$, that is

$$\psi_{soil} = \psi_m + \psi_z = 0. \tag{2.40}$$

The gravitational potential can be expressed in terms of the difference in elevation between the soil and the reservoir (L) and solving for ψ_m:

$$\psi_m = -\rho_w g L. \tag{2.41}$$

To get pressure head (h), divide by $\rho_w g$ so that $h = -L$. Thus, the pressure head in the soil core after water stops flowing is simply the negative of the difference in height between the soil core and the reservoir. The height of the soil sample is measured at the center of the core as an average height, but matric potential at any point in the core varies according to the height difference. Note that the pressure head is the effective length of the hanging water column L (the left and right side of the U-shaped portion of the hanging column below the reservoir cancel out).

Once a static equilibrium is reached (water stops flowing) for a given value of L, the soil sample is removed and weighed so that the water content that corresponds to that value of h can be calculated. The sample is returned to the plate and the reservoir is lowered to a new height. This procedure is repeated to get pairs of values of θ and h in the wet range of the retention curve. Since a hanging water column cannot be maintained for lengths beyond about 100 cm, this approach is limited to the range of $h = 0$ cm to about -100 cm. In this range, very precise measurements of h can be made, since the height of the reservoir can be measured to within 0.1 cm.

2.8.6.2 Pressure Chambers

Pressure chambers are used to get the dryer portion of the retention curve. If very high accuracy is not needed (within a centimeter for h), then pressure chambers can be used to get the entire retention curve (dispensing with the tension plates) and this is the most common practice. Small pressure chambers that hold a single intact soil core are commonly called *pressure cells* (Figure 2.29). These are usually used for $h = 0$ to about -1000 or -3000 cm (or $\psi_m = 0$ to about -100 or -300 kPa).

The cells consist of top and bottom plastic housing pieces, a porous ceramic plate in the bottom housing, a metal ring to hold the soil sample, and rubber sealing rings between the housing pieces and the ring. The ring is usually 3–6 cm in height. Instead of exerting tension on the soil sample through the use of a hanging water column, a positive air pressure is applied to the cell.

The ceramic plates have small enough pores that they will allow water to pass through, but not air, at the pressures applied to the cell. The pressure beyond which a plate will no longer hold water is called the *air-entry pressure*. For example, a plate with an air-entry pressure of 100 kPa, will not allow air to enter for pressures less than 100 kPa. Porous plates with air-entry pressures slightly above the range to be applied to the cell (e.g., 100 kPa) are usually used because the soil samples equilibrate more rapidly than when plates with very high air-entry pressures (such as 1500 kPa) are used.

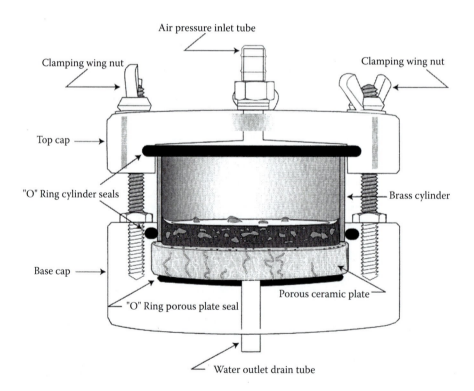

FIGURE 2.29 Cross-sectional view of a pressure cell with soil sample. (Courtesy of Soilmoisture Equipment Corp., Santa Barbara, CA.)

Compressed air is applied through the inlet tube at the top of the cell and water drains from the bottom of the cell. The applied air pressure ($\Delta P = P_a - P_0$) above atmospheric pressure can be measured with water or mercury manometers or with pressure gauges. In this case, the total soil water potential in the core consists of a gravitational, matric, and air potential (since air inside the cell is greater than atmospheric pressure, $P_a > P_0$). One can think of the water flowing out of the cell at the bottom as standard water (free of matric and hydrostatic pressures and at atmospheric pressure) with a total potential of zero. Solute concentrations inside the core and in the water draining from the bottom will be the same, so there is no effect of solute potential. As water drains from the core, the matric potential becomes more negative. Water stops draining (static equilibrium is reached) when the total potential inside the core is the same as the water leaving the core. At that point, all the components in the soil sum to zero:

$$\psi_{soil} = \psi_m + \psi_a + \psi_z = 0. \tag{2.42}$$

The difference in elevation between any point in the core and the drainage point is so small that the gravitational component, ψ_z, can be ignored, therefore:

$$\psi_{soil} = \psi_m + \psi_a = 0. \tag{2.43}$$

Solving for ψ_m and substituting ΔP for ψ_a:

$$\psi_m = -\psi_a = -\Delta P. \tag{2.44}$$

Water stops flowing when the drop in matric potential exactly offsets the increased energy due to the applied air pressure. Under these conditions, the matric potential pressure in the core at equilibrium is simply the negative of the applied air pressure above atmospheric pressure. Pressure head h is calculated in the usual way by dividing the right side of Equation 2.44 by $\rho_w g$.

The usual procedure is to start with a low pressure difference of perhaps $\Delta P = 5$ kPa. The cell is weighed regularly to determine when equilibrium is reached (the core stops losing water and the cell weight becomes constant). Once equilibrium is reached, the weight of the cell is recorded and the pressure is increased to the next value of ΔP. One of the principle advantages of pressure cells is that the soil cores do not have to be removed from the cells to get a weight and therefore the problem of reestablishing water contact between the soil core and the ceramic plate after each measurement is avoided. When all measurements for the desired range in matric potential have been made, the cell is taken apart and the dry weight of the soil and all the cell parts are measured, so that θ can be calculated for each setting of ΔP (and hence ψ_m or h).

2.8.6.3   Pressure Chambers with Disturbed Samples

For matric potential values more negative than about -100 or -300 kPa, pressure cells are usually not used. For several reasons, large pressure chambers and small disturbed samples of soil (not intact cores) are used to get a small number of values beyond this range. A wet ceramic plate with a very high air-entry pressure is placed in the bottom of the chamber. Dried, sieved soil is placed in small rings on top of the plate. Water is added to the samples in the ring and the area around the rings to establish water contact between the plate and soil samples. Then the chamber is closed and the desired air pressure is applied. The principle of operation is the same as the pressure cells and Equation 2.44 applies. There is a drainage port at the bottom of the chamber. Equilibrium is usually reached in a short time (about 3 days), since small soil samples are used.

There are several reasons for using a large chamber, small soil samples, and few pressure readings in the very negative matric potential range. The soil water retention curve changes only gradually for most soils between $\psi_m = -100$ and -1500 kPa (see Figure 2.27), so a few points are sufficient to define this portion of the curve. It will be shown that soil structure is not important in retaining water at these very negative matric potentials, so intact cores are not necessary. In fact, if large intact cores and pressure cells are used in this range of matric potential, it can take weeks for the core to come to equilibrium. If only a single measurement is made, the problem of reestablishing contact is avoided. If measurements are made at several pressures, it is a simple matter to prepare another disturbed soil sample and repeat the measurement at the new pressure.

Once all the measurements of θ have been made at different values of h or ψ_m, the data can be fit using a retention curve equation (more on this in Section 2.9.4). Then, any measurement of θ, obtained in the field using any of the methods discussed in the first part of this chapter, can be converted to matric potential. The disadvantages of this approach are that it is a time-consuming process to obtain all the pairs of values of θ and h (or ψ_m) to construct a retention curve and that the curve may not apply to each point in a field where θ is measured. The advantages of this approach are that it is reasonably accurate and one can learn a great deal about the soil, beyond the ability to convert θ to h, as will be seen in the next section.

The relationship between θ and h can also be obtained in the field by making simultaneous measurements of water content and matric potential on the same soil. For example, tensiometers can be used to measure matric potential and TDR to measure water content. This approach would be limited by the range of operation of tensiometers, 0 to about -80 kPa. When field retention curves are measured, one usually finds that θ in the field is less than in the laboratory, especially near the soil surface and this is assumed to be due to entrapped air (more on this in Chapter 3).

2.9   THE SOIL WATER RETENTION CURVE

The soil water retention curve is used for many purposes in addition to serving as an indirect method for determining matric potential.

2.9.1   TEXTURE AND STRUCTURE EFFECTS

The retention curve varies with texture in a predictable way and can be divided into three regions of matric potential (Jury and Horton 2004) (Figure 2.30). The *adsorption region* occurs at very negative matric potentials where water content is low and

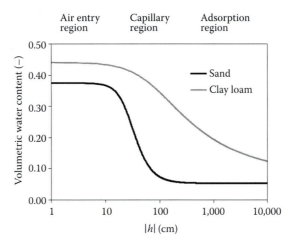

FIGURE 2.30 Typical soil water retention curves for a sand and a clay loam showing three regions: air-entry, capillary, and adsorption.

nearly constant. Water is held in films on clay surfaces and not by capillarity. This is why intact cores that retain the structure of the soil are not needed for measuring the retention curve in this region. The clay loam in Figure 2.30 has higher water content in this region than the sand due to greater specific surface.

The *capillary region* occurs in the middle range of *h*. As matric potential becomes less negative, a succession of larger pores fills as capillarity pulls water into these pores. The shape of this region of the curve reflects the *pore-size distribution*. If the curve changes significantly in a part of this region, then there are many pores in this size range. The sand curve changes over a narrow range, so it tends toward pores of a uniform size, or a narrow pore-size distribution. The clay loam changes over a wide range in *h*, so it tends toward pores of many sizes, or a broad pore-size distribution. Since capillarity dominates this portion of the curve, it is critical that intact cores that retain the soil's structure be used for determining the retention curve in this region. This also means that a retention curve is a way to quantify structure.

The *air-entry region* occurs at values of *h* near zero. In this range, θ is high and again nearly constant. Starting at the *y*-intercept of the graph, these values represent the *saturated water content* (θ_s) where there is no soil air. As matric potential decreases, there is no decrease in water content until a tension is reached that exceeds the force of capillarity in the largest pores holding water. At that point, air enters these pores. This matric potential is called the *air-entry value* (h_a) and it usually occurs between *h* = −10 to −100 cm. In Figure 2.30, the maximum pore size is approximately the same in the sand and clay loam. In a clay loam soil with a lot of structure, the maximum pore size might be larger and the air-entry value could be less than the sand.

2.9.2 PLANT AVAILABLE WATER

The shape of the soil water retention curve affects the *plant available water* (*PAW*) [$L^3 L^{-3}$] in a soil. This is the maximum amount of water held by a soil that would be available to plants and is defined as the difference between the soil *field capacity* (θ_{fc}) and *permanent wilting point* (θ_{wp}) water contents [L^3L^{-3}]:

$$PAW = \theta_{fc} - \theta_{wp}. \tag{2.45}$$

Permanent wilting point is the water content at which most agronomic plants can no longer extract water from soil and consequently reach a state of permanent wilt. This water content corresponds to the soil water content at a matric potential pressure of about −1500 kPa ($h = -15{,}330$ cm).

Field capacity, also called the *drained upper limit*, is the water content a thoroughly wetted soil reaches after free drainage becomes negligible, usually 1–2 days after wetting. This drainage process is discussed further in Chapter 5, which covers transient water flow. Romano and Santini (2002) suggest that θ_{fc} corresponds to the soil water content at a pressure head of about −100 cm for sandy soils, −350 cm for medium-textured soils, and −500 cm for clayey soils.

PAW varies with texture. It is low in coarse-textured soils, such as sands, because θ_{fc} is low since most of the water is held in large pores that drain before field capacity is reached (Figure 2.31, top graph). It is also low for fine-textured soils, such as clays, because θ_{wp} is high due to the large specific surface in these soils, which retain water at very low matric potentials (Figure 2.31, bottom graph). *PAW* is greatest for medium-textured soils, such as silts (which have intermediate-size primary particles) and loams (which are mixtures of all sizes). In these soils, θ_{fc} is high and θ_{wp} is low (Figure 2.31, middle graph).

2.9.3   HYSTERESIS IN THE SOIL WATER RETENTION CURVE

Soil water retention curves are usually developed by going from high to low water contents, often called a *moisture release curve* or *drying curve*. However, for the tension plate (Figure 2.28), one could start with the reservoir well below the sample and raise the reservoir. This would produce a *wetting curve*. Similarly, one could connect a water source to the drain tube at the bottom of a pressure cell (Figure 2.29), start with a high pressure, reduce the pressure, and let the sample imbibe water. The wetting and drying curves will not be the same. The wetting curve will often plot below the drying curve (Figure 2.32). This phenomenon is called *hysteresis* (from the Greek word to be late or fall short). If the drying process is reversed, then the water retention curve will follow a *scanning curve* from the point of reversal. Another scanning curve is followed if the wetting process is reversed.

Hysteresis is due in part to the *ink-bottle effect* caused by soil pores that do not have a uniform radius (Figure 2.33). The capillary rise law (Equation 2.10) seen earlier in this chapter showed that water will rise to a height h in a pore with radius r. This same equation can be used to determine at what pressure head (h) a pore of radius r will fill, by adding a negative sign to the right side of the equation:

$$h = -\frac{2\sigma\cos(\alpha)}{\rho_w g r}. \tag{2.46}$$

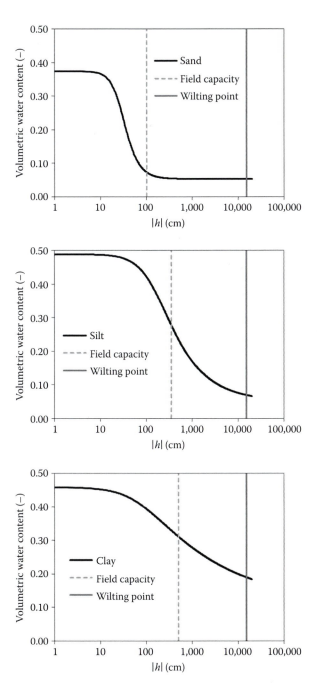

FIGURE 2.31 Typical soil water retention curves for a sand (top graph), silt (middle graph), and clay (bottom graph). The matric potential that corresponds to field capacity in each soil (and varies with texture) and the matric potential that corresponds to wilting point in each soil (the same for all soils) are shown by vertical lines. The silt has the largest *PAW*.

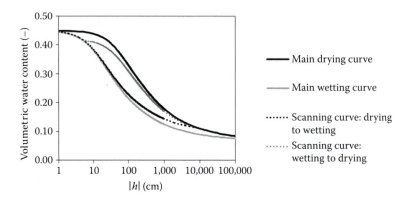

FIGURE 2.32 Main drying and wetting soil water retention curves are not the same because of hysteresis. Scanning curves show the path water retention curves will follow if the wetting or drying process is reversed.

The larger the radius, the closer pressure head has to be to zero for a pore to fill. In Figure 2.33, a pore with a narrow neck, r_1, empties on the drying curve, but cannot fill at the same tension on the wetting curve because water cannot rise past the wide part of the pore. The tension must decrease further and then water will rise past the widest part of the pore all the way to the narrow neck again. In essence, draining depends on the smallest radius and filling depends on the largest radius, so the drying and wetting curves are not the same.

Other mechanisms that contribute to hysteresis include the way water condenses inside a pore and the effect of hydrophobic surfaces during wetting (Jury and Horton 2004). In theory, one should use a drying cycle curve to convert θ to h in a soil where θ is decreasing. If the water content is increasing, then a wetting curve should be used. In practice, one usually measures only the drying retention curve and uses this to convert θ to h, and vice versa.

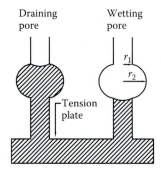

FIGURE 2.33 The ink-bottle effect causes irregularly shaped soil pores to empty and fill at different pressure heads.

2.9.4 Soil Water Retention Curve Equations

When the data for a soil water retention curve have been collected by one of the methods discussed above, there are pairs of values for θ and h (or ψ_m) such as in Table 2.4. These data were measured on six horizons in a Cecil soil in Georgia (Bruce et al. 1983). To use this tabular data to convert measured values of θ to h, one would be forced to interpolate for any values of θ that fell between the tabular data. Since the data are not linear, even when plotted on a log scale for pressure head, this might not give very accurate values between the measured points. Also, there is probably some experimental error in the measurements, so the objective is to fit a smooth curve to the data that will be a "best fit," but will not necessarily go through every point. One of the main reasons for using soil water retention curve equations is that numerical models, such as HYDRUS, work best with smooth functions of soil hydraulic properties.

A number of equations have been developed to describe the soil water retention curve (Leij et al. 1997; Kosugi et al. 2002). In this chapter, we will give only those that are widely used or have some special features that are useful for particular applications.

One of the most widely used water retention functions is that developed by van Genuchten (1980a):

$$S_e(h) = \frac{1}{\left[1 + \left(-\alpha h\right)^n\right]^m},$$ (2.47)

TABLE 2.4

Soil water retention curve data for a Cecil loamy sand

Horizon	Ap	BA	Bt1	Bt2	BC	C
Pressure head (cm)	Volumetric water content (cm³ cm⁻³)					
−5	0.376	0.324	0.431	0.425	0.424	0.446
−10	0.371	0.316	0.425	0.424	0.424	0.445
−20	0.338	0.289	0.414	0.422	0.423	0.442
−30	0.302	0.275	0.406	0.420	0.420	0.437
−40	0.273	0.267	0.401	0.418	0.417	0.432
−50	0.250	0.262	0.396	0.416	0.415	0.427
−100	0.187	0.246	0.383	0.404	0.404	0.409
−200	0.153	0.234	0.371	0.390	0.388	0.385
−300	0.143	0.217	0.351	0.392	0.382	0.344
−500	0.124	0.210	0.343	0.381	0.367	0.318
−1,000	0.107	0.202	0.332	0.365	0.347	0.284
−2,044	0.068	0.174	0.310	0.333	0.315	0.250
−15,330	0.043	0.172	0.255	0.283	0.236	0.156

Source: Plot 4 from Bruce, R.R., et al., *Physical Characteristics of Soils in the Southern Region: Cecil,* Georgia Agricultural Experiment Stations, Athens, GA.

where $S_e(h)$ is the effective soil water saturation given in Equation 2.13 and α [L^{-1}], n [–], and m [–] are fitting parameters. Effective saturation has been written here as $S_e(h)$ instead of S_e to emphasize that it is a function of h. Substituting Equation 2.13 and solving for $\theta(h)$, the equation is

[handwritten annotations: "saturation" pointing to θ_s, "residual" pointing to θ_r]

$$\theta(h) = \frac{\theta_s - \theta_r}{\left[1 + (-\alpha h)^n\right]^m} + \theta_r. \tag{2.48}$$

The "best fit" of this equation to the soil water retention curve data in Table 2.4 is obtained by optimizing the values of the *parameters* in this equation. For reasons to be seen in Chapter 3, one usually assumes that there is a fixed relationship between m and n:

$$m = 1 - \frac{1}{n}. \tag{2.49}$$

That leaves four parameters to optimize: θ_s, θ_r, α, and n. Since Equation 2.48 is not a linear equation (h is raised to a power), a nonlinear optimization method is required.

The HYDRUS suite of software includes the free RETC program, which is designed to fit a number of different analytical functions to soil water retention data, including Equation 2.48 (van Genuchten et al. 1991). It does this in an iterative manner, starting with initial estimates of the parameters, and seeking new values that will minimize the least squares objective function:

$$O(b) = \sum_{i=1}^{N} \left\{ w_i \left[\theta_i - \hat{\theta}_i (b) \right] \right\}^2, \tag{2.50}$$

where b is the set of parameter estimates (a vector), θ_i and $\hat{\theta}_i$ are the observed and fitted water contents, respectively, and N is the number of retention data points. The weighting coefficients, w_i, may be used to assign more or less weight to a single data point, depending on the level of confidence in the data point. The RETC fit of the van Genuchten (1980a) equation to the Ap horizon water retention curve data in Table 2.4 is shown in Figure 2.34. It is clear that the curve does not pass through each point, but it is a good fit to the data.

The optimum values for the parameters found using RETC are shown in Table 2.5 for the Ap and other horizons from Table 2.4. The fitted value for saturated water contents, θ_s, is lowest in the sandy BA horizon. This is due to the higher bulk density typically found in sand compared to clay, which results in a lower porosity and saturated water content (see Equation 1.9). The Ap horizon is sandy, but the bulk density is not high, probably due to high organic matter content and cultivation of the topsoil, so θ_s is intermediate. The fitted values of the residual water content, θ_r, are lowest in the clay horizons, which might seem surprising, given that clay soils

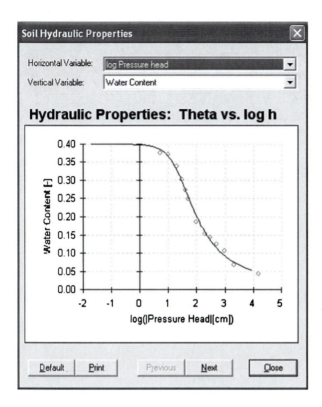

FIGURE 2.34 Output from the RETC program showing the fit of the van Genuchten (1980a) equation to the water retention data for the Ap horizon from Table 2.4.

tend to have high permanent wilting point water contents, θ_{wp}, due to the high specific surface of clays (see Section 2.9.1). The residual water content is clearly not equivalent to θ_{wp} and is best thought of as a fitting parameter (as mentioned earlier in discussing Equation 2.13). For soils such as clays that have a very gradual slope to the retention curve that does not level out at very negative matric potential values, it can be quite low (or zero). The parameter α is related to the air-entry potential, h_a, which is the matric potential where air first enters (θ first decreases). Lower values of α indicate that the air-entry region is broad, as shown in Figure 2.30 for the clay loam. As one might expect, the clayey lower horizons in Table 2.5 have small values of α. The parameter n affects the steepness of the curve. Large values of n result in a steeper curve, such as that shown for the sand in Figure 2.30. The sandy upper horizons in Table 2.5 have higher values of n.

An important function derived from soil water retention equations is the *hydraulic capacity function*, $C_w(h)$, which is the slope of the retention curve or the derivative of θ with respect to h:

$$C_w(h) = \frac{d\theta}{dh}.$$ (2.51)

Soil properties (texture, depth, ρ_b, and ρ_s) and fitted parameters for the van Genuchten equation (θ_r, θ_s, α, and n) for the data in Table 2.4 found using the RETC program

Horizon	Texture	Depth (cm)	ρ_b (g cm⁻³)	ρ_s (g cm⁻³)	θ_r	θ_s	α (cm⁻¹)	n
Ap	Loamy sand	0–21	1.45	2.64	0.032	0.399	0.0495	1.46
BA	Clay loam	21–26	1.69	2.68	0.137	0.346	0.1396	1.25
Bt1	Clay	26–102	1.44	2.72	0.000	0.433	0.0382	1.08
Bt2	Clay	102–131	1.56	2.73	0.000	0.424	0.0058	1.09
BC	Clay loam	131–160	1.53	2.75	0.000	0.423	0.0042	1.14
C	Sandy clay loam	160–250+	1.44	2.73	0.000	0.449	0.0078	1.22

The capacity function is used in the equations that describe nonsteady water movement and these will be seen in Chapter 5. For the moment, this function will be examined. Taking the derivative of the van Genuchten (1980a) soil water retention function (Equation 2.48), one obtains (see derivation in Section 2.12.2):

$$C_w(h) = \frac{\alpha^n \left(\theta_s - \theta_r\right) mn \left(-h\right)^{n-1}}{\left[1 + \left(-\alpha h\right)^n\right]^{m+1}}. \tag{2.52}$$

The RETC program also calculates $C_w(h)$ and it is shown in Figure 2.35 for the soil water retention curve from Figure 2.34 for the Ap horizon data. $C_w(h)$ is near zero in the high and low range of h where the retention curve is nearly level. It reaches a maximum in the capillary region of the retention curve where the slope is steepest.

The hydraulic capacity function can also be used to look at *pore-size distributions*. As shown earlier, the capillary rise law gives the size of a pore that will fill at a given matric potential (or the largest water-filled pore at a given value of h) (Equation 2.46). Using this equation, one can convert values of h on the x-axis of Figure 2.35 to values of soil pore radius r. Since there is an inverse relationship between h and r in Equation 2.46, this flips the capacity curve. This is now a *frequency distribution function* for pore size (or a histogram), where the y-axis is the relative volume of pores for *each value of r* (Figure 2.36). It is clear that most of the pores have a radius slightly larger than 0.01 cm. There are relatively few small pores and large pores (but the large pores may have a very important effect on water and solute movement, as will be seen in Chapters 3, 5, and 6).

The most popular equation for fitting retention curve data, historically, is that developed by Brooks and Corey (1964). It is also given in terms of effective saturation and there are two equations, one for the air-entry region of the curve and the other for the rest of the retention curve:

$$S_e(h) = \frac{\theta - \theta_r}{\theta_s - \theta_r} = 1 \qquad 0 > h \geq h_a$$

$$S_e(h) = \frac{\theta - \theta_r}{\theta_s - \theta_r} = \left(\frac{h_a}{h}\right)^\lambda \qquad h < h_a,$$

$$(2.53)$$

where h_a is the air-entry value and λ is the pore-size distribution index. This equation has four fitting parameters: θ_s, θ_r, h_a, and λ. The value of h_a shifts the retention curve left and right (analogous to α in Equation 2.48) and λ controls the steepness of the curve (analogous to n in Equation 2.48). The Brooks and Corey curve fitted to the Ap horizon data using RETC is shown in Figure 2.37. Note the difference near the air-entry point between the fit of the van Genuchten (1980a) equation (Figure 2.34) and the Brooks and Corey (1964) equation.

Another water retention curve equation is the Haverkamp equation (Haverkamp et al. 1977):

$$S_e(h) = \frac{1}{1 + (\alpha h)^\beta}, \qquad (2.54)$$

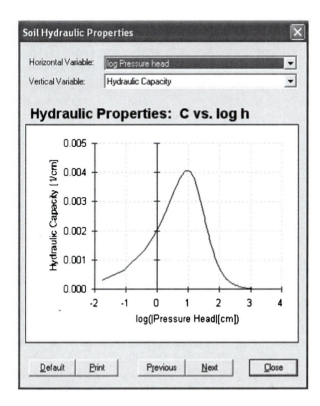

FIGURE 2.35 Output from the RETC program showing the capacity function for the water retention data for the Ap horizon from Table 2.4.

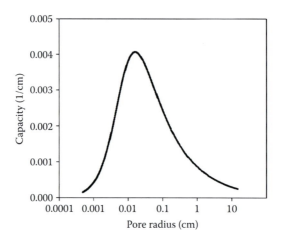

FIGURE 2.36   The capacity function from Figure 2.35 with the *x*-axis converted from the absolute value of pressure head to pore size using the capillary rise law. It is a frequency distribution of pore size.

where the fitting parameters are θ_s, θ_r, α, and β. It is quite similar to the van Genuchten equation 2.47, but simpler.

A fourth equation for fitting retention data is that developed by Durner (1994), who divided soil pores into two (or more) overlapping regions and suggested using a van Genuchten (1980a) type equation for each of these regions. Linear supposition of the functions for each particular region results in the composite equation:

$$S_e(h) = \frac{\theta - \theta_r}{\theta_s - \theta_r} = \frac{w_1}{\left[1 + |\alpha_1 h|^{n_1}\right]^{m_1}} + \frac{w_2}{\left[1 + |\alpha_2 h|^{n_2}\right]^{m_2}} + \cdots$$

$$= w_1 S_{e_1} + w_2 S_{e_2} \ldots,$$

(2.55)

where w_i are the weighting factors for the overlapping regions and α_i, n_i, and m_i are fitting parameters for the separate hydraulic functions, S_{e_i} ($i = 1,2\ldots N$). When $N = 2$, this is known as a *dual-porosity* equation. It is commonly assumed that Equation 2.49 applies, so there are seven fitting parameters with this equation: θ_s, θ_r, α_1, α_2, n_1, n_2, and w_2 (since the weights sum to one, $w_1 = 1 - w_2$ for the dual-porosity equation). This equation is particularly suitable for structured soils with macropores.

When the porosity of a soil is substantially higher that the water content at the pressure head closest to zero on the retention curve (one cannot measure retention at pressures of zero), that is evidence that macropores are present. For example, using the values for ρ_b and ρ_s in Table 2.5, one can calculate the porosity (φ) for each horizon using Equation 1.9. For the Ap horizon, $\varphi = 0.451$ cm³ cm⁻³, which is considerably higher than the first water content in the retention curve data at a pressure head of 5 cm, $\theta = 0.376$ cm³ cm⁻³ (Table 2.4). It is also higher than the fitted value for $\theta_s = 0.399$ found by RETC when fitting the van Genuchten (1980a) equation (Table 2.5). One can use RETC to fit the

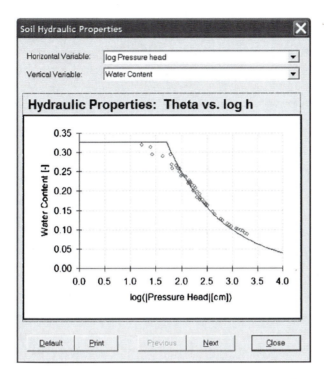

FIGURE 2.37 Output from the RETC program showing the fit of the Brooks and Corey (1964) equation to the water retention data for the Ap horizon from Table 2.4.

dual-porosity equation to the Ap horizon data by setting $\theta_s = \varphi = 0.451$ cm^3 cm^{-3}, $\alpha_2 = 3$ cm^{-1}, $n_2 = 6$, and fitting θ_r, α_1, n_1, and w_2. The fitted dual-porosity equation for the Ap horizon data is shown in Figure 2.38. The macropore effect is clearly present near saturation in this curve (compare to Figure 2.34). The values for α_2 and n_2 were chosen to describe the shape of the curve in the macropore region and, since there are no measurements within this region, the values are arbitrary.

All these retention curve equations can be modified to reflect the effect of hysteresis. For example, with the van Genuchten (1980a) equation 2.48, it is common to assume that the main wetting and drying curves have the same parameters, except for α and that the wetting curve α is twice the drying curve α. Equations for main wetting and drying curves and scanning curves are given by Scott et al. (1983), Kool and Parker (1987), Lenhard et al. (1991), and Lenhard and Parker (1992).

Example 2.10

Use Microsoft Excel to make a bar graph of *PAW* for each horizon of the Cecil soil, using the data in Table 2.4. Use the water content for $h = -15,300$ cm as θ_{wp} for each horizon. For θ_{fc}, use the water content at a pressure head of -100 cm for the Ap (loamy sand) horizon; -300 cm for the BA (sandy clay), BC (clay loam), and C (sandy clay loam) horizons; and -500 for the Bt1 (clay) and Bt2 (clay) horizons.

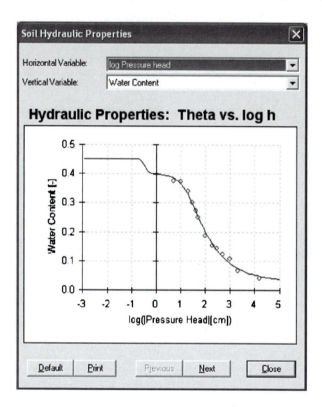

FIGURE 2.38 Output from the RETC program showing the fit of the dual-porosity equation to the water retention data for the Ap horizon from Table 2.4.

What is the total *PAW* in centimeters in the profile to a depth of 160 cm (most annual crops would not have roots below this depth)?

Enter the data as shown in the first three columns of Figure 2.39 and calculate *PAW* using a formula as the difference between θ_{fc} and θ_{wp} in column D. Select the data in column A including the first row and then hold down the Ctrl key and select the data in column D. Then click on the Insert tab on the toolbar. In the Chart section, select Column and then one of the 2-D column formats from the pull-down menu to produce the graph in Figure 2.39.

To calculate the total *PAW* in centimeters for the profile, enter the top and bottom depths for each horizon from Table 2.5 in columns E and F and calculate the thickness using a formula in column G. Then, based on Equation 2.15 for expressing a water content as an equivalent depth, multiply the *PAW* in column D by the thickness in column G and place this in column H. Lastly, calculate the sum of the entries in column H (see cell H8 and the Formula Bar in Figure 2.39).

From the spreadsheet, it can be seen that there are about 17 cm of *PAW* in the Cecil profile to a depth of 160 cm. During the summer in Georgia, plant use of water can approach 1 cm d^{-1}, so there is a little more than a 2-week supply of water in the soil profile. The medium-textured (loam) horizons have the greatest *PAW* and the sand and clay horizons have the least, except for the sandy Ap horizon. The high *PAW* in the Ap may be due to a higher organic matter content in the topsoil.

	A	B	C	D	E	F	G	H
1	Horizon	Theta-fc	Theta-wp	PAW	Top	Bottom	Thickness, cm	PAW, cm
2	Ap	0.187	0.043	0.144	0	21	21	3.024
3	BA	0.217	0.172	0.045	21	26	5	0.225
4	Bt1	0.343	0.255	0.088	26	102	76	6.688
5	Bt2	0.381	0.283	0.098	102	131	29	2.842
6	BC	0.382	0.236	0.146	131	160	29	4.234
7	C	0.344	0.156	0.188				
8							Sum	17.013

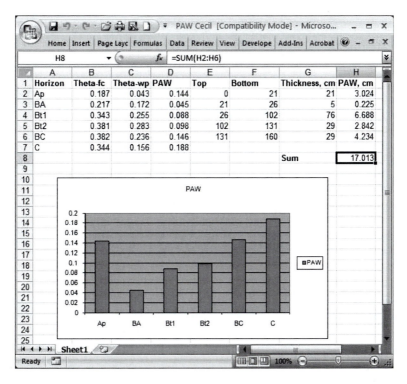

FIGURE 2.39 Excel spreadsheet calculation of *PAW* in each horizon of the Cecil soil using the water retention data from Table 2.4.

2.10 RETC PROGRAM

We will demonstrate here how to use RETC to fit the van Genuchten (1980a) equation to the Ap horizon data from Table 2.4 (RETC Simulation 2.1 – Cecil Ap Horizon).

Once the software is installed, open the program and click on the "New Project" icon at the top left of the toolbar. This will bring up the window in Figure 2.40, where a name and short description can be given for a new project.

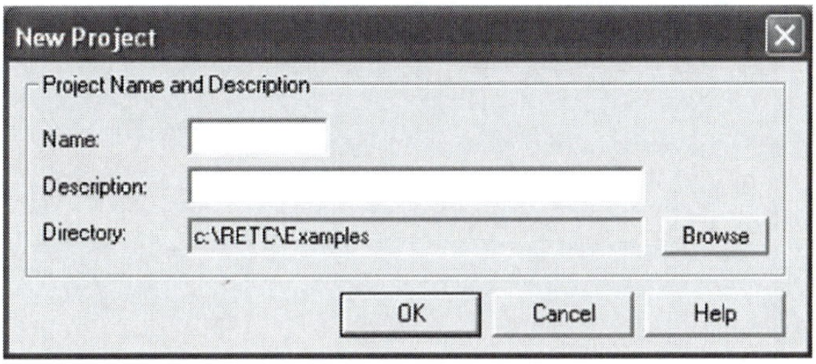

FIGURE 2.40 New Project window records the name and an optional short description for a new project and shows the directory where the project will be stored. This directory can be changed using the Browse button.

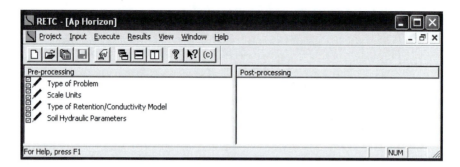

FIGURE 2.41 The main window showing Pre-processing (left) and Post-processing (right) panels.

Once this information is entered, the main RETC window will appear (Figure 2.41) through which you can access the four windows used to set up a program (left panel, "Pre-processing"). After the program has been run, output will appear in the right ("Post-processing") panel.

The easiest way to set up a new project is to go step-by-step through the Pre-processing windows starting at the top. Double-click on "Type of Problem" and you will see the window in Figure 2.42. RETC can be used to fit hydraulic conductivity data (discussed in Chapter 3) as shown by some of the options in this window. Choose the "Retention data only" option for "Type of Fitting" and then click on "Next".

Then, the second dialogue window from the Pre-processing panel will appear where the units for length and time are specified and the maximum number of iterations is set (Figure 2.43). Use the default settings for length (centimeters), time (days), and iterations (50). Time units are irrelevant in this example since hydraulic conductivity data are not being fit. Click on "Next".

This will bring up the "Type of Retention/Conductivity Model" dialogue (Figure 2.44), where the type of retention curve equation to be fitted to the data is identified.

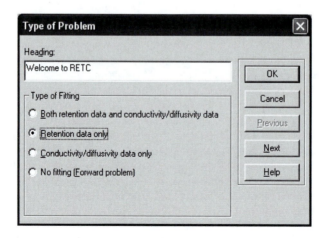

FIGURE 2.42 Type of Problem dialogue window.

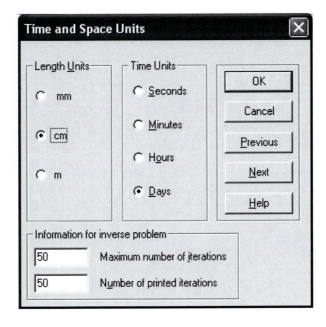

FIGURE 2.43 Time and Space Units window. Use the default settings.

Choose the default "van Genuchten, $m = 1-1/n$" model and change "Number of retention data points" to 13 (there are 13 pairs of θ and h for the Ap horizon in Table 2.4).

In the "Water Flow Parameters" window (Figure 2.45), the parameters to be fitted are identified and initial estimates for the parameter values are provided. From left to right, these are θ_r, θ_s, α, n, and saturated hydraulic conductivity (K_s, which will be

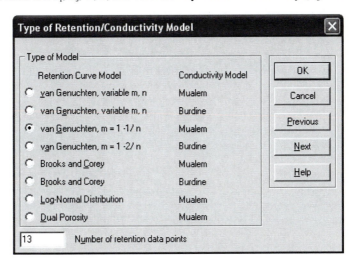

FIGURE 2.44 Type of Retention/Conductivity Model window. Choose the "van Genuchten, $m = 1 - 1/n$" model and enter the number of retention θ and h data pairs.

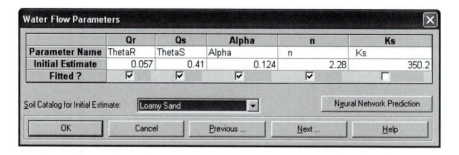

FIGURE 2.45 Water Flow Parameters window. Turn on the check marks below ThetaR, ThetaS, Alpha, and n. Turn off the check mark below K_s. Choose Loamy Sand from the Soil Catalog to get good initial estimates for θ_r, θ_s, α, and n.

discussed in Chapter 3). The check box below each parameter value is used to indicate which parameters are to be fitted. Indicate that θ_s, θ_r, α, and n will be fitted by clicking on these boxes. Saturated hydraulic conductivity will not be fitted, so click on the check mark to toggle it off. To get good initial estimates of the parameters (it is important for the iteration process to start with good initial values), click on the down arrow to the right of the "Soil Catalog for Initial Estimates" entry and select "Loamy Sand", since this is the texture of the Ap horizon (Table 2.5).

In the "Retention Curve Data" window (Figure 2.46), enter the pressures heads (as positive values) and water contents from Table 2.4 for the Ap horizon. If you have these in a spreadsheet, you can use "Ctrl C" to copy and "Ctrl V" to paste each column of data. Also enter a weight of 1 for each entry. Click "Next".

Click "OK" in the next window when you are asked if you want to save the input data and "OK" again when asked if you want to run RETC. Then, you should see a black window that says "Press Enter to continue". This indicates that the program

Retention Curve Data

	Pressure	Theta	Weight
1	5	0.376	1
2	10	0.371	1
3	20	0.338	1
4	30	0.302	1
5	40	0.273	1
6	50	0.25	1
7	100	0.187	1
8	200	0.153	1
9	300	0.143	1
10	500	0.124	1
11	1000	0.107	1
12	2044	0.068	1
13	15330	0.043	1

OK
Cancel
Previous ...
Next ...
Add Line
Delete Line
Help ...

FIGURE 2.46 Retention Curve Data window. Enter the water retention curve data for the Ap horizon from Table 2.4.

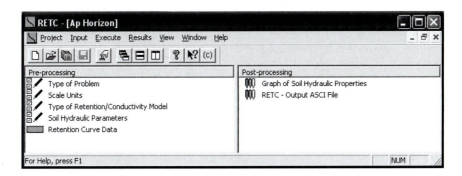

FIGURE 2.47 After the program has run successfully, you should see two new entries in the right panel.

has run successfully. Press "Enter" and you should see the main RETC window with two new entries in the "Post-processing" right panel (Figure 2.47).

Double-click on the "Graph of Soil Hydraulic Properties" and you should see a graph of the water retention curve data (circles) and the fitted water retention equation (black line) (Figure 2.48). This looks different from the earlier graph of the data

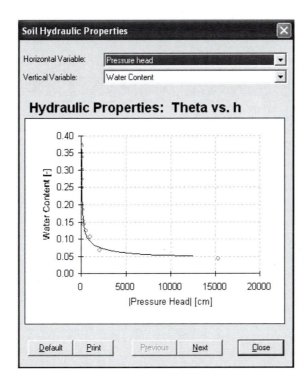

FIGURE 2.48 The water retention curve data and fitted van Genucthen (1980a) retention curve for the Ap horizon data from Table 2.4.

(Figure 2.34) because a log scale has not been used on the *x*-axis for pressure head. If you click on the down arrow to the right of "Pressure head" at the top of the window (List box "Horizontal Variable") and select "log Pressure head", you will get the same graph as in Figure 2.34. To find the values of the fitted parameters, close this window and double-click on the second entry in the right panel of the main RETC window, "RETC – Output ASCI File".

You will see the "RETC – Output File" window. Scroll down until you see the display shown in Figure 2.49. Near the middle of the figure, you can see R^2 for the fit of the van Genuchten (1980a) equation to the data, which is quite good (0.99). In the table at the bottom of the figure, you can see the values of the parameters that produced the best fit to the data. You can also see 95% upper and lower confidence limits for the parameters. If these limits are narrow, then the fit was sensitive to this parameter and other values will not produce as good a fit. At the top of the figure, you can see a correlation matrix that shows all the possible correlations between the parameters. Ideally, the parameters would not be highly correlated (positively or negatively) because this would indicate that each parameter acted independently and was not influenced by the value of other parameters. In this case, there are relatively high correlations between θ_r and *n*, between θ_s and α, and between α and *n*. This indicates that there are probably other combinations of parameter values that will produce a good fit. It is for this reason that good initial estimates of the parameter values (ones that are typical for the soil texture, structure, or some other *a priori* information) are needed.

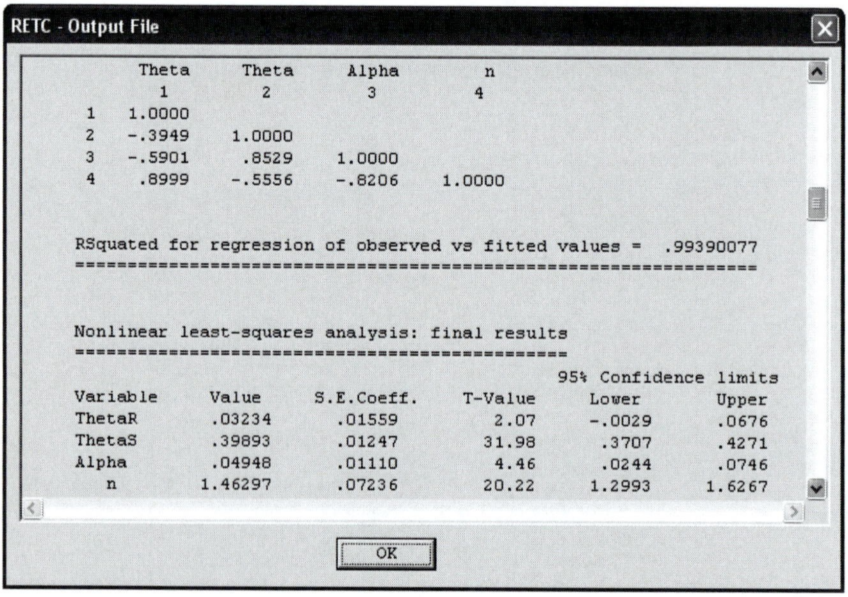

```
RETC - Output File                                                          ☒

        Theta      Theta      Alpha        n
          1          2          3          4
     1   1.0000
     2   -.3949    1.0000
     3   -.5901     .8529    1.0000
     4    .8999    -.5556    -.8206     1.0000


     RSquated for regression of observed vs fitted values =  .99390077
     ================================================================


     Nonlinear least-squares analysis: final results
     =================================================
                                                    95% Confidence limits
     Variable     Value     S.E.Coeff.    T-Value    Lower      Upper
     ThetaR       .03234      .01559        2.07     -.0029      .0676
     ThetaS       .39893      .01247       31.98      .3707      .4271
     Alpha        .04948      .01110        4.46      .0244      .0746
        n        1.46297      .07236       20.22     1.2993     1.6267

                        ┌─────── OK ───────┐
```

FIGURE 2.49   RETC – Output File window showing the correlation matrix for the fitted variables, R^2 values of fitted variables, and the 95% confidence limits.

2.11 SUMMARY

In this chapter, the basic properties of water have been presented and it has been shown how water's unique molecular structure leads to macroscale behavior in soil, such as capillarity. Soil water potential is a useful way to quantify the moisture status of soils. It consists of different components and there are methods for measuring soil water potential. Soil water content and soil water potential are related through a water retention curve. The retention curve provides very useful information about a soil, including an estimate of the *PAW* and the pore-size distribution. The three soil characteristics described in Chapter 1 (texture, structure, and mineralogy) affect soil water content and potential.

Using the concepts of water content and potential, water movement under steady flow conditions (Chapter 3) and transient flow conditions (Chapter 5) can be predicted.

2.12 DERIVATIONS

In the following sections, the capillary rise law (Equation 2.10) and the hydraulic capacity function (Equation 2.52) for the van Genuchten (1980a) soil water retention equation are derived.

2.12.1 CAPILLARY RISE LAW

The decrease in pressure ($\Delta P\!\downarrow$) within the cylinder in Figure 2.9 (causing water to rise) is described by Equation 2.9. An equation is sought that does not involve the radius of the meniscus (R), which is generally unknown. Through trigonometry, it can be shown that the relationship between the radius of the cylinder (r), the radius of the meniscus (R), and the contact angle α in Figure 2.9 is

$$\cos(\alpha) = \frac{r}{R}.$$

Substituting for R in Equation 2.9:

$$\Delta P\!\downarrow = \frac{2\sigma\cos(\alpha)}{r}.$$

The increase in pressure at the entrance to the cylinder is due to the weight of the column of water acting over the cross-sectional area of the cylinder:

$$\Delta P\!\uparrow = \frac{mg}{\pi r^2} = \frac{(\rho_w \pi r^2 h)g}{\pi r^2} = \rho_w gh,$$

where the weight of the water is equal to the mass (m) times the acceleration due to gravity (g). In the second step, a substitution has been made for m using the density of water (ρ_w) times the volume of the column of water, which is the cross-sectional area times the height of the column of water (h).

Water stops rising in the cylinder when the decrease in pressure caused by capillarity is exactly offset by the increase in pressure due to the lengthening column of water:

$$\Delta P \uparrow = \Delta P \downarrow$$

$$\rho_w g h = \frac{2\sigma \cos(\alpha)}{r}$$

$$h = \frac{2\sigma \cos(\alpha)}{\rho_w g r}.$$

2.12.2 HYDRAULIC CAPACITY FUNCTION FOR VAN GENUCHTEN EQUATION

Taking the derivative of the van Genuchten (1980a) soil water retention function (Equation 2.48) with respect to h:

$$C_w(h) = \frac{d\theta}{dh} = (\theta_s - \theta_r)\frac{d}{dh}[1 + (-\alpha h)^n]^{-m}$$

$$= (\theta_s - \theta_r)\frac{-m}{[1 + (-\alpha h)^n]^{m+1}}\frac{d}{dh}[1 + (-\alpha h)^n]$$

$$= (\theta_s - \theta_r)\frac{-m}{[1 + (-\alpha h)^n]^{m+1}}(-1)\alpha^n n(-h)^{n-1}$$

$$= \frac{\alpha^n(\theta_s - \theta_r)mn(-h)^{n-1}}{[1 + (-\alpha h)^n]^{m+1}}.$$

2.13 PROBLEMS

1. Derive Equation 2.12 using the definitions for θ, ρ_b, and ρ_w.
2. If the water contents measured with two sets of vertically installed TDR waveguides (one 15 cm long and the other 30 cm long) are 0.15 cm^3 cm^{-3} in the top 15 cm of soil and 0.24 cm^3 cm^{-3} in the top 30 cm of soil, what is the water content of the 15–30 cm depth increment? You can assume that the 0–30 cm depth reading is an average of the 0–15 and 15–30 cm water content.
3. See the last equation in Example 2.3. If you had longer rods, what would be the equation for the water content of the 45–60 cm depth (θ_{45-60})?
4. Soil swelling is caused by the osmotic pressures that are created as water is attracted to exchangeable ions on surfaces of 2:1 type clays. What acts as a semipermeable membrane (confines the hydrated ions) in this system?

5. The relative permittivity for air is nearly one. What does that imply about the velocity of light in air? See Equation 2.18.

6. What is the matric potential in kilopascals at the cup of a 0.30 m long tensiometer buried to a depth of 0.20 m in a soil, if the gauge reading is −160 hPa? What is the total potential in kilopascals using the soil surface as the reference elevation for gravitational potential? Ignore solute and soil air pressure effects. Use a temperature of 25°C.

7. What is the total potential in meters of water for a soil at a depth of 4.39 m with a water table at a depth of 2.53 m and a total ion concentration of 0.005 mol L^{-1}? Use the soil surface as the reference elevation for gravitational potential. The temperature is 25°C.

8. Piezometers are installed at a location to two depths to determine the vertical direction of water flow. The shallow piezometer extends to a depth of 100 cm and the water level in the piezometer is 92 cm below the soil surface. The deep piezometer extends to a depth of 300 cm and the water level in the piezometer is 79 cm below the soil surface. What is the total potential head in centimeters (use the soil surface as the reference elevation and ignore solute potential)? What is the direction of water flow?

9. What is the *PAW* in centimeters in the top 50 cm of a sandy soil that has a soil water characteristic curve that can be described by the Brooks and Corey (1964) equation where $\lambda = 2$, $h_a = -50.4$ cm, $\theta_s = 0.43$ cm^3 cm^{-3}, and $\theta_r = 0.05$ cm^3 cm^{-3}?

10. Use Microsoft Excel to calculate the water contents for pressure heads from 0 to 1000 cm in 10 cm increments for a loam soil with van Genuchten (1980a) parameters shown in Figure 2.50. Make a graph of the water retention curve using a log scale for the *x*-axis (pressure head). Click on Layout under Chart Tools on the toolbar, select an axis in the Current Selection box, and use Format Selection to get options for changing axes, including using a log scale.

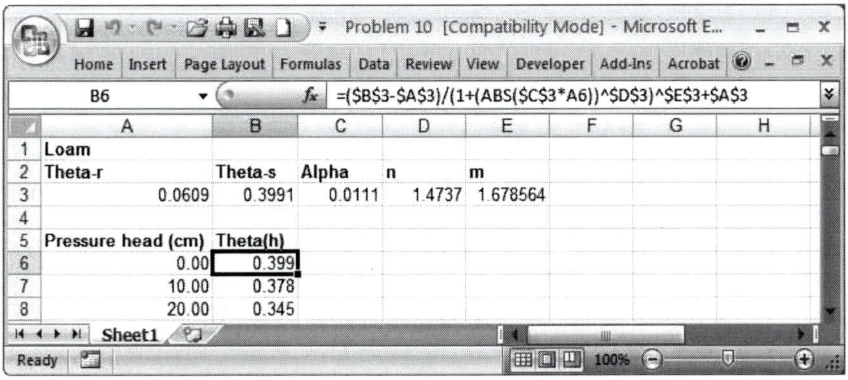

FIGURE 2.50 Excel spreadsheet for calculating θ(h) using the van Genuchten (1980a) equation.

TABLE 2.6
Hypothetical data for a neutron probe calibration equation

Count ratio	θ (h)	Count ratio	θ (h)	Count ratio	θ (h)
0.77	0.112	0.90	0.140	1.38	0.264
0.81	0.131	0.94	0.163	1.40	0.276
0.83	0.125	1.02	0.177	1.41	0.283
0.85	0.138	1.34	0.259	1.42	0.295
0.88	0.151	1.35	0.255	1.43	0.280

11. Show that if the soil water characteristic curve is described by the Haverkamp
equation 2.54, then the capacity function has the form:

$$C_w(h) = -(\theta_s - \theta_r) \cdot \frac{\beta \alpha^\beta h^{\beta-1}}{\left[1+(\alpha h)^\beta\right]^2}.$$

TABLE 2.7
Water retention data for the Bandalier Tuff

Pressure head (cm)	θ (h)	Pressure head (cm)	θ (h)	Pressure head (cm)	θ (h)
Laboratory data		Field data		Field data	
293.9	0.165	16.9	0.319	142.6	0.226
322.5	0.162	25.1	0.313	142.0	0.222
409.6	0.147	27.2	0.294	150.2	0.222
453.2	0.139	43.7	0.289	160.7	0.215
596.6	0.127	61.5	0.294	169.9	0.200
641.5	0.125	64.0	0.268	180.9	0.212
801.5	0.116	66.4	0.257	190.9	0.208
860.1	0.113	79.8	0.264	201.8	0.197
949.7	0.109	85.2	0.257	204.1	0.183
1192.0	0.103	91.6	0.257	228.1	0.191
1298.0	0.101	91.2	0.248	232.6	0.184
1445.0	0.0988	98.6	0.239	234.0	0.182
1594.0	0.0963	104.8	0.241	229.3	0.177
1760.0	0.0915	118.2	0.237	263.2	0.177
1980.0	0.0875	122.4	0.236	287.3	0.170
		136.2	0.219	307.7	0.165

Source: Abeele, W.V., Report No. LA-10037-MS, Los Alamos National Laboratory, Los Alamos, NM, 1984; van Genuchten, M.Th., et al., EPA/600/2-91/065, www.ars.usda.gov/Main/docs.htm?docid=15992, 1991.

12. Use Microsoft Excel to develop a calibration equation using the hypothetical data in Table 2.6 for a neutron probe. Make a graph of θ on the y-axis and count ratio on the x-axis between x-axis limits of 0.7 and 1.5. Click on Layout under Chart Tools on the toolbar and select an axis in the Current Selection box to get options for changing the minimum and maximum values. To get the calibration equation, select Trendline from the Layout toolbar. Select "More Trendline options", then select the Linear trendline and check the options to "Display equation on chart" and "Display R-squared value on chart." You should see the equation for converting count ratios to θ (the calibration equation) and the R^2 for the fit of the equation to the data.

13. Use Microsoft Excel to make plots of TDR volumetric water content vs. apparent permittivity, using Equations 2.20 and 2.21 on the same graph. Cover the range of ε_{ra} from 1 to 80.

14. Use the RETC program to fit retention curves to the observed data for the crushed Bandalier Tuff in Table 2.7 (Abeele 1984; van Genuchten et al. 1991). Use the Brooks and Corey (1964) and van Genuchten (1980a) parametric models to represent the soil water retention curve. Use the Mualem version of both models. The observed data were obtained in the laboratory and in the field. Use the Alt-key and the print screen key to copy the water retention curve graph (see Figure 2.48) and the output file (see Figure 2.49) into a Word document. Which model fits the curve best?

15. Substitute Equation 2.13 into Equation 2.54 and solve for h. Show the intermediary steps.

3 Steady Water Flow in Soils

3.1 INTRODUCTION

Water movement in soils is a key process that affects water quantity and quality in the environment. Since the transport of solutes is closely linked with the soil water flux, any quantitative analysis of solute transport must first evaluate water fluxes into and through soil. Soil water movement occurs under both saturated and unsaturated conditions. Saturated conditions occur below the water table, where water movement is predominately horizontal, with lesser components of flow in the vertical direction. While unsaturated conditions generally predominate above the water table (the *vadose zone*), localized zones of saturation can exist, especially following precipitation or irrigation or above an impermeable zone. As a general rule, water movement in the unsaturated zone is vertical, but can also have large lateral components.

Saturated soils occur when soil pores are entirely filled with water. In this case, a saturated water content, θ_s, prevails, which is equal to the porosity, φ, if there is no entrapped air. While soil pores can be assumed to be fully saturated below the water table, even so-called saturated soils above the water table may retain some residual entrapped air, especially near the soil surface.

Water flow can be either steady or transient. Under steady water flow, variables do not change as a function of time: the water contents and fluxes are constant in time. Water contents and fluxes may change as a function of position, however. For example, under steady flow conditions, water contents may change with depth. On the other hand, under transient conditions, at least one variable characterizing flow changes as a function of time.

This chapter describes the conditions and equations that govern steady water flow, the parameters that quantify these relationships, and several examples of steady water flow. Nonsteady or transient water flow is described in Chapter 5.

3.2 STEADY FLOW IN SATURATED SOIL

The Poiseuille equation, which applies to steady flow at the pore scale, is described first. Then the Darcy equation, which is used to describe steady flow at the macro scale, is presented.

3.2.1 POISEUILLE EQUATION

At the *microscopic scale* of an individual pore, approximated as a water-filled cylinder of a given radius (R) [L], the volumetric flow rate (Q) [L³T⁻¹] can be described by the Poiseuille equation:

$$Q = \frac{\pi R^4 \rho_w g \Delta H}{8\eta L}, \tag{3.1}$$

where η is the coefficient of dynamic water viscosity [$ML^{-1}T^{-1}$], ρ_w is the water density [ML^{-3}], g is the acceleration due to gravity [LT^{-2}], and ΔH is the difference in total head [L] between two points along the cylinder separated by a distance L [L]. Poiseuille developed this equation in 1842 to describe blood flow in arteries. The equation has been used to show the disproportionate effect that large-diameter pores (i.e., macropores) have on transmitting water. Flow is proportional to R^4, whereas the cross-sectional area of a pore is proportional to R^2. Therefore, one large pore with the same cross-sectional area as several smaller pores will transmit considerably more water due to less viscous drag along the pore wall. See Section 3.6.1 for derivation of the Poiseuille equation.

3.2.2 DARCY EQUATION

Since it is usually not feasible to determine the size distribution and interconnectivity of pores that would be required to use the Poiseuille equation, a *macroscopic approach* is normally used to describe flow through soils. This approach was first taken by Henry Darcy, an engineer working on sand filters used to purify the drinking water in Dijon, France. Recently, the contribution of his colleague, Jules Dupuit, has been recognized (Ritzi and Bobeck 2008). Through experimentation, Darcy found that the volumetric flow rate (Q) [L^3T^{-1}] per unit cross-sectional area (A) [L^2] through a sand filter of a given thickness (L) [L] was proportional to the total soil water potential head gradient across the sand ($\Delta H/L$). He called the proportionality constant *saturated hydraulic conductivity* (K_s) [LT^{-1}] (Darcy 1856; Dupuit 1857; Hubbert 1956):

$$\frac{Q}{A} = -K_s \frac{\Delta H}{L}. \tag{3.2}$$

This is known as *Darcy's law*. With the Darcy equation written in this form, it is assumed that $\Delta H = H_2 - H_1$, where H_2 is above H_1 for vertical flow and H_2 is to the right of H_1 for horizontal flow. The term on the left side (Q/A) has units of length per unit time or velocity. Since the gradient term $\Delta H/L$ is unitless, K_s must have the same units as Q/A, length divided by time, or velocity.

If J_w is defined as the volumetric flow rate of water per unit cross-sectional area, Q/A, or the *water flux*, then Darcy's law can be written:

$$J_w = -K_s \frac{\Delta H}{L}, \tag{3.3}$$

where J_w has units of length per unit time [LT^{-1}]. Using derivative notation, Darcy's law for vertical flux can be written:

$$J_w = -K_s \frac{dH}{dz}. \tag{3.4}$$

For vertical movement, a positive J_w indicates upward flow. For horizontal movement, a positive J_w indicates flow to the right. The negative sign in Equation 3.4 indicates that water will flow from areas of high soil water potential head to areas of low soil water potential head.

The way the derivative is written in Equation 3.4 is correct if H is only a function of z. In other words, when taking the derivative, it can only be taken with respect to one independent variable, z. This would be true for steady flow in the vertical direction. If water movement occurs in two (x and z) or three (x, y, and z) directions, H may be a function of two or three space variables. In Chapter 5, under transient flow conditions, water content will be a function of time as well as position. Under these circumstances, one can take the derivative of H with respect to any of these variables. When a dependent variable is a function of more than one independent variable, then *partial derivative* notation is used and Darcy's law is written (note the slight change in how the derivative is written):

$$J_w = -K_s \frac{\partial H}{\partial z}. \tag{3.5}$$

Darcy's law can be easily extended to two- and three-dimensional flows.

Darcy's equation takes the same form as several other important laws in science, among them, Fick's law for molecular diffusion, Ohm's law for electric current flow, and Fourier's law for heat conduction. It clearly shows that for water to move, there must be a difference in water potential, but the rate of water flow will depend on the hydraulic gradient and the hydraulic conductivity of the soil.

Example 3.1

What is the steady water flux through a vertical column of soil 40 cm in length with 5 cm of water ponded on the surface and water allowed to drip from the bottom, if the soil saturated hydraulic conductivity is 10 cm h^{-1} (Figure 3.1)?

Set the reference elevation for zero gravitational head at the bottom of the column.

$$J_w = -K_s \frac{\Delta H}{L} = -K_s \frac{H_2 - H_1}{L} = -K_s \frac{(h+z)_2 - (h+z)_1}{L}$$

[handwritten annotations: "pressure head", "with respect to reference elevation", "Distance between Hs", "It is freely draining so we are at atmospheric pressure"]

$$= -\left(10 \frac{cm}{h}\right) \frac{(5\,cm + 40\,cm) - 0}{40\,cm} = -11.25 \frac{cm}{h}.$$

[handwritten: $= -k_s \frac{(h_2 - h_1) + (z_2 - z_1)}{L} = -(10) \frac{(5-0) + (40-0)}{L}$]

Flow is downward since the sign is negative.

In a uniform soil under saturated conditions, Darcy's law can be used to show that pressure head (h from Chapter 2) varies linearly with depth (Jury and Horton 2004). Typically, a *column* of soil is used as a system where flow is predominately in one direction (along the axis of the column). In a vertical soil column of length L and saturated hydraulic conductivity K_s, water is ponded to a depth b at the top

Constant head condition

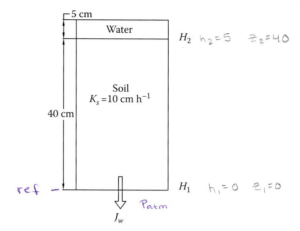

H_2 $h_2 = 5$ $z_2 = 40$

$H = z + h$

ref

H_1 $h_1 = 0$ $z_1 = 0$

P_{atm}

J_w

FIGURE 3.1 A vertical column of soil 40 cm in length with 5 cm of water ponded on the surface and water allowed to drip from the bottom. The soil saturated hydraulic conductivity is 10 cm h⁻¹.

of the column and water discharges freely from the bottom at atmospheric pressure (Figure 3.2). Once steady flow has been achieved in the column, the flux can be described with Darcy's law, setting the reference height for zero gravitational head at the bottom of the column:

$$J_w = -K_s \frac{H_2 - H_1}{L} = -K_s \frac{b + L}{L}.$$
(3.6)

Darcy's law for this system can also be written using an arbitrary point at height z inside the column, instead of the top of the column (Figure 3.3). This is done to introduce the pressure head, h, at an arbitrary depth. H_1 remains the same (zero). Darcy's law for flux between these two points is

$$J_w = -K_s \frac{H_2 - H_1}{L} = -K_s \frac{h + z}{z}.$$
(3.7)

Note that the denominator is the distance between the two points where H is given, not the total length of the column in this case. Once steady flow is achieved, the flux between any two points in the column has to be the same (water contents are not changing in the column, so there is no increase or decrease in storage). Thus, the equations can be set equal and then solved for h:

$$-K_s \frac{h + z}{z} = -K_s \frac{b + L}{L}$$

$$h = \frac{b}{L} z.$$
(3.8)

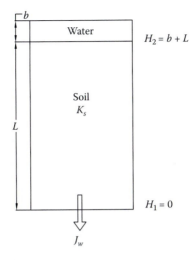

FIGURE 3.2 A vertical soil column of length L and saturated hydraulic conductivity K_s. Water is ponded to a depth b at the top of the column and water discharges freely from the bottom at atmospheric pressure.

Since b and L are constants, the pressure head varies linearly with z from a value of b at the top (equal to the depth of ponding) to a value of zero at the bottom. Thus, the pressure head varies linearly with depth in a saturated, uniform soil under conditions of steady flow. If the pressures at any two points in a soil under these conditions are known, one can draw a straight line between the values to get the pressures at all other points. Later in the chapter, nonuniform soils and unsaturated soils will be discussed to see how these conditions affect linearity in the distribution of pressure

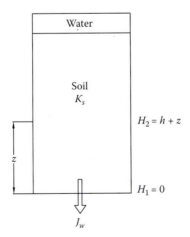

FIGURE 3.3 A vertical soil column with an arbitrary point at height z inside the column, instead of the top of the column.

th. In Chapter 5, the effect of transient flow and unsaturated conditions on head distribution linearity will be discussed.

3.2.3 SATURATED FLOW PARAMETERS

For saturated flow, the most important soil parameter is saturated hydraulic conductivity, which is a function of the fluid and soil properties:

$$K_s = \frac{k\rho_l g}{\eta},$$

(3.9)

where k is the *intrinsic permeability* of the soil [L^2], ρ_l is the density of the fluid [ML^{-3}], g is the gravitational acceleration [LT^{-2}], and η is the coefficient of dynamic viscosity of the fluid [$ML^{-1}T^{-1}$]. From this equation, it is apparent that temperature has an effect on K_s, since it will affect ρ_l and η. It is generally true that, whereas a given soil will have a constant k as long as the soil structure is not affected by the fluid, it will have a different K_s for different liquids.

Soils with low porosity, few large pores, and poor interconnectivity between pores, have low values of K_s. Rawls et al. (1982) compiled values of K_s from 1323 soils collected over 32 states of the United States. Average K_s for 11 of the 12 U.S. Department of Agriculture (USDA) soil textural classes (the silt class is missing) from Rawls et al. (1982) are shown in Table 3.1. The Brooks and Corey (1964) water retention curve parameters (Equation 2.53) are also shown. Saturated hydraulic conductivities were highest in coarse-textured soils (sand, loamy sand, etc.) and

---

TABLE 3.1
Brooks and Corey soil hydraulic parameters (K_s, θ_r, θ_s, h_a, and λ) for 11 textural classes

Textural class	K_s (cm d^{-1})	θ_r (cm^3 cm^{-3})	θ_s (cm^3 cm^{-3})	h_a (cm)	λ (–)	λ_c (cm)
Sand	504.0	0.020	0.417	– 7.26	0.592	9.88
Loamy sand	146.6	0.035	0.401	– 8.69	0.474	12.28
Sandy loam	62.16	0.041	0.412	– 14.7	0.322	22.18
Loam	31.68	0.027	0.434	– 11.1	0.220	17.79
Silt loam	16.32	0.015	0.486	– 20.7	0.211	33.38
Sandy clay loam	10.32	0.068	0.330	– 28.1	0.250	44.16
Clay loam	5.52	0.075	0.390	– 25.9	0.194	42.27
Silty clay loam	3.60	0.040	0.432	– 32.6	0.151	55.03
Sandy clay	2.88	0.109	0.321	– 29.2	0.168	48.61
Silty clay	2.16	0.056	0.423	– 34.2	0.127	58.96
Clay	1.44	0.090	0.385	– 37.3	0.131	64.07

Source: Rawls, W.J., et al., *Trans. ASAE* 1328, 1316, 1982.

Note: Macroscopic capillary length (λ_c) calculated using Equations 3.21 and 3.25.

lower in fine-textured soils (clay, silty clay, etc.), because of larger pores in coarse-textured soils. This can be corroborated by examining air-entry pressure heads, h_a (the pressure head at which the largest capillary pore empties), which are lower in fine-textured soils, indicating that the largest capillary pore is relatively small. In Table 3.1, θ_r and θ_s denote the residual and saturated water contents [$L^3 L^{-3}$], respectively, and λ is a pore-size distribution index that characterizes the width of the pore-size distribution or the steepness of the retention function (see Chapter 2, Section 2.9.4). These parameters will be discussed again in this chapter in the section on unsaturated water flow.

Saturated hydraulic conductivity varies widely among soils within a textural class, especially in the more fine-textured classes in Table 3.1. In Figure 3.4, K_s for 324 soil samples from the UNSODA database (Nemes et al. 2001) are plotted as a function of clay content. Saturated hydraulic conductivity tends to decline as clay content increases, but there is considerable scatter at higher clay contents. The clayey soils with high K_s are probably soils with strong structure and low shrink-swell clay minerals. Hence, soil structure, mineralogy, and texture affect water flow in all but very sandy soils.

Bruce et al. (1983) measured K_s in the Cecil loamy sand profile described in Tables 1.5 and 2.4. The values are given in Table 3.2. The effect of texture and structure on K_s can be seen in these data. The horizons high in sand (Ap and BA) have higher K_s than the horizons high in clay (Bt2 and BC). The exceptions are the Bt1 and C horizons. The Bt1 horizon has the highest clay content, but it also has strong structure and relatively large aggregates (medium-size class, see Table 1.5). Strong structure gives this horizon a high K_s. The BC horizon has a relatively high sand content, but weak structure. The lack of structure results in a low K_s. Research in North Carolina also showed that in a Cecil series soil, K_s reached a minimum in the BC horizon (Schoneberger and Amoozegar 1990) (Figure 3.5).

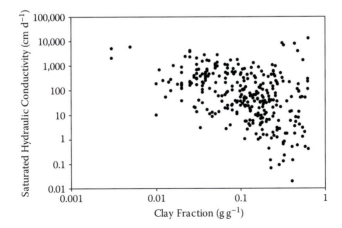

FIGURE 3.4 Saturated hydraulic conductivity as a function of clay fraction for 324 soils from the UNSODA database. (Courtesy of Nemes, A., et al., The UNSODA unsaturated soil hydraulic database, Version 2.0, http://ars.usda.gov/Services/docs.htm?docid=8967, 2001.)

2

Soil properties (texture, sand, silt, clay, and K_s) for the Cecil loamy sand

Horizon	Texture	Clay (%)	Silt (%)	Sand (%)	K_s (cm h⁻¹)
Ap	Loamy sand	7	15	78	19.19
BA	Clay loam	37	20	43	7.69
Bt1	Clay	50	20	30	10.73
Bt2	Clay	41	25	34	0.206
BC	Clay loam	36	27	37	0.035
C	Sandy clay loam	24	24	52	0.467 *low permeability*

Source: Plot 4 from Bruce, R.R., et al., *Physical Characteristics of Soils in the Southern Region: Cecil.*
Georgia Agricultural Experiment Stations, Athens, GA, 1983.

Saturated hydraulic conductivity is highly variable in space, not only in the vertical direction as horizons change (Figure 3.5), but also in the horizontal direction within the same horizon (Nielsen et al. 1973; Russo and Bresler 1980; Russo and Bouton 1992). Saturated hydraulic conductivity also varies in time at much shorter scales than that of pedogenesis. This is especially true near the soil surface, where land use, biology, and weather have more of an effect. Some authors have suggested that soil properties should be divided into static and dynamic properties (Grossman et al. 2001; Tugel et al.

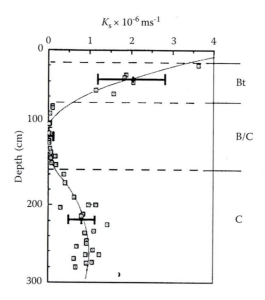

FIGURE 3.5 Saturated hydraulic conductivity as a function of depth in a Cecil series soil from North Carolina Piedmont. (From Schoeneberger, P., and Amoozegar, A., *Geoderma*, 46, 31, 1990. With permission.)

2005). Hence, the effect of texture and mineralogy on K_s might be considered static, whereas other effects might be considered dynamic.

3.2.4 FLOW PERPENDICULAR TO LAYERS

Most soils, other than Entisols and Inceptisols, have well-developed horizons or layers that have different hydrologic characteristics, including K_s, as shown in Table 3.2. Above the water table, water usually flows vertically, so that flow is perpendicular to these layers. Below the water table, flow can occur parallel to the layers.

Steady flow through a layered soil can be described by making an analogy between water flow and flow of electric current (Jury and Horton 2004). These equations were first developed by Darcy and Dupuit in the 1850s (Ritzi and Bobeck 2008). It would be useful to know the flux through a soil column with N layers of thickness $L_1 \ldots L_N$ and saturated hydraulic conductivity $K_1 \ldots K_N$ (left side of Figure 3.6). Using an analogy between Darcy's and Ohm's laws (and between hydraulic and electric resistances, respectively), it can be shown that the *effective* saturated hydraulic conductivity (K_{eff}) [LT^{-1}], a single value that will have the same effect on flow as all layers combined, is:

$$K_{eff} = \frac{\sum_{j=1}^{N} L_j}{\sum_{j=1}^{N} \frac{L_j}{K_j}}. \tag{3.10}$$

The effective hydraulic conductivity is not a simple arithmetic average of conductivities of individual layers, or their limiting value, but a weighted average that

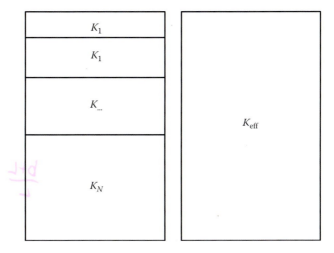

FIGURE 3.6 A soil column with N layers of thickness $L_1, \ldots, L_N$ and saturated hydraulic conductivities $K_1, \ldots, K_N$ on the left side. An equivalent uniform column on the right with a saturated hydraulic conductivity, K_{eff}.

considers the layer thicknesses as well as their conductivities. Thus, the thickness of layers is important.

Steady vertical flow through the layered soil profile is then described with Darcy's equation:

$$J_w = -K_{eff} \frac{\Delta H}{L},$$ (3.11)

where potentials are measured at the top and bottom of the profile and L is the total thickness of the profile.

Because the flux across each layer is the same under steady flow conditions, Equation 3.11 can be used (once J_w is known) to determine the matric potential at each layer interface, as shown in the following example.

Example 3.2

Determine the effective saturated hydraulic conductivity, steady flux, and pressure head at the layer interface in a column consisting of a sand layer 30 cm thick overlying a loam layer 20 cm thick. The saturated hydraulic conductivities of the sand and loam layers are 21 and 1.32 cm h^{-1}, respectively. Water is ponded at the top of the column to a depth of 5 cm and allowed to drip from the bottom of the column (Figure 3.7).

First, calculate the effective saturated hydraulic conductivity of the soil column:

$$K_{eff} = \frac{\sum_{j=1}^{N} L_j}{\sum_{j=1}^{N} \frac{L_j}{K_j}} = \frac{20\,cm + 30\,cm}{\dfrac{20\,cm}{1.32\,cm\,h^{-1}} + \dfrac{30\,cm}{21\,cm\,h^{-1}}} = 3.02\,cm\,h^{-1}.$$

Then calculate the steady flux using K_{eff} and the conditions at the top and bottom of the column:

$$J_w = -K_{eff} \frac{H_2 - H_1}{L}$$

$$= -3.02\,cm\,h^{-1} \frac{(5 + 50)\,cm - 0\,cm}{50\,cm}$$

$$\frac{b + L}{L}$$

$$= -3.32\,cm\,h^{-1}.$$

Note that the flux is negative, indicating downward flow.

Now that the flux is known, set up Darcy's equation using conditions at the bottom of the column and the interface to find the pressure head at the interface h_3:

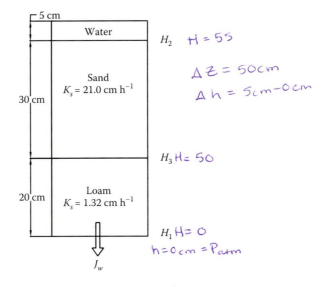

FIGURE 3.7 A column consisting of a sand layer 30 cm thick overlying a loam layer 20 cm thick. The saturated hydraulic conductivities of the sand and loam layers are 21 and 1.32 cm h⁻¹, respectively. Water is ponded at the top of the column to a depth of 5 cm and allowed to drip from the bottom of the column.

$$J_w = -K_{\text{loam}} \frac{H_3 - H_1}{L} \qquad J = -K \left(\frac{h_3 + z_3 - h_1 - z_1}{L} \right)$$

$$-3.32\,\text{cm h}^{-1} = -1.32\,\text{cm h}^{-1} \frac{(h_3 + 20) - 0\,\text{cm}}{20\,\text{cm}}$$

$$h_3 + 20\,\text{cm} = -3.32\,\text{cm h}^{-1} \frac{20\,\text{cm}}{-1.32\,\text{cm h}^{-1}}$$

$$h_3 = 30\,\text{cm}.$$

Note that a positive pressure builds up at the interface above the low-conductivity layer.

In this manner, one can show that when water flows down through a high-conductivity to a low-conductivity layer (e.g., a sand horizon over a clay horizon), a positive pressure head occurs at the interface that is greater than would be present in a uniform soil. This "back pressure" slows flow through the upper layer and speeds flow through the lower layer until there is an equivalent flow through both layers. In the reverse situation (e.g., a clay over a sand), the interface pressure head is less than that in a uniform soil, creating a "suction" that speeds flow through the upper layer and slows flow through the lower layer.

In a saturated layered soil under steady flow, the pressure head distribution is linear within each layer, but with a different slope in each layer. In reality, an abrupt change in slope at boundaries is not usually observed in natural soils because K_s usually varies gradually, leading to a more curvilinear change in pressure head.

For flow from low- to high-conductivity layers, where a large difference in conductivity is present, negative pressure heads can occur at the interface that exceed the air-entry potential of soil in the lower layer. This causes unsaturated conditions in the lower layer, which greatly reduce flow through the profile. This often occurs when a *surface seal* or *crust* is present. Another way to equalize flow through a layered soil with a low-conductivity layer over a high-conductivity layer is for water to flow through only part of the less restrictive layer. This is a form of preferential flow called *fingering*. Both surface seals and fingering are discussed in Chapter 5.

3.2.5   FLOW PARALLEL TO LAYERS

For flow parallel to layers (which usually occurs below the water table), the resistances of particular layers can be considered in parallel (more water flows through high- than low-conductivity layers). If the soil layers are of equal thickness, then K_{eff} is the arithmetic average of the hydraulic conductivities. For layers of different thicknesses, K_{eff} is a weighted average where the layer thicknesses are the weights:

$$K_{eff} = \frac{\sum_{j=1}^{N} L_j K_j}{\sum_{j=1}^{N} L_j}. \tag{3.12}$$

Note the difference between the effective conductivity for flow perpendicular to layers (Equation 3.10) and that for flow parallel to layers (Equation 3.12).

3.2.6   FLOW TO A DRAIN

Flow to a drain provides another example of steady saturated flow (Dupuit 1863; Forchheimer 1930). An equation can be developed for the maximum height of the water table between two drains or ditches under steady flow conditions (Figure 3.8) (Warrick 2003). For a steady rainfall or irrigation rate R [LT^{-1}] and a drain separation distance $2s$ [L] with the water level in the ditch or drain maintained at a height z_0 [L], the assumption is that water in the saturated zone will flow only in the horizontal direction and the gradient will be the change in height of the water table:

$$J_x = -K_s \frac{dz}{dx}, \tag{3.13}$$

where z is the height of the water table [L]. Under these conditions, it can be shown that the maximum height of the water table, z_{max} [L], between the drains or ditches is

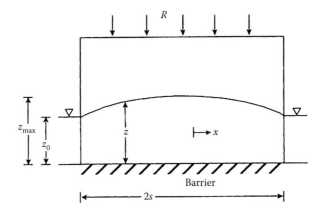

FIGURE 3.8　Geometry considered for defining the drainage equation. Water height in the ditch or drain is z_0. (From Warrick, A.W., and Broadbridge, P., *Water Resour. Res.* 28, 427, 1992. With permission.)

$$z_{max} = \sqrt{\frac{Rs^2}{K_s} + z_0^2}.$$　　　　　　　(3.14)

From this equation, it can be seen that the water table will not rise as high between the drains or ditches when K_s is large. Also, the greater the irrigation rate, the higher the water table will rise. The larger the spacing between drains, the higher the water table will rise. Equation 3.14 can be used to determine the distance for separating drains, in order to maintain the water table within a given range from the soil surface. See Section 3.6.2 for derivation of Equation 3.14.

3.3　STEADY FLOW IN UNSATURATED SOIL

The unsaturated zone, also called the vadose zone, lies between the water table and the soil surface. In this region, the water content of the soil is less than saturation ($\theta < \theta_s$), many pores are air-filled ($a > 0$), and pressure heads are negative ($h < 0$). The water table separates the saturated zone from the unsaturated zone.

Rising above the water table is the *capillary fringe*, a zone where water is under tension, but is very near saturation. The capillary fringe is important for water flow because only a slight change in water content or total head can cause a sharp change in the water table position (Sklash et al. 1986). Also, lateral flow increases in this zone (Liu and Dane 1996; Abit et al. 2008).

The thickness of the capillary fringe depends on the water retention curve and can be approximated by the air-entry pressure head (h_a). From Table 3.1, one can see that sands have thinner capillary fringes than unstructured clays. Capillary fringes are also found around zones of saturation in the unsaturated zone, such as around injection boreholes, surface seepage basins, and losing streams.

3.3.1 BUCKINGHAM–DARCY EQUATION

In the unsaturated zone, larger pores drain more readily than smaller ones, as can be noted using the capillary rise equation (Equation 2.10). Therefore, the hydraulic conductivity is much less under unsaturated than saturated conditions because of water moving through smaller pores or as films along the walls of larger pores. At very low water contents, continuous fluid paths may not exist and water may move in the vapor phase. The unsaturated hydraulic conductivity is therefore represented as a function of negative pressure head [$K(h)$] or as a function of water content [$K(\theta)$]. Buckingham (1907) modified Darcy's equation for unsaturated flow as follows:

$$
\begin{aligned}
J_w &= -K(h)\frac{\partial H}{\partial z} = -K(h)\frac{\partial(h+z)}{\partial z} = -K(h)\left(\frac{\partial h}{\partial z} + \frac{\partial z}{\partial z}\right) \\
&= -K(h)\left(\frac{\partial h}{\partial z} + 1\right),
\end{aligned}
\tag{3.15}
$$

where the total head is the sum of the pressure and gravitational heads, $H = h + z$.

The most important parameters in unsaturated flow are parameters of the unsaturated hydraulic conductivity function $K(h)$, the retention curve $\theta(h)$, and (to a lesser extent) the hydraulic capacity function $C_w(h)$ (Equation 2.51). All depend on pressure head and the pore-size distribution. Values of $K(h)$ vary by orders of magnitude with pressure head and are sensitive to the same variables that affect K_s: texture, structure, and mineralogy (Grossman et al. 2001; Tugel et al. 2005).

3.3.2 UNSATURATED HYDRAULIC CONDUCTIVITY

As is the case with the water retention functions, equations have been developed to describe the hydraulic conductivity function. One of the simplest equations is the Gardner (1958) equation:

$$
K(h) = K_s e^{\alpha h},
\tag{3.16}
$$

where α is a positive constant dependent on the soil. Since h is negative, the exponential term causes $K(h)$ to decrease rapidly as h becomes more negative. A coarse-textured soil (sand), which has a rapidly decreasing hydraulic conductivity function compared to a fine-textured soil (clay), would be expected to have a larger value of α (and a larger K_s).

Other simple equations include the Campbell (1974) equation:

$$
K(\theta) = K_s \left(\frac{\theta}{\theta_s}\right)^m,
\tag{3.17}
$$

where m is a fitting parameter [−] and the Haverkamp equation (Haverkamp et al. 1977):

$$K(h) = \frac{K_s}{1 + \left(\dfrac{h}{a}\right)^N},$$ (3.18)

where a [L^{-1}] and N [−] are fitting parameters.

Many hydraulic conductivity functions have been derived using the pore-size distribution models of Burdine (1953) or Mualem (1976) in combination with various retention functions (see Section 2.9.4). Burdine (1953) derived his equation by considering the soil as a bundle of capillary tubes of different radii:

$$K(S_e) = K_s S_e^l \frac{\displaystyle\int_0^{S_e} \frac{dS_e}{h^2(S_e)}}{\displaystyle\int_0^1 \frac{dS_e}{h^2(S_e)}},$$ (3.19)

where K_s is the saturated hydraulic conductivity [LT^{-1}], l is a pore-connectivity parameter [−], S_e is effective saturation (Equation 2.13) [−], and $h(S_e)$ is the water retention curve relationship [L]. Mualem (1976) developed a slightly different equation by considering the soil as two bundles of capillary tubes connected in series:

$$K(S_e) = K_s S_e^l \frac{\left[\displaystyle\int_0^{S_e} \frac{dS_e}{h(S_e)}\right]^2}{\left[\displaystyle\int_0^1 \frac{dS_e}{h(S_e)}\right]^2}.$$ (3.20)

The Brooks and Corey (1964) retention equation (Equation 2.53) is commonly associated with Burdine's pore-size distribution model (Equation 3.19), leading to the hydraulic conductivity function:

$$K(S_e) = K_s S_e^{1+l+2/\lambda},$$ (3.21)

with l assumed to be 2.0 in the original study of Brooks and Corey (1964).

The van Genuchten (1980a) retention function (Equation 2.47) is similarly coupled most often with the model of Mualem (Equation 3.20) to give (see Section 3.6.3 for derivation):

$$K(S_e) = K_s S_e^l \left[1 - (1 - S_e^{1/m})^m\right]^2,$$ (3.22)

where

$$m = 1 - 1/n, \quad n > 1. \tag{3.23}$$

The pore-connectivity parameter l in Equation 3.22 was estimated by Mualem (1976) to be about 0.5 as an average for many soils. However, Schaap and Leij (2000) recently recommended using $l = -1$ as an appropriate value for most soil textures.

Finally, Durner (1994) associated his dual porosity retention model (Equation 2.55) with Mualem's pore-size distribution model (Equation 3.20) to give:

$$K(S_e) = K_s \frac{(w_1 S_{e_1} + w_2 S_{e_2})^l \left\{ w_1 \alpha_1 \left[1 - (1 - S_{e_1}^{1/m_1})^{m_1} \right] + w_2 \alpha_2 \left[1 - (1 - S_{e_2}^{1/m_2})^{m_2} \right] \right\}^2}{(w_1 \alpha_1 + w_2 \alpha_2)^2}, \tag{3.24}$$

where w_i are the weighting factors for the overlapping regions of the water retention functions, and α_i, and m_i are fitting parameters for the separate functions, S_{e_i} ($i = 1, 2, \ldots$).

Figure 3.9 presents examples of hydraulic conductivity functions for the sand, loam, and clay textural classes, using the Brooks and Corey (1964) unsaturated hydraulic conductivity function (Equation 3.21) and the parameters given in Table 3.1. The hydraulic conductivity curves are presented as functions of both the negative of pressure head (top) and water content (bottom). It is apparent that Equation 3.21 results in a linear relationship between $\ln K(h)$ and $\ln h$. Notice that the hydraulic conductivity at saturation is significantly larger for coarse-textured soils than for loams and clays. This difference is often several orders of magnitude. Also notice that the hydraulic conductivity decreases several orders of magnitude as the soil becomes unsaturated. This decrease, when expressed as a function of the pressure head (Figure 3.9, top), is much more significant for sands than for loams or clays. The decrease for coarse-textured soils is so dramatic that, eventually, the hydraulic conductivity becomes smaller than for the loam or clay. These properties of the hydraulic conductivity function are often used in the design of engineered structures, such as capillary barriers (finer-textured soils above coarser-textured soils) in landfill covers to divert water from flowing through the underlying waste (see Section 5.8.3 in Chapter 5), or in thin sand or gravel layers at the soil surface to prevent or limit evaporation.

3.3.3 Macroscopic Capillary Length

A useful way to quantify the unsaturated hydraulic conductivity function in a single parameter is in terms of the macroscopic capillary length (λ_c), defined as:

$$\lambda_c = \frac{\int_{h_i}^{h_0} K(h) dh}{K(h_0) - K(h_i)}, \tag{3.25}$$

where h_0 is a matric potential at or near saturation and h_i is a more negative matric potential such that $h_0 > h_i$ (White and Sully 1987).

There are a number of ways to interpret λ_c. One way is a $K(h)$-weighted average pressure head over the range from h_0 to h_i (except that it is a positive number). If λ_c is small, this indicates that the *effective pressure head for unsaturated flow* is near saturation. In other words, $K(h)$ is near zero except for matric potentials near saturation. This would be the case for a sand where $K(h)$ rises sharply at the wet end of the curve. An unstructured clay might have a larger λ_c because the effective pressure head for unsaturated flow is more negative and $K(h)$ does not approach zero as fast at the dry end of the curve. Typical values for λ_c in four broad groups of soils are shown in Table 3.3. The effect of structure in fine-textured soils is apparent in that an unstructured clay may have a λ_c of 25 cm, whereas a structured clay may have a λ_c of 2.8 cm. Values for λ_c are also shown in Table 3.1 for each of the textural classes. These values were computed using Equation 3.25 with $h_0 = 0$ cm, $h_i = -15,000$ cm, and the Brooks and Corey (1964) equations for $K(h)$ (Equation 3.2.1). In both tables, leaving aside the effect of structure, λ_c increases as the clay content increases.

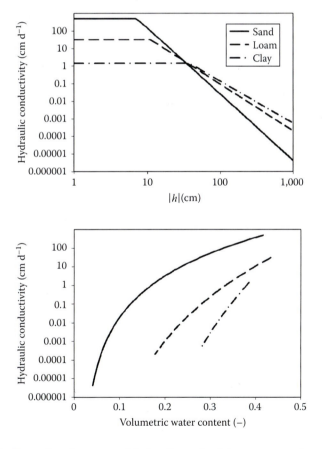

FIGURE 3.9 Examples of unsaturated hydraulic conductivity curves as a function of pressure head (top) and volumetric water content (bottom) for the sand, loam, and clay soil textural classes based on parameters in Table 3.1.

:/structure categories for estimation of macroscopic capillary length (λ_c)

Soil texture/structure category	λ_c (cm)
Coarse and gravelly sands; may also include some highly structured soils with large cracks and/or macropores	2.8
Most structured soils from clays through loams; also includes unstructured medium and fine sands	8.3
Soils that are both fine textured (clayey) and unstructured	25
Compacted, structureless, clayey materials, such as landfill caps and liners, lacustrine or marine sediments, etc.	100

Source: Adapted from Elrick, D.E., and Reynolds, W.D., *Advances in Measurement of Soil Physical Properties: Bringing Theory into Practice*, SSSA, Madison, WI, 1992.

Another way to interpret λ_c is as a measure of the effect of capillarity (the attraction of water to dry soil) as opposed to gravity on water movement. Water flow from a point source such as a borehole into a soil with a large value of λ_c (unstructured clay) will have more lateral flow than into a soil with a small value of λ_c (sand or well-structured clay) (Figure 3.10).

If the Gardner equation 3.16 is used to describe $K(h)$ in Equation 3.25, one gets another interesting interpretation of λ_c. It can be shown (see Section 3.6.4) in this case that:

$$\lambda_c = \frac{1}{\alpha}. \tag{3.26}$$

This makes it much easier to understand that a small value of λ_c indicates a rapid exponential decrease of the $K(h)$ function. It will be seen that λ_c appears in a number of equations later in this chapter, where it represents the effect of capillarity on unsaturated water flow.

In many cases, the $K(h)$ function is such that the integral in Equation 3.25 cannot be evaluated analytically. In this case, "numerical integration" can be used (see Example 3.3).

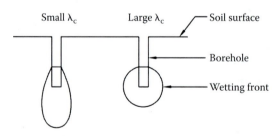

FIGURE 3.10 Boreholes and wetting fronts in a soil with a small λ_c and a soil with a large λ_c.

Example 3.3

Use numerical integration and Excel to estimate λ_c for the loamy sand textural class in Table 3.1. Use the Brooks and Corey equation for $K(h)$. Use $h_0 = 0$, and to simplify the calculation, $h_i = -100$ (normally a more negative pressure head would be used to cover the range of unsaturated hydraulic conductivity).

Simpson's rule is a common method used to evaluate integrals numerically (Gottfried 2007). To integrate an equation $y(x)$ over an interval from $x = a$ to $x = b$, divide the interval into equally spaced segments: $x_1, x_2, ..., x_n$, where Δx is the interval and n is an odd number ($x_1 = a$, $x_n = b$). Evaluate the function at each data point to obtain the values $y_1, y_2, ..., y_n$. The integral is then:

$$\int_a^b y(x)dx = \frac{\Delta x}{3}\left(y_1 + 4y_2 + 2y_3 + 4y_4 + 2y_5 + \cdots + 2y_{n-2} + 4y_{n-1} + y_n\right). \quad (3.27)$$

Starting with the Brooks and Corey hydraulic conductivity function (Equation 3.21) for $K(S_e)$:

$$K(S_e) = K_s S_e^{1+l+2/\lambda},$$

substitute the Brooks and Corey Equation 2.53 for $S_e(h)$:

$$S_e(h) = \left(\frac{h_a}{h}\right)^\lambda \quad h < h_a,$$

to obtain $K(h)$ with l assumed to be 2 as in the original study of Brooks and Corey (1964):

$$K(h) = K_s \qquad\qquad h \geq h_a$$
$$K(h) = K_s \left(\frac{h_a}{h}\right)^{2+3\lambda} \quad h < h_a \cdot \qquad (3.28)$$

In an Excel spreadsheet, enter the Brooks and Corey parameters from Table 3.1 for a loamy sand: λ, h_a, and K_s (Figure 3.11). Set up a column of values for h that go from 0 to −100 cm in steps of 10 cm. In the next column, calculate $K(h)$ for each value of h using Equation 3.28 and an IF statement that returns K_s if $h > h_a$. This column represents the values $y_1, y_2, ..., y_n$.

In the third column, multiply the values in the second column by the appropriate coefficient for the series in Equation 3.27. The first value is multiplied by one, the second value by four, the third value by two, etc. The last value is multiplied by one. Calculate the sum of the Simpson terms in column C and place this in cell C18 (Figure 3.12). The integral of $K(h)$ is calculated in cell C19 by multiplying the sum by 10/3 ($\Delta h/3$) (see Equation 3.27). In cell C20, calculate λ_c by dividing the integral from cell C19 by the difference between K_s and $K(-100)$ (see formula bar and Equation 3.25).

The value calculated using the Excel spreadsheet is 12.26 cm, which is very close to the value of 12.28 cm in Table 3.1 calculated using numerical integration in the Mathcad software (Mathsoft, Inc.). A more accurate estimate of the value could be obtained using the Excel spreadsheet by dividing the interval into more than 11 data points.

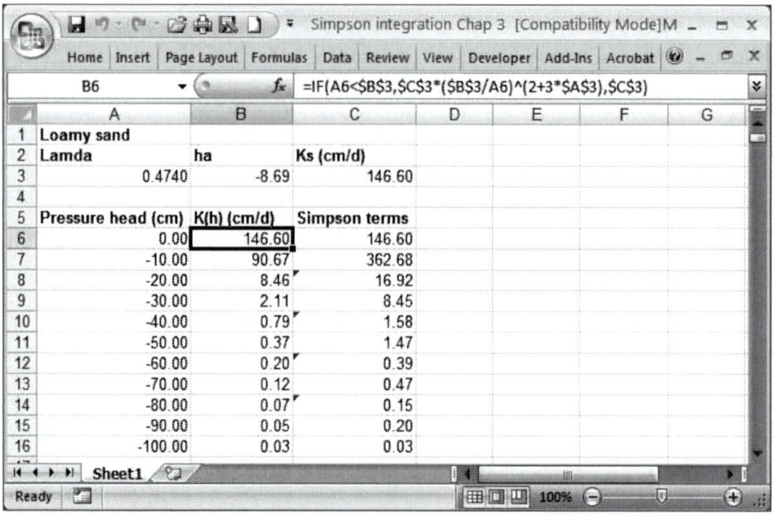

FIGURE 3.11 Excel spreadsheet with Brooks and Corey parameters for a loamy sand from Table 3.1.

3.3.4 ANALYTICAL SOLUTIONS FOR STEADY FLOW

It is useful to consider several cases of steady unsaturated flow. Although true steady flow may occur only rarely under unsaturated conditions, transient flow (discussed in Chapter 5) usually approaches steady flow over long time periods and the steady flow equations help to show what factors may be important under transient conditions.

As noted before, steady one-dimensional flow in a uniform soil can be described by the Buckingham-Darcy equation 3.15. If the water content is uniform with depth, then $\partial h/\partial z = 0$, only gravity causes flow, and the flux is equal to the hydraulic conductivity:

$$J_w = -K(h). \tag{3.29}$$

In this case, the distribution of h with depth is a vertical line (the same at all depths). This is called *gravity flow* or *unit-gradient flow*.

Analytical solutions to the Buckingham-Darcy differential equation (Equation 3.15) have been developed for a number of steady flow conditions. The usual procedure is to rewrite the equation in an integral form using the separation of variables approach (Jury and Horton 2004). All the terms in the independent variable (z in this case) are brought to one side of the equation and all the terms in the dependent variable (h in this case) are brought to the other side. If this can be done, then both sides are integrated over the range of h and z particular to the problem:

$$\int_{h_1}^{h_2} \frac{dh}{1 + \dfrac{J_w}{K(h)}} = z_1 - z_2. \tag{3.30}$$

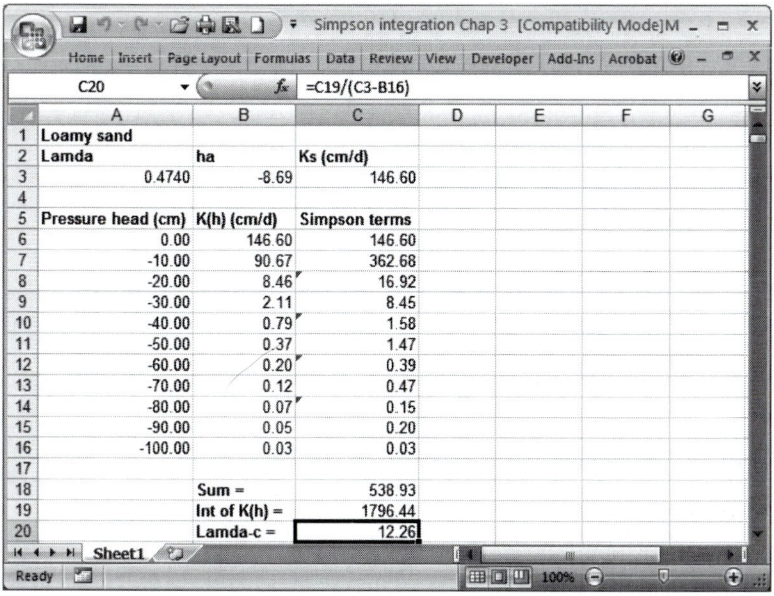

	A	B	C	D	E	F	G
1	Loamy sand						
2	Lamda	ha	Ks (cm/d)				
3	0.4740	-8.69	146.60				
4							
5	Pressure head (cm)	K(h) (cm/d)	Simpson terms				
6	0.00	146.60	146.60				
7	-10.00	90.67	362.68				
8	-20.00	8.46	16.92				
9	-30.00	2.11	8.45				
10	-40.00	0.79	1.58				
11	-50.00	0.37	1.47				
12	-60.00	0.20	0.39				
13	-70.00	0.12	0.47				
14	-80.00	0.07	0.15				
15	-90.00	0.05	0.20				
16	-100.00	0.03	0.03				
17							
18		Sum =	538.93				
19		Int of K(h) =	1796.44				
20		Lamda-c =	12.26				

Cell C20 formula: =C19/(C3-B16)

FIGURE 3.12 The sum of the Simpson terms in column C is calculated in cell C18. The integral of $K(h)$ is calculated in cell C19 by multiplying the sum by $10/3$ $(\Delta h/3)$ (see Equation 3.27). In cell C20, λ_c is calculated by dividing the integral by the difference between K_s and $K(-100)$ (see Equation 3.25).

If the equation for $K(h)$ is such that the integral on the left-hand side can be evaluated analytically (has an *analytical solution*), then the differential equation can be solved.

3.3.4.1 Infiltration to a Water Table

An example is steady infiltration ($J_w = -i$) in a uniform soil from the surface to a water table at depth L (Jury and Horton 2004). This might occur in a soil remediation effort designed to flush contaminants from the unsaturated zone down into the capture zone of a well. In this case, $h_1 = 0$ at the water table where $z_1 = -L$. The other limit is an arbitrary depth below the surface ($h_2 = h$ and $z_2 = z$), so that integration will yield an equation for the distribution of h with depth under steady flow conditions. The $K(h)$ function is described using the Haverkamp equation 3.18. To get an integral that can be evaluated, it is assumed that $N = 2$ in the Haverkamp equation. This produces a differential equation, the solution to which gives the distribution of h as a function of depth (z). It can be shown (see Section 3.6.5) that the solution is (Jury and Horton 2004):

$$h = a\sqrt{\frac{K_s}{i} - 1} \cdot \tanh\left[-\frac{\sqrt{(1 - i/K_s)i/K_s}}{a}(L+z)\right], \qquad (3.31)$$

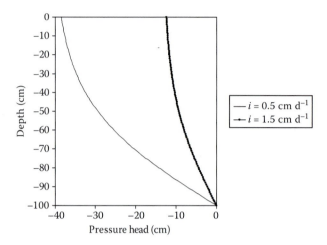

FIGURE 3.13 Distribution of h as a function of z for a steady low infiltration rate (0.5 cm d^{-1}) and a high infiltration rate (1.5 cm d^{-1}) based on Equation 3.31. (Adapted from Jury, W.A., and Horton, R., *Soil Physics*, John Wiley, Hoboken, NJ, 2004. With permission.)

where tanh(x) is the hyperbolic tangent function.

The distribution of h vs. z predicted by Equation 3.31 for a Chino clay with $K_s = 1.95$ cm day^{-1}, $a = -23.8$ cm, and $N = 2$ (Gardner and Fireman 1958) is shown for a low ($i = 0.5$ cm day^{-1}) and a high ($i = 1.5$ cm day^{-1}) infiltration rate in Figure 3.13. The distribution of h is curvilinear, contrary to the distribution for a uniform (Equation 3.8) or layered saturated soil, especially with the lower infiltration rate, which causes a more negative h at the surface. At the water table, the pressure head is zero, of course. Near the surface, especially at the high infiltration rate, the curve is nearly vertical, indicating that gravity flow is the dominant form of flow (Equation 3.29).

3.3.4.2   Infiltration from a Ring

Another example of steady downward flow is three-dimensional water flow from a point source such as a ring placed at the soil surface (Figure 3.14). A solution to the integral form of the Buckingham-Darcy equation is used to find the steady infiltration rate i_s, which will be approached after a period of time. Flow into the soil is considered a positive flux in this case. Wooding (1968) developed an equation for these circumstances:

$$i_s = -J_w \big|_{z=0}$$

$$= K(h_0)\left(1 + \frac{4\lambda_c}{\pi r}\right), \tag{3.32}$$

where i_s is the steady infiltration rate [LT^{-1}], r is the radius of the ring [L], and $K(h_0)$ is the hydraulic conductivity [LT^{-1}] corresponding to the pressure head of the water supply at the soil surface ($z = 0$).

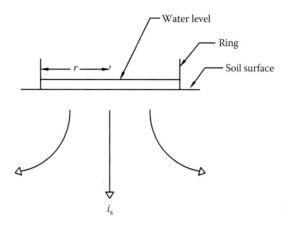

FIGURE 3.14   Steady 3-D infiltration of water from a ring of radius *r* at the soil surface.

By comparing Equation 3.32 and Equation 3.29, it is apparent that the second term in Equation 3.32 accounts for the increased infiltration rate from a ring due to lateral flow into dry soil (Figure 3.14). It is no surprise that this term includes λ_c, since this represents the effect of capillarity, which will vary with soil texture and the shape of the $K(h)$ function. The Wooding equation assumes that the unsaturated hydraulic conductivity function can be described by the Gardner equation 3.16.

If water is ponded at the surface, $h_0 > 0$ and $K(h_0)$ should be what corresponds to K_s in the field. This brings up an important point. Experiments have shown that the *field-saturated hydraulic conductivity* (K_{fs}) is less than the saturated hydraulic conductivity that would be measured on a carefully saturated, intact core in the laboratory (K_s). This has been attributed to the effect of entrapped air under field conditions that reduces the cross-sectional area available for water flow and the water potential gradient. Bouwer (1969) recommended using a value for field-saturated hydraulic conductivity, K_{fs}, equal to one-half of the true saturated hydraulic conductivity, K_s, for predicting the steady state infiltration rate.

 Example 3.4

What is the field-saturated hydraulic conductivity for a well-structured soil with macropores that has a steady infiltration rate of 12.3 cm h^{-1} from a ring with shallow-ponded water. The radius of the ring is 30 cm.

From Table 3.3, choose the appropriate value of λ_c:

$$\lambda_c = 2.8\,\text{cm}.$$

Since water is ponded in the ring, $K(h_0) = K_s$ in Equation 3.32. Solve for K_s and substitute values:

$$K_s = \frac{i_s}{1+\dfrac{4\lambda_c}{\pi r}} = \frac{12.3\dfrac{cm}{h}}{1+\dfrac{4\cdot 2.8\,cm}{\pi\cdot 30\,cm}} = 11\frac{cm}{h}.$$

3.3.4.3 Infiltration from a Borehole

The Buckingham-Darcy differential equation can also be solved for steady flow from a borehole (Figure 3.15).

$$Q_s = K_s\left(\pi r^2 + \frac{H\lambda_c}{G} + \frac{H^2}{G}\right), \tag{3.33}$$

where Q_s is the steady volumetric flow rate [L^{3}T^{-1}], H is the height of water ponded in the borehole [L], r is the radius of the hole [L], and G is a dimensionless geometric factor that depends primarily on the ratio of H/r (Elrick and Reynolds 1992). If the steady infiltration flux, i_s, is defined as the volumetric flow rate divided by the cross-sectional area of the borehole, $Q/\pi r^2$, the equation can be expressed in a form comparable to the Wooding equation 3.32:

$$i_s = K_s\left(1+\frac{H\lambda_c}{G\pi r^2} + \frac{H^2}{G\pi r^2}\right). \tag{3.34}$$

The similarity between Equation 3.34 and the Wooding equation 3.32 is apparent. In sequence, the terms inside the brackets in Equation 3.34 account for the effect of gravity, capillarity (note the λ_c term), and hydrostatic pressure in the borehole.

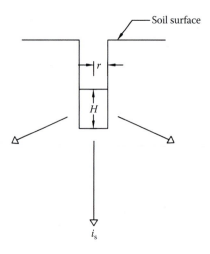

FIGURE 3.15 Steady 3-D infiltration of water from a borehole.

TABLE 3.4

Dimensionless coefficients for the polynomial (Equation 3.35) describing the geometric factor G, valid for H/r < 10

Soil texture/structure	A_1	A_2	A_3	A_4
Sand	0.079	0.516	− 0.048	0.002
Structured loams and clays	0.083	0.514	− 0.053	0.002
Unstructured clays	0.094	0.489	− 0.053	0.002

Source: Bosch, D.D., and West, L.T., *SSSAJ*, 62, 90, 1998.

Bosch and West (1998) fitted a polynomial equation to the data of Elrick and Reynolds (1992) to determine the value of G:

$$G = \frac{1}{2\pi}\left[A_1 + A_2\frac{H}{r} + A_3\left(\frac{H}{r}\right)^2 + A_4\left(\frac{H}{r}\right)^3 \right].$$
(3.35)

The values of the coefficients $A_1 \ldots A_4$ in this polynomial depend on texture and structure (Table 3.4).

Equation 3.34 is used to measure K_s using a borehole permeameter, described later in this chapter.

3.4 MEASUREMENTS OF HYDRAULIC PROPERTIES

Accurate measurement of hydraulic parameters by both laboratory and field methods is essential for predicting water movement in soils. However, uncertainties in parameter estimates arise because of spatial variation, measurement accuracy, and scale effects. Many of these methods use steady flow conditions, but some recent advances have been made in using transient conditions and these are discussed in Chapter 5. The important laboratory and field methods commonly employed in estimating hydraulic properties using steady flow conditions are discussed in the following sections with additional procedures available in Dane and Topp (2002). Pedotransfer functions are also discussed as an alternative to direct measurement of hydraulic properties.

3.4.1 LABORATORY METHODS

The constant-head and falling-head methods are common methods for measuring K_s in the laboratory. In the former method, a constant head of water is maintained at the top and bottom of a soil core and the steady water flux, J_w, is recorded (Reynolds and Elrick 2002a). Solving Darcy's equation for K_s:

$$K_s = \frac{J_w L}{H_1 - H_2},$$
(3.36)

where L is the length of the soil core, and H_1 and H_2 are total potential heads at the bottom and top of the core, respectively.

In the falling-head method, a stand-pipe is attached to the top of the core and the head at the top is allowed to drop over time (Reynolds and Elrick 2002c). The depth of water ponded at the top of the core changes with time, so the pressure head at the top of the core is $h = b(t)$. If the water at the top is contained in a cylinder of the same diameter as the soil, then the rate at which the level drops $[LT^{-1}]$ is the same as the Darcy flux $[LT^{-1}]$:

$$J_w = \frac{db}{dt} = -K_s \frac{H_2 - H_1}{L}. \tag{3.37}$$

Integrating this equation and solving for K_s (see Section 3.6.6 for derivation):

$$K_s = \frac{L}{t_1} \ln\left(\frac{b_0 + L}{b_1 + L}\right), \tag{3.38}$$

where b_0 is the height of water ponded [L] at the top of the core at time zero and b_1 is the height of water ponded [L] at a later time, t_1 [T]. The falling-head method is useful in soils with very low K_s, where it is difficult to collect sufficient drainage from a core to measure J_w accurately within an interval of several hours. The initial reading (b_0) can be recorded and the core left overnight before recording b_1, and there's no concern about the effect of evaporation on the drainage sample collected.

Values of K_s measured on cores in the laboratory will depend on what measures are used to remove entrapped air from the soil before making the measurement. Flushing the core with CO_2, wetting the core slowly from the bottom, and measuring K_s under conditions of an upward flux are commonly used.

A number of methods have been developed for measuring unsaturated hydraulic conductivity on cores in the laboratory. Unsaturated hydraulic conductivity can be measured on cores in a manner similar to the constant-head method for K_s by dripping water onto the top of the core at a rate that does not cause ponding (Booltink and Bouma 2002a). The total potential within the core is measured at two heights using miniature tensiometers and the steady flux is recorded. The Buckingham-Darcy equation is solved for $K(h)$, where h in this case is the average pressure head within the core. Steady evaporation from the top of a core can also be used to measure $K(h)$ by placing tensiometers at several heights, recording the steady evaporation rate, and calculating the gradient between each pair of tensiometers (Arya 2002).

Steady unsaturated flow can be imposed on a long soil core by supplying water continuously at a rate less than K_s. This can be done by applying a crust made of a mixture of gypsum and sand or cement and sand to the soil surface (Booltink and Bouma 2002b). The steady water flux is measured with a constant positive head above the crust. Because of the low K_s of the crust, flow in the core is unsaturated (as discussed in Section 3.2.4). A tensiometer in the core records h, and if a unit-gradient is assumed, then the measured flux is $K(h)$. Successive flux measurements with different crusts, each with a different conductivity, provide measurements of $K(h)$ over the range of pressures achieved. Bouma and Denning (1972) used this method in the field also.

In all laboratory methods of measuring K_s and $K(h)$, the effect of the chemistry of the added water on clay dispersion must be considered. The goal is to prevent dispersion so that pore-size distribution and connectivity do not change during the measurement, which can be achieved by using a dilute solution with a divalent cation (Chiang et al. 1987).

3.4.2 FIELD METHODS

Common field methods for measuring K_s are use of the *ring infiltrometer* and *borehole permeameter*. Ring infiltrometers can be used to pond water on the soil surface and measure the soil infiltration rate (Reynolds et al. 2002c). It is usually assumed that the Wooding equation 3.32 describes the three-dimensional steady infiltration rate. Using this equation, typical values of λ_c (Tables 3.1 or 3.3) can be used to convert steady flow rates to K_s (see Example 3.4). Concentric double rings have been used to force one-dimensional flow in the interior ring. In this case, the steady infiltration rate is an estimate of K_s in a uniform soil. However, Bouwer (1986) found that flow, for the typical dimensions used in a double ring (20 cm diameter for inner ring and 30 cm diameter for outer ring), was not one-dimensional.

Borehole permeameters are used to measure saturated hydraulic conductivity below the soil surface in the unsaturated zone (Reynolds and Elrick 2002d). They consist of a Mariotte siphon (McCarthy 1934) that maintains water at a constant level in a borehole and allows measurement of the steady flow rate into the soil. Then Equation 3.33 applies, but there are two unknowns in this equation: K_s and λ_c. One approach is to measure Q_s at two values of H in the same borehole and solve simultaneous equations for K_s and λ_c. Since changing the level of H in the borehole necessarily changes the region of soil that is being sampled, soil heterogeneity in the form of layering or macropores can result in unrealistic and invalid (i.e., negative) K_s and λ_c. According to Elrick and Reynolds (1992), as many as 30%–80% of measurements of K_s and λ_c in structured soils may be invalid (both negative). Alternatively, a single measurement of Q_s may be used and λ_c estimated using Table 3.1 or Table 3.3 (see Example 3.5). Studies suggest that this method yields values of K_s that are usually accurate to within a factor of two (Reynolds et al. 1992).

Another approach that does not require multiple measurements in the same borehole is based on the Glover solution (Zangar 1953):

$$i_s = K_s \frac{H^2}{G_G \pi r^2},$$

(3.39)

where G_G is the dimensionless geometric factor for the Glover analysis:

$$G_G = \frac{\sinh^{-1}\left(\dfrac{H}{r}\right) - \sqrt{\left(\dfrac{r}{H}\right)^2 + 1} + \dfrac{r}{H}}{2\pi}.$$

(3.40)

This approach only considers the effect of hydrostatic pressure in the borehole and ignores the effect of gravity and capillarity (Elrick and Reynolds 1992). It can overestimate K_s by an order of magnitude or more in dry, fine-textured, structure-less soils (i.e., soils where capillarity is most important). On the other hand, the Glover solution can provide good estimates of K_s in wet, coarse-textured, or structured soils when the ratio H/r is kept high (> 10) and hydrostatic pressure dominates flow.

Example 3.5

What is the saturated hydraulic conductivity for a structured loam soil if the steady state percolation rate measured with a borehole permeameter is 100 cm³ min⁻¹? The radius of the borehole is 3 cm and the depth of ponding is 15 cm.

Calculate the geometric factor using Equation 3.35 and the coefficients $A_1 \ldots A_4$ for a structured loam from Table 3.4:

$$G = \frac{1}{2\pi}\left[0.083 + 0.514\left(\frac{15\,cm}{3\,cm}\right) - 0.053\left(\frac{15\,cm}{3\,cm}\right)^2 + 0.002\left(\frac{15\,cm}{3\,cm}\right)^3\right]$$

$$= 0.251.$$

Solve Equation 3.33 for K_s and, using $\lambda_c = 8.3$ cm from Table 3.3, calculate K_s:

$$K_s = \frac{Q_s}{\pi r^2 + \dfrac{H\lambda_c}{G} + \dfrac{H^2}{G}}$$

$$= \frac{100\,\dfrac{cm^3}{min}}{\pi(3\,cm)^2 + \dfrac{(15\,cm)(8.3\,cm)}{0.251} + \dfrac{(15\,cm)^2}{0.251}}$$

$$= 0.07\,\frac{cm}{min}.$$

Tension infiltrometers (also called *disc permeameters*) are a popular method for determining unsaturated hydraulic conductivity and other hydraulic parameters in field soils. They consist of a circular porous plate or membrane, which is placed on the soil surface or an excavated soil surface. Water is supplied to the plate under tension using a Mariotte bottle arrangement and the rate of water entry into the soil can be measured on a graduated cylinder or with a pressure transducer (Clothier and Scotter 2002). In some cases, a ring is attached to the tension infiltrometer to allow ponded infiltration measurements at the same location (Perroux and White 1988). Two approaches are common, one in which the infiltration rate is measured after it attains a steady rate, and the other in which the early nonsteady infiltration rate is measured (Warrick 1992).

3.4.3 PEDOTRANSFER FUNCTIONS

Considering that direct measurements of the unsaturated soil hydraulic properties are relatively tedious, difficult, and time consuming, many have attempted to predict soil hydraulic properties from more easily measurable surrogate properties, such as soil texture and other readily available soil information. Relationships between the soil hydraulic and other (textural) properties are commonly called *pedotransfer functions* (PTFs). PTFs are usually obtained using various mathematical and statistical approaches, such as regression (Rawls and Brakensiek 1985; Rawls et al. 1991; Vereecken et al. 1989, 1990) or neural network analysis (Schaap et al. 1998; Schaap and Leij 2000). PTFs can be used to predict either the soil hydraulic properties directly, such as the water content at specified pressure heads or the saturated hydraulic conductivity, or parameters in the analytical models used for the soil hydraulic properties (Rawls and Brakensiek 1985; Vereecken et al. 1989, 1990). Recent reviews of PTFs are given by Leij et al. (2002) and Pachepsky et al. (2006).

The HYDRUS numerical models offer two PTF options. The first option is a soil catalog based on data from Carsel and Parish (1988). The second option is the Rosetta-Lite module of Schaap et al. (2001) to predict soil hydraulic parameters using five different levels of input data. Schaap et al. (1998) calibrated neural network PTFs on a large database of soil hydraulic and related properties. The simplest model (Model 1) uses the average of fitted hydraulic parameters within a textural class in the USDA textural triangle. The four other models in Rosetta use progressively more detailed input data, starting with the sand, silt, and clay fractions (Model 2), then adding a measured bulk density (Model 3), and water contents at 33 (Model 4) and 1500 (Model 5) kPa.

Although very convenient to use, PTFs can produce relatively large errors in the predicted soil hydraulic properties. Schaap et al. (1998) and Schaap and Leij (1998) reported that the root mean square errors (RMSE) for water content and $\log(K_s)$ predictions were 0.108 and 0.741, respectively, using the simplest Rosetta model (Model 1). They still obtained relatively large RMSE values of 0.063 and 0.610 for water contents and $\log(K_s)$, respectively, with Model 5, which requires the most detailed information. Other studies found that the accuracy for predicting water content using various PTFs is relatively good (RMSE of 0.02–0.11 cm^3 cm^{-3}) compared to predicting $\log(K_s)$ (RMSE no better than 0.5) (Jaynes and Tyler 1984; Ahuja et al. 1989; Tietje and Hennings 1996; Donatelli et al. 2004).

Example 3.6

Use the Rosetta-Lite module in HYDRUS-1D to predict K_s and the van Genuchten (1980a) water retention parameters (Equation 2.47) for the Ap horizon of the Cecil sandy loam soil shown in Table 3.2 using Model 1. How does the estimate of K_s compare to the measured value? This is HYDRUS Simulation 3.1–Rosetta-Lite Module.

Open HYDRUS-1D and close any projects that may be currently open. Then, click on the New Project icon in the toolbar at the upper left. You should see the New Project window (Figure 3.16). Type in a suitable name for the project (such as "PTF 1") and a description (such as "Example 3.6"). You can use the Browse button to change the directory. Click OK to close the window.

FIGURE 3.16 New Project window in HYDRUS-1D.

You should now see the Project window with two panels: on the left side a Pre-processing panel and on the right side a Post-processing panel (Figure 3.17). The normal procedure is to go through each of the preprocessing steps in order. Since only the Rosetta-Lite module will be used in this case, a few of the preprocessing steps usually used will be skipped.

Double-click on Geometry Information and the dialog window shown in Figure 3.18 will appear. Select "cm" as the unit of length.

Double-click on Time Information and the Time Information window shown in Figure 3.19 will appear. Select hours as the unit of time. Double-click on Water Flow–Soil Hydraulic Property Model and the dialog window shown in Figure 3.20 will appear. Select the van Genuchten-Mualem model. Note that other hydraulic models including Brooks and Corey (1964) and Durner (1994) are available. Also note the options for incorporating hysteresis. Select no hysteresis for this example.

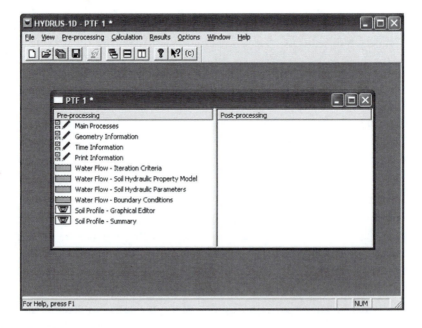

FIGURE 3.17 Project window with Pre-processing and Post-processing panels.

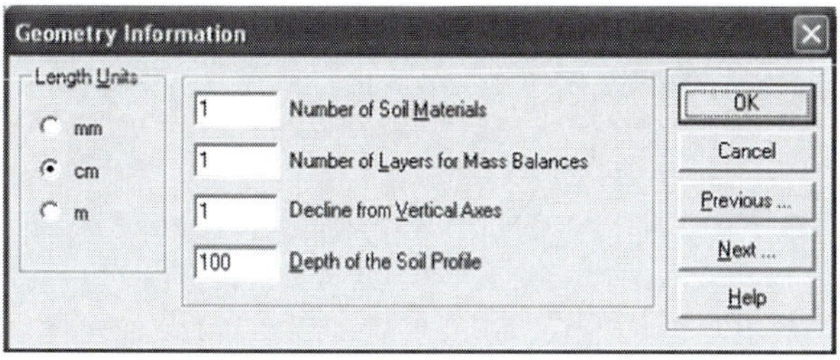

FIGURE 3.18   Geometry Information window.

Double-click on Water Flow–Soil Hydraulic Parameters to see the dialog window in Figure 3.21. The Soil Catalog PTF option appears in the lower left of the window. To use the Rosetta-Lite PTF, click on Neural Network Prediction and the Rosetta-Lite window will appear (Figure 3.22). In the Select Model panel, select Textural classes (Model 1 PTF that uses only textural class). In the Input panel, use the pulldown list to select loamy sand (the textural class of the Ap horizon) and click on "Predict".

You should now see van Genuchten-Mualem parameters, including K_s in the Output panel in units of cm day^{-1} (Figure 3.23). Click Accept and the Water Flow Parameters window will appear again with K_s converted to units of cm h^{-1}.

The predicted K_s is 4.38 cm h^{-1}, which is lower than the measured value in Table 3.2 (19.19 cm h^{-1}). The relatively high value of the measured K_s is probably due to the fact that this is the surface horizon where macropores are common. Structure and macropores are usually not included in PTFs because these variables are difficult to quantify, as discussed in earlier chapters. However, Lilly et al.

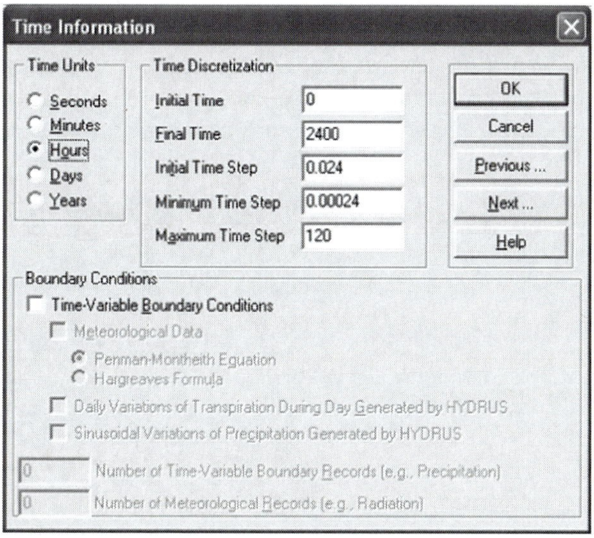

FIGURE 3.19   Time Information window.

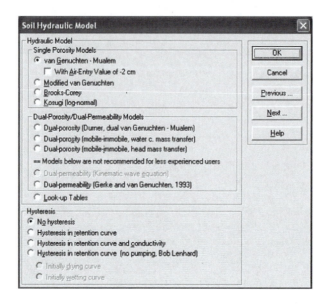

FIGURE 3.20   Soil Hydraulic Property Model window.

(2008) developed PTFs based on soil survey structure data (see Table 1.5) and showed that these inputs were important in predicting K_s.

3.5 SUMMARY

Water movement in soils occurs under both saturated and unsaturated conditions, under steady and transient flow conditions. In this chapter, the equations that govern steady water flow, the parameters that quantify these equations, and several applications to common problems are given. Soil texture, structure, and mineralogy affect saturated and unsaturated hydraulic conductivities. A number of equations are used to describe the unsaturated hydraulic conductivity function. Laboratory and field methods have been developed to measure saturated and unsaturated hydraulic

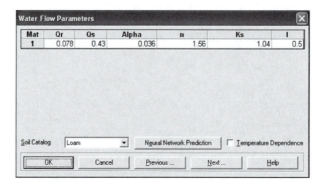

FIGURE 3.21   Water Flow Parameters window.

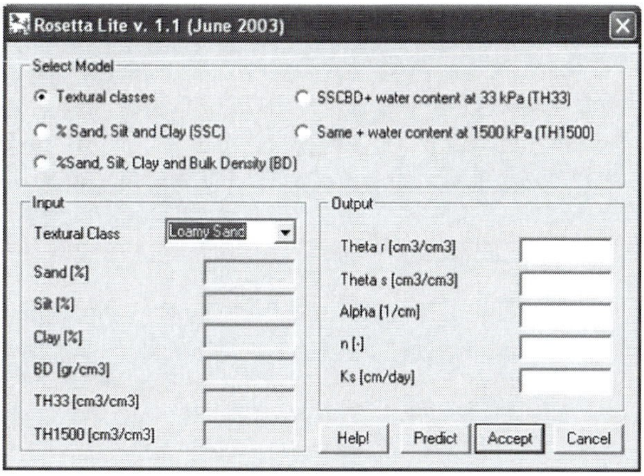

FIGURE 3.22 Rosetta-Lite window.

conductivities. Many of these methods use steady flow conditions. PTFs can be used to predict soil hydraulic properties from more easily measurable surrogate properties, such as soil texture. The HYDRUS-1D software package implements the Rosetta-Lite PTF module. Although true steady flow may occur only rarely, transient flow usually approaches steady flow over long time periods and the steady flow equations help to show what factors may be important under transient conditions.

3.6 DERIVATIONS

In the following sections, the Poiseuille equation 3.1, the drainage ellipse Equation 3.14, the van Genuchten (1980a) Equation 3.22, the relationship between α in the Gardner

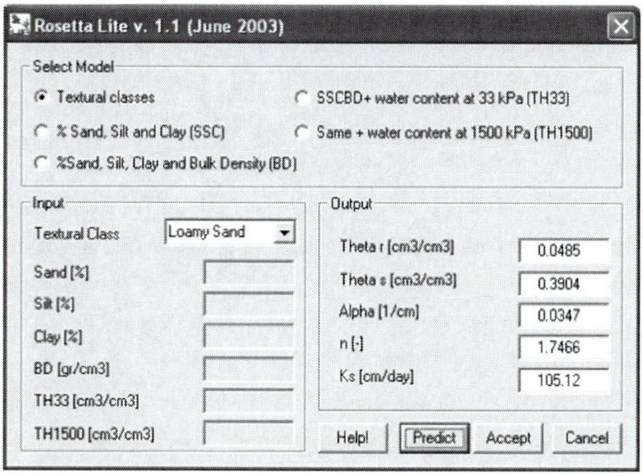

FIGURE 3.23 Predicted parameters for a loamy sand textural class in the output panel.

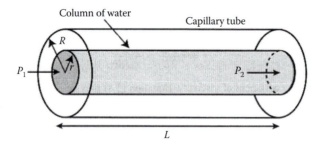

FIGURE 3.24   Section of capillary tube of radius R and length L filled with water flowing in response to a pressure difference $P_2 - P_1$. (Adapted from Jury, W.A., and Horton, R., *Soil Physics*, John Wiley, Hoboken, NJ, 2004. With permission.)

equation 3.16 and λ_c (Equation 3.26), the solution for steady downward flow to a water table (Equation 3.31), and the falling-head method (Equation 3.38) are derived.

3.6.1   POISEUILLE EQUATION

A horizontal section of a cylinder of radius R and length L is filled with water flowing to the right (Figure 3.24).

Hydrostatic pressures of P_1 and P_2 [ML^{-1}T^{-2}] act on the left and right ends of the cylinder. Consider the forces acting on an interior cylinder of water of radius $r < R$ [L]. The net hydrostatic force (F_p) [MLT^{-2}] is the difference in pressure acting on the ends of the interior cylinder, multiplied by the cross-sectional area of the cylinder:

$$F_p = \pi r^2 (P_2 - P_1) = \pi r^2 \Delta P.$$

Opposing this force is the fluid resistance or the shear drag force (F_s) [MLT^{-2}] acting along the length of the surface area of the interior cylinder caused by cohesion with other water molecules. The force is the shear stress (τ) [ML^{-1}T^{-2}] times the area:

$$F_s = \tau (2\pi r L).$$

At steady state, these forces balance each other exactly, such that (setting the forces equal to each other):

$$\tau = r \frac{\Delta P}{2L}.$$

In Chapter 2, the shear stress was shown to be related to the velocity gradient (dV/dy) [T^{-1}] and coefficient of dynamic viscosity of water (η) [ML^{-1}T^{-1}] (Equation 2.8):

$$\tau = -\eta \frac{dV}{dy}.$$

Writing the velocity gradient as dV/dr and substituting the definition for τ into the force balance equation, one obtains:

$$r \frac{\Delta P}{2L} = -\eta \frac{dV}{dr}.$$

To integrate this equation and find an equation for $V(r)$ [LT^{-1}], the equation is rearranged to separate variables. Then, the equation is integrated over the range r (an arbitrary radius less than R) to R and the corresponding range for V, $V(r)$ to $V(R) = 0$ (zero velocity at the wall of the cylinder):

$$r\,dr = -\frac{2L\eta}{\Delta P}\,dV$$

$$\int_r^R r\,dr = -\frac{2L\eta}{\Delta P}\int_{V(r)}^0 dV$$

$$\left.\frac{r^2}{2}\right|_r^R = -\frac{2L\eta}{\Delta P}\left.V\right|_{V(r)}^0$$

$$\frac{R^2}{2} - \frac{r^2}{2} = \frac{2L\eta}{\Delta P}V(r)$$

$$V(r) = \frac{\Delta P}{4L\eta}\left(R^2 - r^2\right).$$

The last equation shows the velocity distribution within a cylinder. It is zero at the wall where $r = R$ and reaches a maximum $V_{max} = \Delta PR^2/(4L\eta)$ at the center of the cylinder where $r = 0$. This distribution has the shape of a parabola, as can be seen in Figure 3.25, where the radius is plotted as a relative radius ($r/R = 0 \rightarrow 1$) and velocity as a relative velocity ($V/V_{max} = 0 \rightarrow 1$).

To find the volume of water that flows past a point in a unit time (Q) [L^{3}T^{-1}], the velocities must be integrated over the area (finding the volume of the parabola):

$$Q = \int V(r)\,dA.$$

Since the area is circular, it is easiest to use cylindrical coordinates where $dA = r\,dr\,d\varphi$ and φ is the angle in radians. For a parabolic cone, r ranges from 0 to R and φ from 0 to 2π radians:

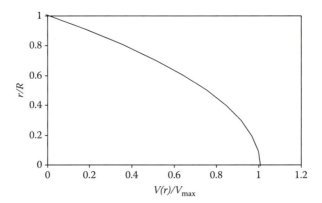

FIGURE 3.25 Parabolic distribution of velocities within a cylinder. The velocity is zero at the wall where $r = R$ and at a maximum $= V_{max}$ at the center of the cylinder where $r = 0$. (Adapted from Jury, W.A., and Horton, R., *Soil Physics*, John Wiley, Hoboken, NJ, 2004. With permission.)

$$Q = \int_0^R \int_0^{2\pi} V(r)\, r\, dr\, d\phi$$

$$= \phi\Big|_0^{2\pi} \int_0^R V(r)\, r\, dr$$

$$= 2\pi \int_0^R V(r)\, r\, dr.$$

Substituting the parabolic equation for $V(r)$:

$$Q = \frac{\Delta P \pi}{2L\eta} \int_0^R (R^2 - r^2)\, r\, dr$$

$$= \frac{\Delta P \pi}{2L\eta} \left[R^2 \int_0^R r\, dr - \int_0^R r^3\, dr \right]$$

$$= \frac{\Delta P \pi}{2L\eta} \left[R^2 \frac{r^2}{2}\Big|_0^R - \frac{r^4}{4}\Big|_0^R \right]$$

$$= \frac{\Delta P \pi}{2L\eta} \left[\frac{R^4}{2} - \frac{R^4}{4} \right]$$

$$= \frac{\Delta P \pi R^4}{8L\eta}.$$

For pressure (P), a substitution of $\rho_w g H$ is made, where ρ_w is the density of water $[ML^{-3}]$, g is the acceleration due to gravity $[LT^{-2}]$, and H is the total soil water potential head $[L]$. This results in the Poiseuille equation 3.1:

$$Q = \frac{\pi R^4 \rho_w g \Delta H}{8 L \eta}.$$

3.6.2 THE DRAINAGE ELLIPSE EQUATION

Geometry considered for defining the drainage equation (Equation 3.14) is given in Figure 3.8. The Dupuit-Forchheimer theory assumes that at steady state, the amount of water infiltrating the soil over an incremental distance x is equal to Rx, where R is the precipitation or irrigation rate $[LT^{-1}]$. This is equal to the water leaving the increment due to lateral flow:

$$Rx = -K_s \frac{dz}{dx} z.$$

To get the total inflow and outflow at steady state, indefinite integrals are used after separating variables:

$$Rx\,dx = -K_s z\,dz$$

$$R \int x\,dx = -K_s \int z\,dz$$

$$R \frac{x^2}{2} + C_1 = -K_s \frac{z^2}{2} + C_2$$

$$Rx^2 + K_s z^2 = C,$$

where C_1, C_2, and C are unknown constants that can be obtained by specifying boundary conditions. This equation gives the height of the water table z as a function of x. When $x = 0$, then $z = z_{max}$. These conditions can be used to find the value of the integration constant C:

$$K_s z_{max}^2 = C.$$

Substituting for C in the equation and rearranging:

$$Rx^2 + K_s z^2 = K_s z_{max}^2$$

$$\frac{Rx^2}{K_s z_{max}^2} + \frac{z^2}{z_{max}^2} = 1.$$

The last equation is that of an ellipse in the form:

$$\frac{x^2}{a^2} + \frac{y^2}{b^2} = 1,$$

centered around the origin with x-axis intercepts of $(\pm a, 0)$ and y-axis intercepts of $(0, \pm b)$. The water table in Figure 3.8 forms the upper part of this ellipse. One point on this curve is known where it intercepts the ditch or drain. At that point, $x = s$ and $z = z_0$. Solving the equation for z_{max} and then substituting these values into the equation produces Equation 3.14:

$$z_{max} = \sqrt{\frac{Rx^2}{K_s} + z^2}$$

$$= \sqrt{\frac{Rs^2}{K_s} + z_0^2}.$$

3.6.3 The van Genuchten $K(S_e)$ Function

The van Genuchten (1980a) water retention function (Equation 2.47):

$$S_e = \frac{1}{\left[1 + \left(-\alpha h\right)^n\right]^m},$$

is solved for h:

$$h(S_e) = -\frac{1}{\alpha}\left(\frac{1 - S_e^{1/m}}{S_e^{1/m}}\right)^{1/n}.$$

This equation is substituted into the Mualem equation 3.20:

$$K(S_e) = K_s S_e^l \frac{\left[\displaystyle\int_0^{S_e} \frac{dS_e}{h(S_e)}\right]^2}{\left[\displaystyle\int_0^1 \frac{dS_e}{h(S_e)}\right]^2}$$

$$= K_s S_e^l \frac{\left[-\alpha \displaystyle\int_0^{S_e} \left(\frac{S_e^{1/m}}{1 - S_e^{1/m}}\right)^{1/n} dS_e\right]^2}{\left[-\alpha \displaystyle\int_0^1 \left(\frac{S_e^{1/m}}{1 - S_e^{1/m}}\right)^{1/n} dS_e\right]^2} = K_s S_e^l \left[\frac{f(S_e)}{f(1)}\right]^2,$$

where:

$$f(S_e) = \int_0^{S_e} \left(\frac{S_e^{1/m}}{1 - S_e^{1/m}} \right)^{1/n} dS_e,$$

and $f(1)$ is similarly defined. To integrate this equation, the substitution $y = S_e^{1/m}$ is used, such that $dS_e = m \cdot y^{m-1} \, dy$ (the substitution will also cause the upper limit to change):

$$f(S_e) = \int_0^{S_e^{1/m}} \left(\frac{y}{1-y} \right)^{1/n} m \, y^{m-1} dy$$

$$= m \int_0^{S_e^{1/m}} (1-y)^{-1/n} y^{m-1+(1/n)} dy.$$

This integral cannot be solved unless the power of y is an integer. Let $k = m-1 + 1/n$. The simplest case is for $k = 0$, which leads to the assumption in Equation 3.23:

$$m = 1 - 1/n.$$

Substituting this equation for the exponents:

$$f(S_e) = m \int_0^{S_e^{1/m}} (1-y)^{-1/n} \, dy.$$

To simplify this integral, another substitution of $z = 1-y$ is made, such that $dy = -dz$. The lower and upper limits change to $z = 1$ and $z = 1-S_e^{1/m}$:

$$f(S_e) = m \int_1^{1-S_e^{1/m}} -z^{-1/n} dz$$

$$= -\frac{m}{-\dfrac{1}{n}+1} z^{-(1/n)+1} \Big|_1^{1-S_e^{1/m}}$$

$$= -\frac{m}{m} z^m \Big|_1^{1-S_e^{1/m}}$$

$$= -\left[\left(1 - S_e^{1/m}\right)^m - 1 \right]$$

$$= 1 - \left(1 - S_e^{1/m}\right)^m.$$

In a similar manner, it can be shown that $f(1) = 1$. Substituting $f(S_e)$ and $f(1)$ into the equation for $K(S_e)$, Equation 3.22 is produced:

$$K(S_e) = K_s S_e^l \left[1 - \left(1 - S_e^{1/m} \right)^m \right]^2.$$

3.6.4 Relationship between the Gardner Exponent and Macroscopic Capillary Length

Starting with Equation 3.25, substitute Equation 3.16 for $K(h)$ and then factor out the constant terms:

$$\lambda_c = \frac{\displaystyle\int_{h_i}^{h_0} K_s e^{\alpha h} dh}{K(h_0) - K(h_i)}$$

$$= \frac{K_s}{K(h_0) - K(h_i)} \int_{h_i}^{h_0} e^{\alpha h} dh.$$

To simplify the integral, make a substitution of $u = \alpha h$, such that $dh = du/\alpha$. The lower and upper limits will change to $u = \alpha h_i$ and $u = \alpha h_0$:

$$\lambda_c = \frac{K_s}{K(h_0) - K(h_i)} \int_{\alpha h_i}^{\alpha h_0} e^u \frac{du}{\alpha}$$

$$= \frac{K_s}{K(h_0) - K(h_i)} \cdot \frac{1}{\alpha} e^u \Big|_{\alpha h_i}^{\alpha h_0}$$

$$= \frac{K_s}{K(h_0) - K(h_i)} \cdot \frac{1}{\alpha} \left(e^{\alpha h_0} - e^{\alpha h_i} \right)$$

$$= \left[\frac{K_s e^{\alpha h_0}}{K(h_0) - K(h_i)} \cdot \frac{1}{\alpha} \right] - \left[\frac{K_s e^{\alpha h_i}}{K(h_0) - K(h_i)} \cdot \frac{1}{\alpha} \right]$$

$$= \left[\frac{K(h_0)}{K(h_0) - K(h_i)} \cdot \frac{1}{\alpha} \right] - \left[\frac{K(h_i)}{K(h_0) - K(h_i)} \cdot \frac{1}{\alpha} \right].$$

If h_0 is at or near saturation ($h_0 \approx 0$) and h_i is reasonably negative, then $K(h_0)$ will be much larger than $K(h_i)$. In that case, subtracting $K(h_i)$ from $K(h_0)$ has almost no effect and these terms can be dropped:

$$\lambda_c = \left[\frac{K(h_0)}{K(h_0)} \cdot \frac{1}{\alpha} \right] - \left[\frac{K(h_i)}{K(h_0)} \cdot \frac{1}{\alpha} \right] = \frac{1}{\alpha} - \left[\frac{K(h_i)}{K(h_0)} \cdot \frac{1}{\alpha} \right].$$

Also, if $K(h_0) >> K(h_i)$, then the second term is negligible. Thus, the final result leads to Equation (3.26):

$$\lambda_c = \frac{1}{\alpha}.$$

3.6.5 STEADY INFILTRATION TO A WATER TABLE

In accordance with Jury and Horton (2004), start with Equation 3.30, make the substitutions described in Section 3.3.4.1, and rearrange the denominator:

$$\int_0^h \frac{dh}{1 - \dfrac{\dfrac{i}{K_s}}{1 + \left(\dfrac{h}{a}\right)^2}} = -L - z$$

$$\int_0^h \frac{dh}{1 - \dfrac{i}{K_s}\left[1 + \left(\dfrac{h}{a}\right)^2\right]} = -L - z.$$

To simplify the integral, make a substitution of $y = h/a$, such that $dh = a\,dy$. The upper limit changes to $y = h/a$:

$$\int_0^{h/a} \frac{a\,dy}{1 - \dfrac{i}{K_s}\left[1 + y^2\right]} = -L - z$$

$$a \int_0^{h/a} \frac{dy}{1 - \dfrac{i}{K_s} - \dfrac{i}{K_s}y^2} = -L - z.$$

This equation is a standard form that may be looked up in integral tables (Abramowitz and Stegan 1970):

$$\int \frac{dy}{\alpha - \beta y^2} = \frac{1}{\sqrt{\alpha\beta}} \tanh^{-1}\left(\frac{y\sqrt{\alpha\beta}}{\alpha}\right),$$

where $\tanh^{-1}$ is the inverse hyperbolic tangent function. Let $\alpha = 1 - i/K_s$ and $\beta = i/K_s$:

$$a \int_0^{h/a} \frac{dy}{\alpha - \beta y^2} = -L - z$$

$$\frac{a}{\sqrt{\left(1 - \dfrac{i}{K_s}\right)\dfrac{i}{K_s}}} \tanh^{-1} \left[\frac{y\sqrt{\left(1 - \dfrac{i}{K_s}\right)\dfrac{i}{K_s}}}{1 - \dfrac{i}{K_s}} \right]_{y=0}^{y=h/a} = -(L + z)$$

$$\frac{a}{\sqrt{\left(1 - \dfrac{i}{K_s}\right)\dfrac{i}{K_s}}} \tanh^{-1} \left[\frac{h\sqrt{\left(1 - \dfrac{i}{K_s}\right)\dfrac{i}{K_s}}}{a\left(1 - \dfrac{i}{K_s}\right)} \right] - 0 = -(L + z).$$

The $\tanh^{-1}(y)$ function has a value of zero for $y = 0$, so the lower limit of integration produces a value of zero.

The above equation must be solved for h, which is inside the argument of the inverse hyperbolic tangent function. This can be done using the definition of the inverse hyperbolic tangent function:

$$\tanh^{-1}(y) = x \quad \text{if} \quad y = \tanh(x).$$

Solving for $\tanh^{-1}(y)$ and then using the above definition, one obtains:

$$\tanh^{-1} \left[\frac{h\sqrt{\left(1 - \dfrac{i}{K_s}\right)\dfrac{i}{K_s}}}{a\left(1 - \dfrac{i}{K_s}\right)} \right] = -\frac{\sqrt{\left(1 - \dfrac{i}{K_s}\right)\dfrac{i}{K_s}}}{a}(L + z)$$

$$\frac{h\sqrt{\left(1 - \dfrac{i}{K_s}\right)\dfrac{i}{K_s}}}{a\left(1 - \dfrac{i}{K_s}\right)} = \tanh \left[-\frac{\sqrt{\left(1 - \dfrac{i}{K_s}\right)\dfrac{i}{K_s}}}{a}(L + z) \right].$$

Solving for h and simplifying produces Equation 3.31:

$$h = a \frac{\left(1 - \dfrac{i}{K_s}\right)}{\sqrt{\left(1 - \dfrac{i}{K_s}\right)\dfrac{i}{K_s}}} \tanh\left[-\frac{\sqrt{\left(1 - \dfrac{i}{K_s}\right)\dfrac{i}{K_s}}}{a}(L+z)\right]$$

$$h = a\sqrt{\left(1 - \frac{i}{K_s}\right)\frac{K_s}{i}} \tanh\left[-\frac{\sqrt{\left(1 - \dfrac{i}{K_s}\right)\dfrac{i}{K_s}}}{a}(L+z)\right]$$

$$h = a\sqrt{\frac{K_s}{i} - 1} \tanh\left[-\frac{\sqrt{\left(1 - \dfrac{i}{K_s}\right)\dfrac{i}{K_s}}}{a}(L+z)\right].$$

3.6.6 THE FALLING-HEAD METHOD

Starting with Equation 3.37 and using separation of variables, the equation is

$$\frac{db}{dt} = -K_s \frac{b+L}{L}$$

$$\frac{db}{b+L} = -\frac{K_s}{L} dt.$$

Then, the equation is integrated from the starting time, $t = 0$, to a later time, $t = t_1$, when the depth of ponding is measured. The corresponding depths of ponding for these times are b_0 and b_1:

$$\int_{b_0}^{b_1} \frac{db}{b+L} = -\frac{K_s}{L} \int_0^{t_1} dt$$

$$\ln(b+L)\Big|_{b_0}^{b_1} = -\frac{K_s}{L} t\Big|_0^{t_1}$$

$$\ln(b_1 + L) - \ln(b_0 + L) = -\frac{K_s}{L} t_1.$$

Solving for K_s and simplifying produces Equation 3.38:

$$K_s = -\frac{L}{t_1}[\ln(b_1 + L) - \ln(b_0 + L)]$$

$$= \frac{L}{t_1}[-\ln(b_1 + L) + \ln(b_0 + L)]$$

$$= \frac{L}{t_1}\ln\left(\frac{b_0 + L}{b_1 + L}\right).$$

3.7 PROBLEMS

1. A soil column consists of a 5.0-cm thick clay layer underlain by a 25.0-cm layer of sand. Water is ponded on the surface to a depth of 2.0-cm and allowed to drip from the bottom of the column. K_s is 0.04 cm h^{-1} for the clay and 3.4 cm h^{-1} for the sand. What is the effective saturated hydraulic conductivity in centimeters per hour for the column? What will the steady flux be in centimeters per hour? What will the pressure head (cm) be at the interface between the clay and sand?

2. A clay liner at the bottom of a swine lagoon is 30 cm thick and has a saturated hydraulic conductivity of 10^{-7} cm s^{-1}. If the lagoon is filled to an average depth of 3.0 m and the pressure head immediately below the clay liner is −50 cm, what can you expect the steady seepage rate through the liner to be?

3. Calculate the discharge rate in cubic centimeters per hour through a cylindrical vertical macropore with a radius of 100 μm. The macropore is 30 cm in length and only gravity is causing flow. Assume a temperature of 20°C. Show that your units cancel.

4. Do Problem 10 in Chapter 2 using Excel. Add a new column that calculates S_e and a column that calculates $K(S_e)$ using the van Genuchten (1980a) Equation 3.22. Use $K_s = 30$ cm d^{-1} and $l = 0.5$. Make a graph of K vs. pressure head by plotting the first column on the x-axis and the last column on the y-axis. Show two graphs: one with the normal scale for the x- and y-axes and another using the log scale for both. Click on Layout under Chart Tools on the tool bar, select an axis in the Current Selection box, and use Format Selection to get options for changing axes including using a log scale.

5. Use the RETC program to fit a retention curve to the water retention data for the Ap horizon from Table 2.4 following the directions in Section 2.10. Include the value of K_s from Table 3.2 in the Water Flow Parameters window (Figure 2.45), but do not check it as a fitted value (since it is known). Print out the predicted log $K(h)$ vs. log h curve.

6. Follow Example 3.3 and use an Excel spreadsheet to calculate λ_c for a clay loam soil in Table 3.1.

7. Use an Excel spreadsheet to calculate the effective saturated hydraulic conductivity (K_{eff}) of the Cecil soil profile to a depth of 250 cm from the data

shown in Tables 2.5 and 3.2. Calculate the steady water flux (J_w) through this profile under saturated conditions if water was ponded at the surface to a depth of 1 mm and tile drains were installed (pressure head = 0) at 250 cm. Calculate the pressure head at each interface by setting up Darcy's equation for each layer (starting at the bottom) and solving for h at the top of the layer:

$$h_{top} = h_{bottom} - L - J_w \frac{L}{K_s},$$

where L is the thickness of the layer and K_s is the saturated hydraulic conductivity of the layer (see formula bar Figure 3.26). Make a graph of h vs. z. Enter the depths as negative values, but to calculate the thickness of each layer, take the absolute value of the difference between the depths. You can

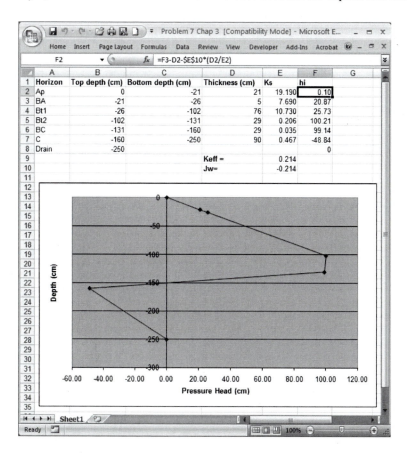

FIGURE 3.26 Excel spreadsheet for calculating pressure head distribution under steady flow in a profile of the Cecil soil with properties from Tables 2.5 and Table 3.2. Water is ponded to a depth of 0.1 cm at the surface and there are tile drains at a depth of 250 cm (pressure head equals zero).

control the position of the x-axis by double-clicking on the y-axis, clicking on the Scale tab, and entering a value of -300 for the point where the x-axis crosses the y-axis. Assume that any negative pressure heads, h, resulting during the solution are larger than the air-entry pressure head, h_a (i.e., that the profile remains saturated with constant hydraulic conductivities).

8. Show that if a soil's unsaturated hydraulic conductivity data fits the Brooks and Corey equation for $K(h)$ (Equation 3.28) with $\lambda = -1/3$, then the macroscopic capillary length (Equation 3.25) $\lambda_c = h_a \ln(h_0/h_i)$. Assume that λ_c is calculated over the range from a pressure head equal to h_a (the air-entry value) to a very negative pressure head where the unsaturated hydraulic conductivity is essentially zero.

9. If a soil has a $K_s = 5$ cm h^{-1} and there is an impermeable layer at a depth of 1.3 m, how far apart should drain tiles be installed if the depth of installation is 80 cm below the soil surface and the water table must not come closer to the surface than 30 cm? The steady irrigation rate is 0.5 cm d^{-1}.

10. What is the field-saturated hydraulic conductivity in centimeters per hour for a coarse sand if the steady percolation rate measured with a borehole permeameter is 1423 cm^3 min^{-1}? The radius of the borehole is 3.8 cm and the depth of ponding is 11 cm.

11. Show that when you substitute the Brooks and Corey water retention curve Equation 2.53 into the Brooks and Corey unsaturated hydraulic conductivity Equation 3.21 and assume that $l = 2$, you obtain (see Example 3.3):

$$K(h) = K_s \qquad\qquad h \geq h_a$$

$$K(h) = K_s \left(\frac{h_a}{h}\right)^{2+3\lambda} \qquad h < h_a$$

12. Solve Equation 3.14 for s. Show the intermediary steps.
13. Solve Equation 3.16 for α. Show the intermediary steps.
14. Show that the Gardner equation for $K(h)$ (Equation 3.16) plots as a straight line in a $\ln K(h)$ vs. h graph.

4 Heat Flow in Soils

4.1 INTRODUCTION

Many soil processes are affected by temperature, so heat transport is an important field of soil physics. In this chapter, the primary equations that apply to heat flow will be examined. The similarity of the heat flux equation to Darcy's law for water flux is emphasized. Heat transport will also be used to introduce nonsteady flow, the continuity equation, and methods of solving the partial differential equation (PDE) that describes nonsteady heat flow. The same methods will be used in Chapters 5 and 6 to solve the transient water flow and solute transport equations.

4.2 SURFACE ENERGY BALANCE

The source of all radiant energy on Earth is radiation from the sun. This represents an *energy flux* in units of energy per area per time [MT^{-3}], such as watts per square meter (W m^{-2}) (a watt is 1 Joule per second). The sun emits shortwave electromagnetic radiation as a consequence of its high temperature (5700 K at the surface). Typically, extraterrestrial solar radiation reaching the earth's atmosphere is about 1350 W m^{-2} (Hillel 2004).

Much of this shortwave radiation that reaches the outer atmosphere of the Earth does not penetrate to the Earth's surface (Figure 4.1) (Jury and Horton 2004). Clouds reflect about 28% of extraterrestrial solar radiation. Some extraterrestrial solar radiation is adsorbed by water vapor, oxygen, ozone, and CO_2 molecules in the atmosphere (about 16%). About 37% of extraterrestrial solar radiation undergoes diffuse scattering by molecules and particles in the air. Of the scattered radiation, about 11% is scattered upward out of the atmosphere and about 26% is scattered downward toward the earth's surface. This results in 19% of extraterrestrial solar radiation reaching the soil surface by direct transmission. The solar radiation reaching the soil surface, called *global solar radiation*, R_S, is the direct transmission component (19% of extraterrestrial solar radiation) and the downward scattering component (26% of extraterrestrial solar radiation) for a total of 45% of extraterrestrial solar radiation, all as shortwave radiation.

Some of the global solar radiation is reflected at the soil surface and this depends on the *albedo* (*a*) of the surface (Figure 4.2). The larger the albedo, the more shortwave radiation that is reflected. Albedo depends on the color, moisture content, and roughness of the soil surface and angle of the sun. It is the ratio of reflected to incident radiation, so it has no units. The average value for the earth surface is about 0.36 and typical values range from <0.10 for dark, wet, bare clays to 0.70 for fresh snow (Evett 2002). The earth, as a consequence of its temperature, emits longwave radiation, R_{earth}. It also receives longwave radiation from the sky, R_{sky}, thus the net downward radiation to the soil, R_n, is

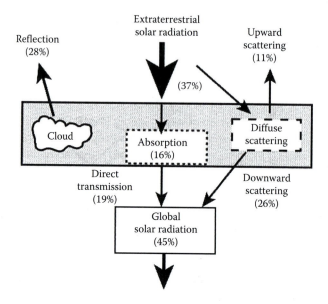

FIGURE 4.1 Typical partitioning of extraterrestrial radiation as it passes through the atmosphere. (From Jury, W.A., and Horton, R., *Soil Physics*, John Wiley, Hoboken, NJ, Inc., 2004. With permission.)

$$R_n = (1 - a)R_S + R_{sky} - R_{earth}, \tag{4.1}$$

where all radiation terms are in units of energy flux, such as watts per square meter. All terms on the right side of the equation are positive. Net radiation may be positive or negative, indicating an addition or loss of radiant energy, respectively, depending on the values of the other terms in Equation 4.1.

At the soil surface, an energy balance equation can be written indicating that the net heat energy penetrating the surface equals the net heat energy leaving the surface:

$$R_n = S + \lambda \cdot ET + J_H, \tag{4.2}$$

where all terms on the right side are positive when energy is moving away from the soil surface (up into the air or down into deeper soil). S is the *sensible* or *convective heat flux*, which represents vertical transport of warm air between the soil surface and the atmosphere above. J_H is the *soil heat flux* between the surface and deeper soil. Both flux terms have units of energy flux, such as watts per square meter, the same as net radiation. The middle term on the right side of Equation 4.2 represents the loss in heat energy due to *evapotranspiration* (evaporation plus plant transpiration). ET is the evapotranspiration in units of mass per area per time (kg m^{-2} s^{-1}, for example) and λ is the latent heat of vaporization in units of heat per mass (J kg^{-1}, for example). Since the heat of vaporization is quite

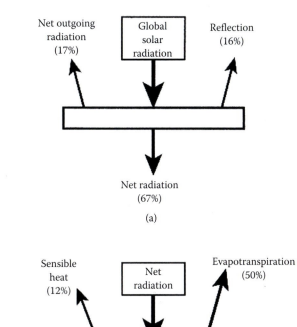

FIGURE 4.2 Typical partitioning of the global solar radiation as it reaches the land surface: (a) contributions to the net radiation; (b) net radiation partitioned into its components. (From Jury, W.A., and Horton, R., *Soil Physics*, John Wiley, Hoboken, NJ, Inc., 2004. With permission.)

large for water, 2441 J g^{-1} at 25°C, this is an important part of the energy balance (Table 2.1).

The relative sizes and signs of the different heat components in Equation 4.2 change from daytime to nighttime (Figure 4.3). During the daytime when the sun is shining, R_n is positive and the *soil surface* gains heat from the sun. It also loses heat as water is lost to evapotranspiration and heat moves away from the surface to the air above and soil below. At night, there is no input from solar radiation. As the surface cools below the temperature of the air above and the soil below, the sensible heat flux and soil heat flux is toward the surface. Some heat is still lost due to evaporation.

The energy balance equation is used to develop an important estimate of *potential* evapotranspiration, ET_0, or the amount of water that would be lost from a well-watered, completely vegetated, surface. This is the Penman-Monteith equation (Monteith 1981; Monteith and Unsworth 1990; FAO 1990):

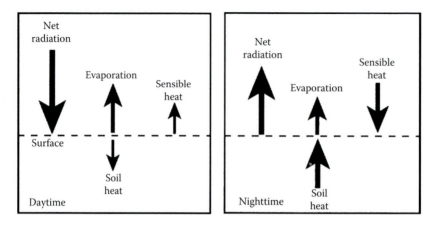

FIGURE 4.3 Components of the surface energy balance during daytime (left) and night-time (right). (From Jury, W.A., and Horton, R., *Soil Physics*, John Wiley, Hoboken, NJ, Inc., 2004. With permission.)

$$ET_0 = \frac{1}{\lambda}\left[\frac{\Delta(R_n - J_w)}{\Delta + \gamma(1 + r_c/r_a)} + \frac{\rho c_p(e_a - e_d)/r_a}{\Delta + \gamma(1 + r_c/r_a)}\right],\tag{4.3}$$

where ET_0 is the evapotranspiration rate (kg m^{-2} s^{-1}), λ is the latent heat of vaporization (J kg^{-1}), R_n is the net radiation at the surface (J m^{-2} s^{-1}), J_w is the soil heat flux (J m^{-2} s^{-1}), ρ is the atmospheric density (kg m^{-3}), c_p is the specific heat of moist air (J kg^{-1} °C^{-1}), e_a is the saturation vapor pressure at temperature T (Pa), e_d is the actual vapor pressure (Pa), r_c is the crop canopy resistance (s m^{-1}), and r_a is the aerodynamic resistance (s m^{-1}). The slope of the vapor pressure curve, Δ (Pa °C^{-1}) (Tetens 1930; Murray 1967), and the psychrometric constant, γ (Pa °C^{-1}) (Brunt 1952), are defined as follows:

$$\Delta = \frac{4098\,e_a}{(T + 237.3)^2},\tag{4.4}$$

$$\gamma = \frac{c_p P}{\varepsilon\lambda} = \frac{1013\,\text{J kg}^{-1}\,°\text{C}^{-1}}{0.622}\frac{P}{\lambda} = 1629\frac{P}{\lambda},\tag{4.5}$$

respectively, where T is the average air temperature (°C), P is the atmospheric pressure (Pa) and ε is the ratio of the molecular weights of water vapor and dry air (i.e., 0.622).

The Penman-Monteith equation is frequently used to estimate potential evapotranspiration based on daily measurements of net radiation, temperature, and wind speed.

Example 4.1

1350 w/m² *(handwritten)*

If extraterrestrial solar radiation is about 1.35 kW m⁻², what sort of upper limit does this put on evapotranspiration in centimeters per day?

From Table 2.1, the heat of vaporization of water at 25°C is 2441 J g⁻¹. Divide extraterrestrial solar radiation by the heat of vaporization:

$$\frac{\left(1.35\dfrac{kW}{m^2}\right)\left(\dfrac{10^3 W}{kW}\right)\left(\dfrac{J/s}{W}\right)}{2441\dfrac{J}{g}} = 0.553\frac{g}{m^2 s}.$$

Convert meters to centimeters and seconds to days:

$$\left(0.553\frac{g}{m^2 s}\right)\left(\frac{m^2}{10^4 cm^2}\right)\left(\frac{8.64 \cdot 10^4 s}{d}\right) = 4.78\frac{g}{cm^2 d}.$$

Divide by the density of water, approximately 1 g cm⁻³:

$$\frac{4.78\dfrac{g}{cm^2 d}}{1\dfrac{g}{cm^3}} = 4.78\frac{cm}{d}.$$

equivalent depth (handwritten)

Hence, the upper limit for evapotranspiration is about 4.78 cm d⁻¹.

4.3 STEADY SOIL HEAT FLUX

The soil heat flux term in Equations 4.2 and 4.3 represents transport of heat in soil by a number of mechanisms. The two most important heat transport mechanisms are *conduction* and *convection*. Conduction is the transfer of heat within the solid and liquid part of the soil and convection is the transfer of heat through the motion of water vapor and liquid water. For most of this chapter, it will be assumed that water is stationary, thus there is no heat movement due to convection. In most soils, convection does not have a large effect because conduction in a wet soil is so high that it dominates heat transport (convection can be ignored). In a dry soil, convection will also be minimal since water movement is very slow. The effect of conduction on vertical heat flux is described using a form of Fourier's law:

$$J_H = -\lambda_e \frac{dT}{dz}, \tag{4.6}$$

where T is temperature and λ_e is the soil effective thermal conductivity. If the temperature gradient is in units of kelvin per meter and the heat flux is in watts per square meter, then the units for λ_e are watts per meter per kelvin. Note the similarity between Fourier's law and Darcy's law (Equation 3.4). Heat flows from locations with high temperature to locations with low temperature, so a negative sign appears on the right side of the equation. Upward fluxes will be considered positive and downward fluxes negative, the same as Darcy's equation (and contrary to the sign convention in the energy balance Equation 4.2).

Typical values for effective thermal conductivity are shown in Figure 4.4. It can be seen that λ_e increases sharply with water content (water is a good conductor of heat). There is not a large difference between a sand and a clay soil at a given water content when the soils have the same porosity (0.60), so the solid phase is not as important as the water content. The peat soil has the lowest conductivity because, with the large value for porosity, much of the pore space is filled with air even at the highest water content. Essentially, thermal conductivity decreases for all soils as volumetric air content (a) increases (Jury and Horton 2004).

Although λ_e is a function of θ, it is *not* a function of the variable in the gradient term, T. This is an important distinction between Fourier's law and the Buckingham-Darcy law (Equation 3.15), as will be seen in Chapter 5.

Example 4.2

Calculate the steady heat flux in a dry sand with a porosity of 0.6, soil surface temperature of 35°C, and a soil temperature at a depth of 1.0 m of 25°C. Do the same calculation for a wet sand.

Estimate the values for λ_e using Figure 4.4 (approximately 0.25 W m^{-1} K^{-1} for a dry sand and 1.4 W m^{-1} K^{-1} for a wet sand). Approximate the gradient derivative in Equation 4.6 by $\Delta T/L$, where $\Delta T = T_2 - T_1$. The temperature at the surface is T_2

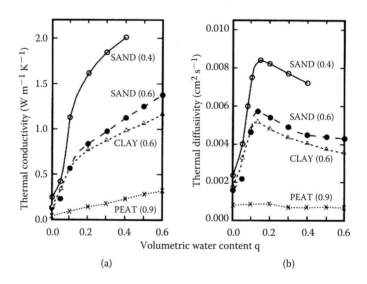

FIGURE 4.4 Effective soil thermal conductivity (a) and diffusivity (b) as a function of water content for various soil types. Numbers in parentheses refer to porosity. (From Jury, W.A., and Horton, R., *Soil Physics*, John Wiley, Hoboken, NJ, Inc., 2004. With permission.)

and the temperature at a depth of 1.0 m is T_1. L is the distance between the two locations, 1.0 m. Since a change in temperature in degree Celsius is the same as a change in temperature in kelvin, it does not matter whether you use degree Celsius or kelvin. For the wet sand:

$$J_H = -\lambda_e \frac{\Delta T}{L}$$

$$= -\lambda_e \frac{T_2 - T_1}{L}$$

$$= -\left(1.4 \frac{W}{m\,K}\right)\left(\frac{35\,°C - 25\,°C}{1.0\,m}\right)$$

$$= -14 \frac{W}{m^2}.$$

For the dry sand:

$$J_H = -\left(0.25 \frac{W}{m\,K}\right)\left(\frac{35\,°C - 25\,°C}{1.0\,m}\right)$$

$$= -2.5 \frac{W}{m^2}.$$

Fluxes are downward since they are negative.

Temperature in soil can be measured using thermocouples (McInnes 2002). As discussed in Chapter 2, a thermocouple consists of an electrical junction between two types of metal wires. As the junction changes temperature, the *Seebeck effect* causes a current to flow. This is detected by a volt meter and a calibration converts voltage to temperature. A thermocouple junction is embedded in a small, aluminum housing attached to a cable so it can be buried in soil.

One might ask what the temperature distribution looks like in soils under steady heat flow conditions. Since Fourier's law and Darcy's law for saturated flow take identical forms, the same procedure employed in Chapter 3 (Equations 3.6 through 3.8) can be used to show that temperature varies linearly with depth in a uniform soil under steady heat flow conditions.

4.4   TRANSIENT SOIL HEAT FLUX

Steady heat flow rarely occurs for very long in soils because temperature is constantly changing at the soil surface (as a function of time of day, season, plant cover, etc.), so the equations that describe *nonsteady* or *transient* heat transport are of interest.

4.4.1   Heat Transport Equation

The *heat conservation equation* is developed (as are conservation equations for other transient soil processes) by considering a small elementary soil volume of dimensions Δx, Δy, and Δz (Figure 4.5). Only heat flow in the vertical direction (z positive up) will be considered for simplification purposes (the same procedure can be used to develop an equation for three-dimensional flow). According to the law of conservation, the heat flux into the volume during a small time interval Δt must be equal to the heat flux out of the volume, plus any change in heat stored within the volume and any loss or gain of heat within the volume due to sources or sinks:

$$\text{flux in} = \text{flux out} + \text{heat stored} + \text{heat source / sinks.}$$

Under steady flow, temperature is a function of position only, $T(z)$, for example, if changes in the vertical position only (not the x- or y-dimensions) are considered. Similarly, for steady flow, heat flux is a function of position only, $J_H(z)$. In nonsteady flow, both of these will be a function of position *and* time: $T(z,t)$ and $J_H(z,t)$.

For transient (nonsteady) flow, one can write the flux up through the bottom of the elementary volume at time t as $J_H(z,t)$ (Figure 4.5). The flux through the top of the elementary volume is $J_H(z + \Delta z,t)$. To get the fluxes during a time interval, Δt, the flux at a time midway through the interval is considered, i.e., $t + \Delta t/2$. The amount of heat crossing the bottom face in the time interval would be the flux (or J m^{-2} s^{-1}) multiplied by the area of the face (m^2) and the time interval (s):

$$J_H\left(z, t + \frac{\Delta t}{2} \right) \cdot \Delta x \cdot \Delta y \cdot \Delta t. \tag{4.7}$$

Similarly, the amount of heat exiting the top face would be

$$J_H\left(z + \Delta z, t + \frac{\Delta t}{2} \right) \cdot \Delta x \cdot \Delta y \cdot \Delta t. \tag{4.8}$$

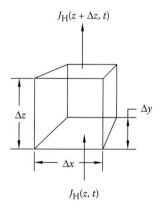

FIGURE 4.5   Elementary volume of soil for calculating the heat conservation equation.

If the heat stored per unit volume (in J m^{-3}) in the element at time t is $H(z + \Delta z/2, t)$, where the midpoint depth of the volume is used to represent the average for the volume, then the change in the amount of heat (in J) during the time interval Δt will be

$$\left[H\left(z + \frac{\Delta z}{2}, t + \Delta t \right) - H\left(z + \frac{\Delta z}{2}, t \right) \right] \cdot \Delta x \cdot \Delta y \cdot \Delta z. \tag{4.9}$$

The amount of heat disappearing from the volume due to heat transformations can be expressed as:

$$s_H \cdot \Delta x \cdot \Delta y \cdot \Delta z \cdot \Delta t, \tag{4.10}$$

where s_H is the heat sink rate (J m^{-3} s^{-1}).

Putting each of these terms (Equations 4.7 through 4.10) into the heat conservation equation for the elementary volume:

$$J_H\left(z, t + \frac{\Delta t}{2} \right) \cdot \Delta x \cdot \Delta y \cdot \Delta t = J_H\left(z + \Delta z, t + \frac{\Delta t}{2} \right) \cdot \Delta x \cdot \Delta y \cdot \Delta t$$

$$+ \left[H\left(z + \frac{\Delta z}{2}, t + \Delta t \right) - H\left(z + \frac{\Delta z}{2}, t \right) \right] \tag{4.11}$$

$$\cdot \Delta x \cdot \Delta y \cdot \Delta z + s_H \cdot \Delta x \cdot \Delta y \cdot \Delta z \cdot \Delta t.$$

Dividing through by $\Delta x \cdot \Delta y \cdot \Delta z \cdot \Delta t$:

$$\frac{J_H\left(z, t + \frac{\Delta t}{2} \right)}{\Delta z} = \frac{J_H\left(z + \Delta z, t + \frac{\Delta t}{2} \right)}{\Delta z}$$

$$+ \frac{H\left(z + \frac{\Delta z}{2}, t + \Delta t \right) - H\left(z + \frac{\Delta z}{2}, t \right)}{\Delta t} + s_H. \tag{4.12}$$

Moving all the terms to the right side:

$$0 = \frac{J_H\left(z + \Delta z, t + \frac{\Delta t}{2} \right) - J_H\left(z, t + \frac{\Delta t}{2} \right)}{\Delta z}$$

$$+ \frac{H\left(z + \frac{\Delta z}{2}, t + \Delta t \right) - H\left(z + \frac{\Delta z}{2}, t \right)}{\Delta t} + s_H. \tag{4.13}$$

If the units of each term are checked, one finds that they are all Joules per cubic meter per second or watts per cubic meter. The definition of the partial derivative of the flux with respect to depth is

$$\frac{\partial J_H}{\partial z} = \lim_{\Delta z \to 0}\left[\frac{J_H(z + \Delta z, t) - J_H(z, t)}{\Delta z}\right], \tag{4.14}$$

and the definition of the partial derivative of stored heat with respect to time is

$$\frac{\partial H}{\partial t} = \lim_{\Delta t \to 0}\left[\frac{H(z, t + \Delta t) - H(z, t)}{\Delta t}\right]. \tag{4.15}$$

If the size of the elementary volume and time interval are reduced ($\Delta z \to 0$ and $\Delta t \to 0$), then one can write the heat conservation equation as:

$$\frac{\partial H}{\partial t} + \frac{\partial J_H}{\partial z} + s_H = 0. \tag{4.16}$$

This is a *partial differential equation* with *independent* variables z and t and *dependent* variables H and J_H. Now, the objective is to reduce the number of dependent variables to just one. For simplification purposes, one can assume that there are no sinks ($s_H = 0$). Also, a definition for stored heat content per unit volume can be used:

$$H = C_p(T - T_{ref}), \tag{4.17}$$

where C_p is the *volumetric heat capacity* of the porous media [$ML^{-1}T^{-2}K^{-1}$] in Joules per cubic meter per degree Celsius. This is the heat that a volume of soil can store for a given rise in temperature. For example, wet soils have a large volumetric heat capacity compared to dry soils, due to the large specific heat of water (Table 2.1). T is temperature (°C) and T_{ref} is an arbitrary reference temperature at which $H = 0$. If one takes the derivative of Equation 4.17 with respect to time and assumes that the heat capacity is constant with time:

$$\frac{\partial H}{\partial t} = C_p \frac{\partial (T - T_{ref})}{\partial t} = C_p \frac{\partial T}{\partial t}. \tag{4.18}$$

Fourier's law (Equation 4.6) can be used for J_H. Substituting Equations 4.6 and 4.18 into Equation 4.16 (and assuming $s_H = 0$) gives:

$$C_p \frac{\partial T}{\partial t} = \frac{\partial}{\partial z}\left(\lambda_e \frac{\partial T}{\partial z}\right). \tag{4.19}$$

When λ_e is not a function of depth, it can be brought outside the derivatives on the right side:

$$C_p \frac{\partial T}{\partial t} = \lambda_e \frac{\partial^2 T}{\partial z^2}.$$ (4.20)

This is now a PDE with only one dependent variable, temperature. If this equation can be *solved*, the result will be an equation for $T(z,t)$, which will predict temperature in a soil profile and how it changes over time for a given situation. There are two *parameters* in Equation 4.20, C_p and λ_e, which must be known if this equation is to be used. If Equation 4.20 is divided by C_p, the result is the *heat flow equation*:

$$\frac{\partial T}{\partial t} = K_T \frac{\partial^2 T}{\partial z^2},$$ (4.21)

where K_T is a new parameter [L^2T^{-1}] defined as the *soil thermal diffusivity*:

$$K_T = \frac{\lambda_e}{C_p}.$$ (4.22)

Soil thermal diffusivity (K_T) in square centimeters per second as a function of θ for several soils is shown in Figure 4.4. Thermal diffusivity increases sharply with water content for low values of θ to a maximum and then stays relatively constant with further increases in θ. The reason for this is that heat capacity rises linearly with θ, whereas thermal conductivity increases most rapidly at low θ.

The heat flow Equation 4.21 takes the form of many other equations that describe nonsteady soil processes (as will be seen in Chapters 5 and 6). The equation shows that the change in temperature with time at a particular depth in soil depends on the second derivative of the distribution of temperature within the profile. If the second derivative of a variable is positive in the local area around a point, this means that the function is *concave*. If z is plotted on the horizontal axis, the distribution of temperatures near the point could appear as the curved line in Figure 4.6. Since heat flows from high to low temperature, the direction of net heat flow will be toward the center of the area. Hence, the heat flow equation shows that if $\partial^2 T/\partial z^2$ is positive in a

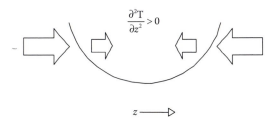

FIGURE 4.6 When the second derivative in T with respect to z is positive in the region of a point, the temperature increases at that point.

region around a point, the temperature at that point will increase ($\partial T/\partial t > 0$) because heat will flow toward that point. Similarly, if the second derivative of the temperature distribution is negative (the curve is *convex*) locally, then heat net flow will be away from the point and the local temperature will decrease ($\partial T/\partial t < 0$). In fact, anywhere along a concave curve temperatures will increase, even if the direction of heat flow is the same. For example, on the left half of Figure 4.6, where temperatures decrease toward the center and heat flow is to the right, inflow from the left to a point is larger (indicated by a larger arrow) than outflow to the right (indicated by a smaller arrow) and the temperature at that point will be rising as well.

The heat flow equation shows that the change in temperature at a point depends on thermal diffusivity, as well as the local distribution of temperature. Thermal diffusivity represents two contrary processes for transmitting heat, conductivity (λ_e) and heat capacity (C_p). If the soil is conductive (as it would be at a high θ), the heat is easily transported so the local temperature is expected to rise or drop, depending on the local temperature distribution. But a high water content would also increase the capacity of the soil to adsorb heat (large C_p) and this would slow the transport of heat.

There is one last question one can answer just by looking at Equation 4.21: what happens if the temperature distribution is locally neither concave nor convex? In that case, it would be a straight line and $\partial^2 T/\partial z^2 = 0$, which means that $\partial T/\partial t = 0$ and the temperature does not change. In other words, it is a steady flow problem and, as seen earlier, the temperature distribution with depth is expected to be a straight line.

4.4.2   ANALYTICAL SOLUTIONS TO THE HEAT TRANSPORT EQUATION

The task now is to *solve* the heat transport Equation 4.21. This means that one must integrate this equation to get an expression for temperature as a function of depth and time: $T(z,t)$. This is analogous to what was done in Chapter 3, where the Buckingham-Darcy integral (Equation 3.15) was solved for a particular set of boundary conditions. The solution was Equation 3.31, which gave pressure head (h) as a function of depth (z). The Buckingham-Darcy equation is an *ordinary differential equation* (ODE) in that there is only one independent variable and that is z (and the differential notation was dz). By contrast, Equation 4.21 is a PDE in that there are two independent variables, z and t (and the differential notation is ∂z and ∂t).

Just as the solution to the Buckingham-Darcy ODE was applicable to a particular set of *boundary conditions* ($h = 0$ at the water table, for example), the solution to the heat flow PDE will be applicable to a particular set of boundary conditions and, since this is a time-dependent equation, one must also specify the *initial conditions*. For example, the boundary conditions will be given at the soil surface where there is a constant temperature (T_0) and at some point deep in the soil where the temperature does not change. The initial temperature between these two points can be specified as being zero, so the temperature deep in the soil is also zero. Hence, the boundary conditions for this particular problem written in terms of $T(z,t)$ are

$$T(0,t) = T_0$$

$$T(\infty,t) = 0,$$

(4.23)

where T_0 is any constant temperature, such as 30°C, and ∞ is infinity (note z is considered positive in the downward direction in this derivation). These are known as *type-1* or *Dirichlet boundary conditions* in that they are given in terms of the dependent variable (T in this case). The initial condition is

$$T(z,0) = 0. \tag{4.24}$$

The first step in solving a PDE is to convert it to an ODE. One of the ways to do this is to use an integral *transform* (Figure 4.7). Then the ODE is solved using one of the standard methods of integration (not that this is a simple task). The final step is to use an *inverse transform* to get the final solution in terms of the variable of interest (T in this case). There are numerous transform types, but the two most common transforms are the *Laplace transform* and the *Fourier transform*. The Laplace transform will be used to change Equation 4.21 to an ODE and then, once the ODE is solved, the inverse Laplace transform will be used to get a solution in $T(z,t)$ (Jury and Roth 1990).

Properties of the Laplace transform are given in Section 4.71. The Laplace transform method of solving the heat flow equation is shown in Section 4.72, where the final solution is

$$T(z,t) = T_0 \cdot \mathrm{erfc}\left(\frac{z}{2\sqrt{K_T t}} \right), \tag{4.25}$$

where "erfc" is the *complementary error function*. Equation 4.25 is the solution to the heat flow equation for the particular set of boundary (Equation 4.23) and initial conditions (Equation 4.24) specified. Properties of the complementary error function are discussed in Section 4.7.3.

Plots of temperature as a function of depth for different times, given a value of $K_T = 4.5 \cdot 10^{-3}$ cm^2 s^{-1} and $T_0 = 30$°C, using Equation 4.25 are shown in Figure 4.8. To make the graphs, z is considered negative in the downward direction so that the

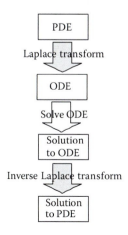

FIGURE 4.7 Using the Laplace transform to solve a PDE.

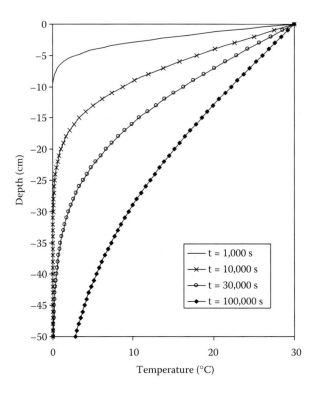

FIGURE 4.8 Temperature (°C) as a function of depth (centimeter) for times of 1,000; 10,000; 30,000; and 100,000 s based on Equation 4.25.

soil surface will appear at the top of the graph. Note that the choice of units for K_T and T_0 determines the units for z, t, and T. Shortly after time zero, the heating front is steep and has penetrated only to about 10 cm. As time increases, heat moves deeper into the soil and the front becomes more diffuse. Note that for all the curves, the boundary condition at the surface ($T_0 = 30°C$) is satisfied. For very large times, the distribution of temperature approaches a straight line, which indicates that it is approaching steady state conditions.

Example 4.3

Use Excel to show the effect of varying K_T using Equation 4.25. Compare the depth of the heating front in a dry soil vs. a wet soil, assuming that the dry soil has a K_T of $2 \cdot 10^{-3}$ cm² s⁻¹ and the wet soil has a $K_T = 4.5 \cdot 10^{-3}$ cm² s⁻¹ based on Figure 4.4. Show the distribution of temperatures after 20,000 s.

See Figure 4.9. Enter depths in column A, starting at zero at the soil surface and going to 50 cm in increments of 5 cm. Enter negative depths for making the graph in column B. Reserve the next two columns (C and D) for the temperatures as a function of depth for the dry and wet soils. In cells E2 through H2, enter the values for the constants: T_0, t, K_T for the dry soil, and K_T for the wet soil. In columns C and D, enter the heat flow equation (Equation 4.25), using the appropriate constant for

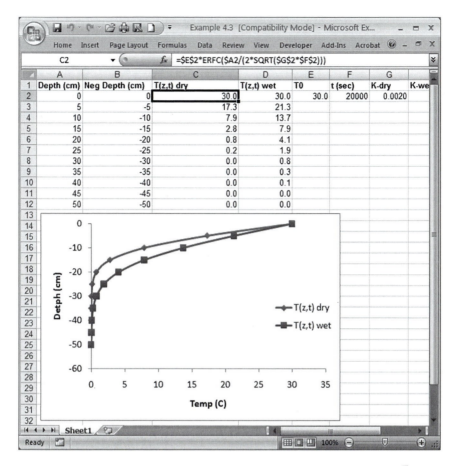

FIGURE 4.9 Excel spreadsheet for Example 4.3. The formula bar shows the equation for temperatures (Equation 4.25).

thermal diffusivity for the dry and wet soils. The complementary error function in Excel is ERFC(X), where X is the argument of the function (see the formula bar in Figure 4.9).

Make a graph using negative depth on the y-axis and temperature on the x-axis. To do this, select the data in columns B, C, and D (don't include the headings) and then click on the Insert tab on the tool bar. Select the Scatter plot with lines connecting points. Click on Select data on the tool bar. Then click on Edit. Use the buttons at the right of each window to select the correct columns for the x and y axes. Click OK and repeat the process for the other data set. Click on the graph and then select Layout under Chart Tools at the top of the tool bar. You can add and edit titles for the axes using icons in the tool bar section for Labels. You can control the format of the axes by clicking on the graph, clicking on the Format tab under Chart Tools, and selecting the horizontal or vertical axes from the pull-down menu at the left of the tool bar. Axis titles can be added in the Labels section of the Layout tab.

The heating front has moved deeper in the wet soil and the effect of K_T is apparent.

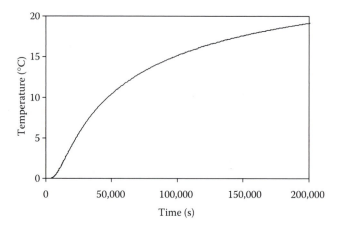

FIGURE 4.10 Temperature as a function of time at a depth of 20 cm using Equation 4.25 and a thermal diffusivity of $4.5 \cdot 10^{-3}$ cm^2 s^{-1}.

Another way to use the solution to the heat flow equation is to look at the temperature at a given depth over time, as might be measured by a thermocouple buried at a given depth. To do this, Equation 4.25 is plotted as a function of time for a fixed value of z. In Figure 4.10, the temperature at a depth of 20 cm is shown, using the same thermal diffusivity used in Figure 4.8 ($4.5 \cdot 10^{-3}$ cm^2 s^{-1}). The temperature starts at zero (the initial condition) and begins to rise when the heating front reaches this depth after about 10^4 s. One might call this a *breakthrough curve* for temperature.

The approach used here can be used to look at the effect of other boundary and initial conditions. The annual distribution of temperature often follows a sinusoidal distribution with time (it is *periodic*). In Figure 4.11, the long-term average monthly soil temperatures at a depth of 10 cm measured at Davis, CA, are shown (Jury and Horton 2004).

The data points can be represented quite well by an equation of the form:

$$T(t) = T_A + A \sin(\omega t + \phi), \tag{4.26}$$

where T_A is the average annual temperature at a depth of 10 cm, A is the *amplitude* of the temperature fluctuations at this depth (half of the difference between the maximum and minimum temperatures), ω is the *angular frequency*, and ϕ is a *phase shift* to ensure that the sine wave starts at the right time. The frequency is how often the periodic function completes a cycle in a given time. The units are inverse time. It is often given in Hertz (Hz), which is cycles per second. A high frequency wave completes many cycles in a given time period.

The angular frequency is inversely related to the *period* of the wave or *wavelength* (τ), which is the time it takes for the wave to complete a cycle:

$$\omega = \frac{2\pi}{\tau}. \tag{4.27}$$

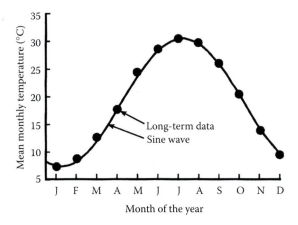

FIGURE 4.11 Long-term mean monthly temperature at a 10-cm depth measured at Davis, CA. (From Jury, W.A., and Horton, R., *Soil Physics*, John Wiley, Hoboken, NJ, Inc., 2004. With permission.)

Obviously, a high frequency wave has a short period, and vice versa.

To model annual fluctuations in soil temperature in response to the changes in mean monthly temperatures at the surface (shown in Figure 4.11), an equation such as Equation 4.26 can be used as the surface boundary condition. Instead of a constant temperature at the surface as in the earlier problem (Equation 4.23), the boundary condition is now:

$$T(0,t) = T_A + A \sin(\omega t + \phi). \tag{4.28}$$

The average temperature (T_A) in Figure 4.11 is about 18°C and the amplitude (A) is about 11°C. To find the value for ϕ, one can use the initial condition that when t equals zero in Equation 4.28, $\sin(\phi)$ should be equal to -1. Thus, $\phi = -\pi/2$. The Laplace transform method can be used to solve the heat flow Equation 4.21 for this surface boundary condition, keeping the boundary condition used earlier for deep in the soil (no change from the initial temperature) and an initial temperature equal to T_A. Again, z is positive in the downward direction in this derivation. The derivation will not be shown here, but the solution is

$$T(z, t) = T_A + Ae^{-(z/d)} \sin\left(\omega t + \phi - \frac{z}{d}\right), \tag{4.29}$$

where d is the *damping depth*:

$$d = \sqrt{\frac{2K_T}{\omega}} = \sqrt{\frac{K_T \tau}{\pi}}. \tag{4.30}$$

The period (τ) of the annual temperatures is 365 days, of course. Converting $K_T = 4.5 \cdot 10^{-3}$ cm^2 s^{-1} to units of square centimeters per day, the result is $K_T = 388.8$ cm^2 d^{-1}. Using these values, $\omega = 0.0172$ d^{-1} (Equation 4.27) and $d = 212$ cm (Equation 4.30). The temperature fluctuations at depths of 0, 100, 200, and 600 cm calculated using Equation 4.29 are shown in Figure 4.12. Time t is allowed to vary over a period that starts when surface temperatures are at a minimum (January) and covers roughly two years. It is apparent that the amplitude of the temperature fluctuations diminishes with depth. This is caused by the exponential decay term that is a function of z in Equation 4.29 and multiplies the amplitude term. In fact, at a depth of 600 cm, there is little change in temperature (the boundary condition at $z = \infty$). Note that all depths have the same average annual temperature. Another interesting feature is that the temperature fluctuations are delayed as depth increases. This is an additional phase shift caused by the third term in the argument of the sine function in Equation 4.29. This phase shift term increases with depth and decreases with d.

The time lag in temperatures at different depths caused by the phase shift can be calculated as:

$$\Delta t = \frac{z}{\omega d}. \qquad (4.31)$$

For example, at a depth of 200 cm, the time lag is 55 days, thus maximum temperatures will occur nearly 2 months later than at the surface.

Both the time lag and the diminishing amplitude depend on the damping depth, d. As d decreases, the temperature changes do not penetrate as deep into the soil. Since d is inversely related to ω, high frequency (large ω) changes in temperature (as might occur on a daily cycle) do not penetrate as deeply as low frequency (annual) changes.

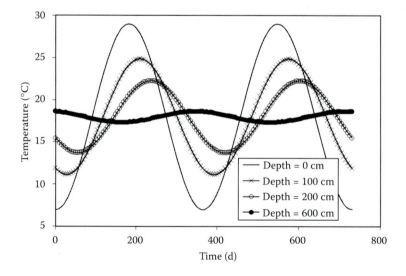

FIGURE 4.12 Temperature at depths of 0, 100, 200, and 600 cm as a function of time, calculated using Equation 4.29 with $K_T = 388.8$ cm^2 d^{-1}, $\tau = 365$ days, $T_A = 18°$C, and $A = 11°$C.

This solution to the heat flow equation can be used to decide how deep water lines must be buried in soil to avoid freeze damage.

Example 4.4

Compare the time lag in annual temperature fluctuations at a depth of 100 cm for a wet soil ($K_T = 4.5 \cdot 10^{-3}$ cm² s⁻¹) and a dry soil ($K_T = 2 \cdot 10^{-3}$ cm² s⁻¹).
 Convert K_T to square centimeter per day:

$$K_T = 4.5 \cdot 10^{-3} \left(\frac{cm^2}{s} \right) \left(\frac{86,400\,s}{day} \right) = 388.8 \frac{cm^2}{day}$$

$$K_T = 2 \cdot 10^{-3} \left(\frac{cm^2}{s} \right) \left(\frac{86,400\,s}{day} \right) = 172.8 \frac{cm^2}{day}.$$

Calculate the damping depth for the wet soil:

$$d = \sqrt{\frac{K_T \tau}{\pi}} = \sqrt{\frac{388.8 \frac{cm^2}{day} \cdot 365\,days}{\pi}} = 212\,cm.$$

Calculate the damping depth for the dry soil:

$$d = \sqrt{\frac{K_T \tau}{\pi}} = \sqrt{\frac{172.8 \frac{cm^2}{day} \cdot 365\,days}{\pi}} = 142\,cm.$$

Calculate the frequency for annual fluctuations:

$$\omega = \frac{2\pi}{365\,days} = 0.0172\ cycles\,per\,day = 0.0172\,days^{-1}.$$

Calculate the time lag for the wet soil:

$$\Delta t = \frac{z}{\omega d} = \frac{100\,cm}{\left(0.0172 \frac{1}{days} \right)(212\,cm)} = 27.4\,days.$$

Calculate the time lag for the dry soil:

$$\Delta t = \frac{z}{\omega d} = \frac{100\,cm}{\left(0.0172 \frac{1}{days} \right)(142\,cm)} = 40.9\,days.$$

In the dry soil, temperature movement is slower so the time lag is greater.

4.4.3 NUMERICAL SOLUTIONS TO THE HEAT FLOW EQUATION

So far, only *analytical solutions* to the heat flow equation have been considered. That is, the heat flow equation (Equation 4.21) could be *solved* using calculus (the Laplace transform approach) because the integrals developed could be evaluated, often by looking them up in integral tables. However, not all integrals have solutions. Many very interesting applications of the heat flow equation, such as for soils with different horizons, result in integrals that have no known solution. In these cases, a *numerical solution* can be developed, but there are drawbacks to this approach as will be seen.

Numerical methods of solving differential equations became feasible with the development of high-speed computers in the early 1960s. The *finite difference* and *finite element* numerical methods are the most commonly used. Despite the similarity in names, the development of both sets of equations is quite different, although the final equations themselves may be similar. The finite element method is more suitable for irregularly shaped flow domains, but the accuracy and speed of the two approaches are nearly the same (McCord and Goodrich 1994). The development of the finite difference equations is much more intuitive and will be discussed here.

The finite difference approach is used to develop a set of algebraic equations from the heat flow Equation 4.21 so that it can be solved numerically. The first step is to divide the soil profile into N layers or depth intervals Δz, such that $z = i \cdot \Delta z$ and $i = 0...N$ (Figure 4.13) (Smith 1985). It is not necessary for the depth intervals to be the same, but for simplicity a constant interval will be assumed here. The point at which intervals join is called a *node*, so there are $N + 1$ nodes in the profile.

The finite difference approximation of the space derivative that is *centered* midway between node $i - 1$ and i can be written as:

$$\left.\frac{\partial T}{\partial z}\right|_{i-1/2} = \frac{T_i - T_{i-1}}{\Delta z}, \tag{4.32}$$

where the vertical bar with subscript $i - 1/2$ indicates that this is the derivative midway between node $i - 1$ and node i. The subscripts on T indicate the node where that

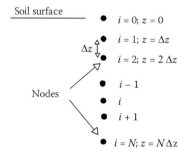

FIGURE 4.13 Dividing the soil profile into N discrete layers of the same thickness Δz results in $N + 1$ nodes. The z-axis is positive downwards in this numerical example.

temperature occurs. Similarly, the derivative centered midway between node i and $i + 1$ is

$$\left. \frac{\partial T}{\partial z} \right|_{i+1/2} = \frac{T_{i+1} - T_i}{\Delta z}. \tag{4.33}$$

In the heat flow Equation 4.21, the second derivative in space appears. A finite difference approximation for the second space derivative can be developed in two steps. First, express the outer derivative as a difference centered on the i node:

$$\left. \frac{\partial^2 T}{\partial z^2} \right|_i = \frac{\partial}{\partial z} \left(\frac{\partial T}{\partial z} \right)$$

$$= \frac{\left. \frac{\partial T}{\partial z} \right|_{i+1/2} - \left. \frac{\partial T}{\partial z} \right|_{i-1/2}}{\Delta z}. \tag{4.34}$$

Next, substitute Equations 4.32 and 4.33 for the derivatives at $i - 1/2$ and $i + 1/2$ distances in Equation 4.34:

$$\left. \frac{\partial^2 T}{\partial z^2} \right|_i = \frac{\left(\dfrac{T_{i+1} - T_i}{\Delta z} \right) - \left(\dfrac{T_i - T_{i-1}}{\Delta z} \right)}{\Delta z}$$

$$= \frac{T_{i+1} - 2T_i + T_{i-1}}{\Delta z^2}. \tag{4.35}$$

Time is also divided into intervals of Δt, such that $t = j \cdot \Delta t$ and $j = 0 \ldots M$. Then the time derivative in Equation 4.21 can be written as a discrete difference divided by the appropriate interval:

$$\left. \frac{\partial T}{\partial t} \right|_i^{j+1/2} = \frac{T_i^{j+1} - T_i^j}{\Delta t} \tag{4.36}$$

where a superscript has been added to T to indicate the time level. The subscript indicates that Equation 4.36 applies at the depth node i.

If Equations 4.35 and 4.36 are substituted for the derivatives in the heat flow Equation 4.21:

$$\frac{T_i^{j+1} - T_i^j}{\Delta t} = K_T \frac{T_{i+1} - 2T_i + T_{i-1}}{\Delta z^2}. \tag{4.37}$$

At this point, it is clear that the PDE has been changed to a simple algebraic equation. Numerical approaches consist of starting with the initial conditions, which give the values of T at all nodes at $t = 0$, then finding T at each node at the next time level ($t = \Delta t$), and repeating the process until the full time period of interest has been covered. Therefore, the crux of the problem is finding the temperature at the next time level (T_i^{j+1}) given the value at the current time level (T_i^j) at each interior node ($i = 1$ to $N - 1$). The boundary conditions give the temperature at the first and last node for the next time level (T_0^{j+1} and T_N^{j+1}). The time level for evaluating the terms on the right side of Equation 4.37 is unspecified at this point since the superscript on T is not given. Depending on the time level assigned to the terms on the right side of Equation 4.37, different numerical approaches are possible.

If all the terms on the right side of Equation 4.37 are evaluated at the known (j) time level, then:

$$\frac{T_i^{j+1} - T_i^j}{\Delta t} = K_T \frac{T_{i+1}^j - 2T_i^j + T_{i-1}^j}{\Delta z^2}. \tag{4.38}$$

Now there is only one unknown temperature at the new time level ($j + 1$). Solving for this temperature, the equation is

$$T_i^{j+1} = \frac{\Delta t}{\Delta z^2} K_T \left(T_{i+1}^j - 2T_i^j + T_{i-1}^j \right) + T_i^j$$

$$= \frac{\Delta t}{\Delta z^2} K_T T_{i+1}^j + \left(1 - 2\frac{\Delta t}{\Delta z^2} K_T \right) T_i^j + \frac{\Delta t}{\Delta z^2} K_T T_{i-1}^j. \tag{4.39}$$

Let:

$$r = \frac{\Delta t}{\Delta z^2}, \tag{4.40}$$

then the finite difference approximation to the heat flow equation can be written as:

$$T_i^{j+1} = rK_T T_{i+1}^j + \left(1 - 2rK_T \right) T_i^j + rK_T T_{i-1}^j. \tag{4.41}$$

This is known as the *explicit* finite difference method. The initial conditions are the temperatures T_i^0 for $i = 0...N$, so to find the temperatures at the first time level, $t = \Delta t$, one can write:

$$T_i^1 = rK_T T_{i+1}^0 + \left(1 - 2rK_T \right) T_i^0 + rK_T T_{i-1}^0. \tag{4.42}$$

Note that all the temperatures on the right side are known as the initial temperatures. Equation 4.42 must be evaluated for $i = 1$ to $N - 1$. For example, the equation for $i = 1$ is

$$T_1^1 = rK_T T_2^0 + \left(1 - 2rK_T\right)T_1^0 + rK_T T_0^0. \tag{4.43}$$

The temperatures for $i = 0$ and $i = N$ at the unknown time level are given by the boundary conditions. Once all the temperatures at the $j = 1$ time level are known, the temperatures at the next time level ($j = 2$) can be found in the same manner:

$$T_i^2 = rK_T T_{i+1}^1 + \left(1 - 2rK_T\right)T_i^1 + rK_T T_{i-1}^1, \tag{4.44}$$

where all the temperatures on the right side are now known.

Before Equation 4.41 can be used, one must decide on the value of r by choosing the distance between nodes (Δz) and the time step (Δt). The choice of Δz depends on how much *resolution* is desired. Using Equation 4.41, one can calculate the temperature at each node. To make a smooth curve with this data, a small Δz is needed. The disadvantage of a small Δz, however, is that a large number of nodes (large N) will be needed to extend the solution to the deepest depth of interest, and this will take more computer time. Once a decision has been made on the value for Δz, the length of the time step can be chosen. Large time steps will require less computer time, but can lead to *inaccurate solutions*. For the *explicit* finite difference numerical method, Δt should be chosen such that $r \cdot K_T$ is less than about 0.5. It will be seen why that is so and what is meant by inaccurate solutions, shortly.

Now the numerical approach (Equation 4.41) will be compared with the analytical solution developed earlier (Equation 4.25). For the numerical solution, $\Delta z = 2$ cm and $\Delta t = 1$ s are chosen so that $r = 0.25$ scm^{-2} and $r \cdot K_T = 0.001125$. The number of nodes ($N + 1$) is set to 16 so that the deepest node occurs at a depth of 30 cm. The numerical solution is simply values of temperature at each node, at each time level, T_i^j. This is a matrix T where the first column ($j = 0$) represents the initial conditions and each subsequent column represents the temperatures at the next time level (Figure 4.14). Each row in the matrix represents a node, where the first row ($i = 0$) is the boundary condition at the surface and the last row ($i = N$) is the boundary condition at the bottom. To go to large times, many columns are needed. To avoid this, a macro can be used to copy the values from the "$j + 1$" column back into the "j" column (see Problem 11).

Hence, the numerical solution is not an equation, but simply columns of numbers. Equation 4.41 is the *algorithm* used in a computer to generate the temperatures. For a numerical approach, no solution can be expressed as an equation. This is one of the drawbacks of numerical approaches. For example, one could look at the analytical solution to the periodic boundary condition (Equation 4.29) and understand why damping would occur with depth (the exponential decay term) and a time lag (the phase shift term) would be present. This can't be done with the numerical approach because the solution is simply numbers. The only way to look at the solution is to graph the results.

The results of the numerical solution are graphed in Figure 4.15, where temperatures after 5000 s are shown as a function of depth. The values from the numerical

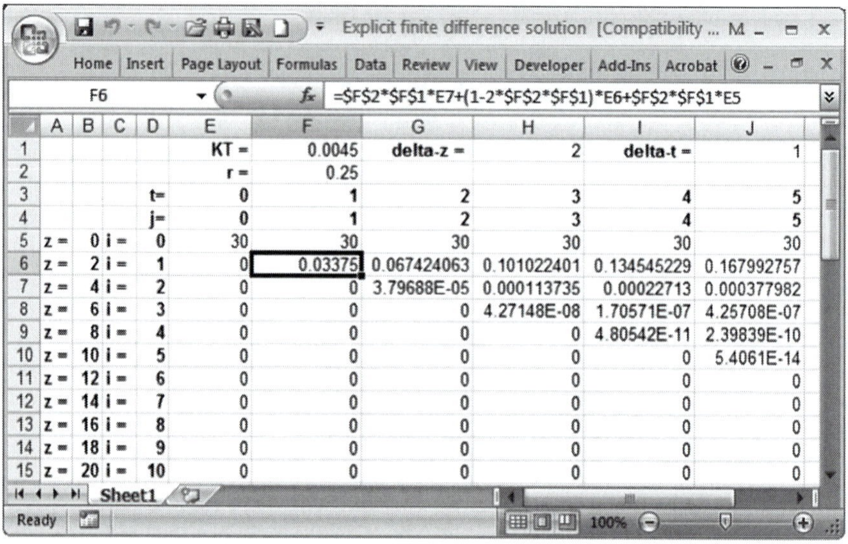

	A	B	C	D	E	F	G	H	I	J
1					KT =	0.0045	delta-z =	2	delta-t =	1
2					r =	0.25				
3				t=	0	1	2	3	4	5
4				j=	0	1	2	3	4	5
5	z =	0	i =	0	30	30	30	30	30	30
6	z =	2	i =	1	0	0.03375	0.067424063	0.101022401	0.134545229	0.167992757
7	z =	4	i =	2	0	0	3.79688E-05	0.000113735	0.00022713	0.000377982
8	z =	6	i =	3	0	0	0	4.27148E-08	1.70571E-07	4.25708E-07
9	z =	8	i =	4	0	0	0	0	4.80542E-11	2.39839E-10
10	z =	10	i =	5	0	0	0	0	0	5.4061E-14
11	z =	12	i =	6	0	0	0	0	0	0
12	z =	14	i =	7	0	0	0	0	0	0
13	z =	16	i =	8	0	0	0	0	0	0
14	z =	18	i =	9	0	0	0	0	0	0
15	z =	20	i =	10	0	0	0	0	0	0

FIGURE 4.14 The numerical *solution* is a matrix where the first column (column E) represents the initial temperatures at each node. The next column (column F) represents the temperatures at the next time level. The rows represent nodes with the surface node in the first row (row 5) (boundary condition of $T = 30$°C). The formula bar shows Equation 4.41.

solution are the square symbols and the smooth curve is the analytical solution (Equation 4.25) seen earlier. There is excellent agreement between the numerical and analytical approaches.

One can get a better idea of what is meant by the *semi-infinite* boundary condition used for the lower boundary (Equation 4.23) by looking at Figure 4.15. For the analytical solution, it was specified in Equation 4.23 that T was zero (same as the initial temperature for the profile) at a depth of $z = \infty$. For the numerical approach, since computers only deal with finite numbers, one can't set a condition at infinity. The boundary condition could only be set at a very deep finite depth, but that will require a very large number of nodes (N very large). In the numerical problem above, only 16 nodes were used, so the boundary condition of $T = 0$ was specified at a depth of 30 cm ($\Delta z = 2$ cm). And yet, there was excellent agreement with the analytical solution. That is because the heating front has not yet reached the lower boundary at 5000 s. If the numerical problem was run any longer, the numerical and analytical results would start to deviate. The semi-infinite boundary condition for numerical solutions simply means that the problem cannot be run beyond the time when temperatures start to change just above the boundary (when the heating front arrives, in this problem).

It was mentioned earlier that inaccurate numerical solutions can occur with the explicit finite difference method if a large enough Δt is chosen such that $r \cdot K_T > 0.5$. To see this, the time step will be increased 500-fold to $\Delta t = 500$ s (leaving Δz the same), so that $r \cdot K_T = 0.563$. The numerical and analytical solutions are compared in Figure 4.16. After only 10 time steps at $t = 5000$ s, the numerical solution diverges

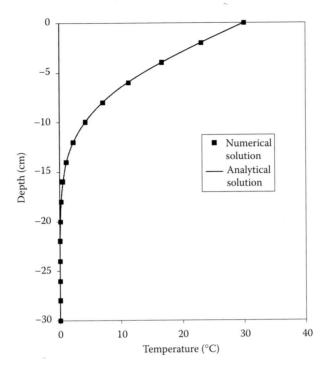

FIGURE 4.15 Temperature as a function of depth after 5000 s based on a numerical solution (symbols) and an analytical solution (smooth curve) for the heat flow problem with a constant surface boundary condition of $T = 30°C$ and a thermal diffusivity of $4.5 \cdot 10^{-3}$ cm^2 s^{-1}. For the numerical solution, the time step is $\Delta t = 1$ s so that $r \cdot K_T = 0.001$.

sharply from the analytical solution in the middle region of the profile. The numerical solution has become *unstable* for this value of r and K_T. Unstable means that values of T will fluctuate sharply between time intervals and diverge from the true solution.

One can see why this happens by returning to Equation 4.41 using the values for r and K_T in Figure 4.15. In that case, $r \cdot K_T = 0.001$ cm^2 s^{-1}. Equation 4.41 is therefore:

$$T_i^{j+1} = 0.001 T_{i+1}^j + (1 - 0.002) T_i^j + 0.001 T_{i-1}^j$$

$$= \frac{1}{1000} T_{i+1}^j + \frac{998}{1000} T_i^j + \frac{1}{1000} T_{i-1}^j. \tag{4.45}$$

The temperature at the i node for the next time level is estimated by taking a *weighted average* of the temperatures at the $i - 1$, i, and $i + 1$ nodes at the previous time level. Note that the weights all add to one. The heaviest weight is assigned to the temperature at the same node (probably the best information on what the temperature is likely to be at the next time level), but some of the weights also go to the neighboring nodes. In this way, temperature changes can be propagated along the profile, from a high temperature at the boundary for example.

Now look at what the weights are when the larger time step in Figure 4.16 is used, such that $r \cdot K_T = 0.563$:

$$T_i^{j+1} = 0.563 \cdot T_{i+1}^j + (1 - 1.126) \cdot T_i^j + 0.563 \cdot T_{i-1}^j$$

$$= \frac{563}{1000} \cdot T_{i+1}^j - \frac{126}{1000} \cdot T_i^j + \frac{563}{1000} \cdot T_{i-1}^j. \tag{4.46}$$

The weights still sum to one, but this time proportionately heavier weights are given to the neighboring nodes and a *negative* weight to the temperature at the same node, even though it is usually the best predictor of what the temperature will be at the next time level. This can cause the temperatures to fluctuate sharply and is a poor algorithm for predicting the temperature at the next time level.

The restrictions on r and consequent requirement for small time steps can be avoided by devising a different finite difference scheme from the explicit finite difference approach examined so far. For the explicit scheme, all the terms on the right

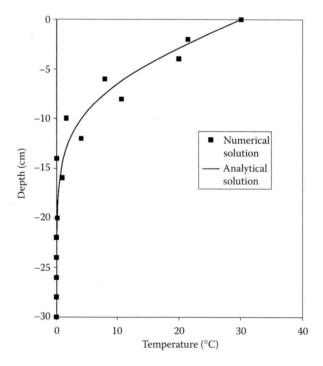

FIGURE 4.16 Temperature (°C) as a function of depth for $t = 5000$ s based on an analytical solution (curve) and a numerical solution (symbols) for the heat flow problem in Figure 4.15. For the numerical solution, the time step is $\Delta t = 500$ s so that $r \cdot K_T = 0.563$.

side of Equation 4.37 were evaluated at the known (j) time level. If, instead, these terms are evaluated at the unknown ($j + 1$) time level, then:

$$\frac{T_i^{j+1} - T_i^{j}}{\Delta t} = K_T \frac{T_{i+1}^{j+1} - 2T_i^{j+1} + T_{i-1}^{j+1}}{\Delta z^2}. \tag{4.47}$$

Solving for the temperatures at the unknown time level ($j + 1$) and using r as $\Delta t/\Delta z^2$:

$$-rK_T T_{i+1}^{j+1} + 2rK_T T_i^{j+1} - rK_T T_{i-1}^{j+1} + T_i^{j+1} = T_i^{j}$$

$$-rK_T T_{i+1}^{j+1} + \left(1 + 2rK_T\right)T_i^{j+1} - rK_T T_{i-1}^{j+1} = T_i^{j}. \tag{4.48}$$

There are now three unknowns in Equation 4.48 instead of the single unknown in the explicit finite difference approach (Equation 4.41). This approach is known as the *fully implicit* finite difference scheme, since the equation can't be solved explicitly for a single unknown temperature. However, one can write Equation 4.48 for each internal node. Hence, there are $N - 1$ equations with $N - 1$ unknowns ($T_1^{j+1}...T_{N-1}^{j+1}$). The boundary conditions give the temperatures at the end nodes. Since there are as many equations as unknowns, this is a *system of simultaneous equations* that can be solved. Equation 4.48 can be written:

$$aT_{i+1}^{j+1} + bT_i^{j+1} + cT_{i-1}^{j+1} = T_i^{j}, \tag{4.49}$$

where:

$$a = c = -rK_T$$

$$b = 1 + 2rK_T. \tag{4.50}$$

Then one can write the system of $N + 1$ equations in matrix (set of rows and columns) and vector (column) notation. Using a very small value for $N = 3$ to save space:

$$\begin{bmatrix} 1 & 0 & 0 & 0 \\ a & b & c & 0 \\ 0 & a & b & c \\ 0 & 0 & 0 & 1 \end{bmatrix} \begin{bmatrix} T_0^{j+1} \\ T_1^{j+1} \\ T_2^{j+1} \\ T_3^{j+1} \end{bmatrix} = \begin{bmatrix} T_0^{j} \\ T_1^{j} \\ T_2^{j} \\ T_3^{j} \end{bmatrix}. \tag{4.51}$$

For matrix and vector multiplication, the elements in each row of the matrix are multiplied by the elements in the vector and the sum is taken. This sum is equal to the element in the vector on the right side of Equation 4.51 for the corresponding row. The result is

$$T_0^{j+1} = T_0^j$$

$$aT_0^{j+1} + bT_1^{j+1} + cT_2^{j+1} = T_1^j$$

$$aT_1^{j+1} + bT_2^{j+1} + cT_3^{j+1} = T_2^j \qquad (4.52)$$

$$T_3^{j+1} = T_3^j,$$

which is the system of equations that must be solved (the first and last equation account for the boundaries). This system of equations can be written as:

$$[A] \cdot \{T\}^{j+1} = \{T\}^j, \qquad (4.53)$$

where the square brackets [–] indicate a matrix and curved brackets {–} indicate a vector. [A] is the matrix shown on the left side of Equation 4.51. It has $N+1$ rows and $N+1$ columns. The *elements* of the matrix are zero unless they are on the three diagonals going from the top left to the bottom right. In other words, [A] is an $N+1$ *square tridiagonal matrix*; $\{T\}^{j+1}$ is a vector with $N+1$ elements, the first and last of which are the boundary temperatures. The other $N-1$ elements are the unknown temperatures at the next time level. $\{T\}^j$ is also a vector with $N+1$ elements and it represents the known temperatures at the current time level (including the boundary values).

To solve the system of equations, both sides of Equation 4.53 are multiplied by the *inverse* of the [A] matrix, written as $[A]^{-1}$:

$$[A]^{-1} \cdot [A] \cdot \{T\}^{j+1} = [A]^{-1} \{T\}^j. \qquad (4.54)$$

A matrix times its inverse is equal to one so:

$$\{T\}^{j+1} = [A]^{-1} \{T\}^j. \qquad (4.55)$$

Inverting large matrices (e.g., $N = 100$) involves many computations, but computer algorithms have been developed to do this efficiently. Many other problems lead to a system of equations similar to Equation 4.53 and they are solved on computers using Equation 4.55. Examples are chemical equilibrium problems where the concentrations of different chemical compounds in a solution are unknown. Each of the compounds can undergo different reactions with other compounds, which can be written as equations. As long as there are at least as many equations as unknown concentrations, the system of equations can be solved for the concentration of each compound at equilibrium.

If the *determinant* of the A matrix is near zero it is impossible to invert the matrix. In this case, the numerical approach will fail. The determinant depends on the values of a, b, and c in Equation 4.50, which depend in turn on the choice of values for Δt and Δz and the value of K_T.

Matrices can be entered into Excel and Excel can invert these matrices. Solving the heat flow equation using the implicit finite difference scheme consists of defining the elements of the [A] matrix using Equations 4.49 and 4.50 for $N + 1$ rows and columns, defining the elements of the vector $\{T\}^0$ as the initial temperatures, and then using Equation 4.55 to find the new temperature vector. A macro can be used to store the new temperature vector $\{T\}^{j+1}$ back into the previous temperature vector $\{T\}^j$ until the final time of interest is reached. See Problem 12 at the end of this chapter.

One other finite difference scheme that is commonly used is the *Crank-Nicolson implicit method* (Smith 1985). This is obtained by evaluating the temperatures on the right side of Equation 4.37 at the $j + 1/2$ time level. To do this, each temperature on the right-hand side is written as the arithmetic average of the temperatures at the j and $j + 1$ time levels. This approach also leads to a set of simultaneous equations that must be solved using matrices, but the elements of the matrices and vectors differ from the fully implicit method (Equations 4.49 and 4.50).

Before leaving this section, the numerical solution will be used for a soil where K_T is not a constant, to show the advantage of the numerical approach. It was seen earlier that K_T was most affected by changes in water content (compared to changes in texture, for example). Therefore, one must assume that water contents are relatively uniform in order to use Equation 4.25. That is not the case with the numerical approach, since different values of K_T can be assigned to nodes. In Figure 4.17, the profile temperatures after 5000 s in a uniformly wet soil ($K_T = 4.5 \cdot 10^{-3}$ cm^2 s^{-1}) are compared to temperatures in a nonuniform soil with a wet layer ($K_T = 4.5 \cdot 10^{-3}$ cm^2 s^{-1}) in the top 5 cm, underlain by a dry layer ($K_T = 2.3 \cdot 10^{-3}$ cm^2 s^{-1}). The analytical solution (Equation 4.25) was used for the uniform wet soil and the explicit finite difference numerical scheme (Equation 4.41) was used for the nonuniform soil. Above about 3 cm, the temperatures are about the same. But below this depth, temperatures in the dry layer are lower than they would be in a uniformly wet soil due to the slower movement of heat in the dry soil.

4.5   SOIL HEAT FLOW WITH HYDRUS-1D

HYDRUS-1D uses a finite element in space and finite difference in time approach to find a numerical solution to the heat flow equation. The user specifies the maximum depth of the soil profile (L) and the grid spacing (Δz, which may be variable) and then HYDRUS-1D adjusts the time step (Δt) to maintain an accurate solution.

HYDRUS-1D solves a form of the heat transport equation that includes conduction *and* convection and accounts for the effect of water content on soil heat capacity and apparent thermal conductivity (compare this to Equation 4.19, which did not include convection):

$$C_p(\theta) \cdot \frac{\partial T}{\partial t} = \frac{\partial}{\partial z}\left(\lambda(\theta) \cdot \frac{\partial T}{\partial z}\right) - C_w J_w \frac{\partial T}{\partial z}, \tag{4.56}$$

where C_w is the volumetric heat capacity [ML^{-1}T^{-2}K^{-1}] of water and J_w is the Darcian flow of water [LT^{-1}]. The first term on the right side of the equation represents heat

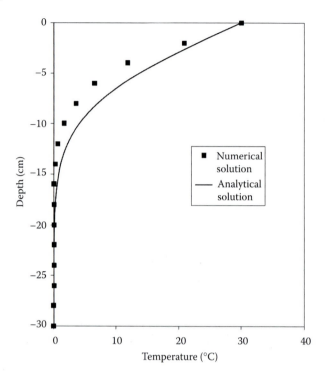

FIGURE 4.17 Temperature as a function of depth after 5000 s in a uniform wet soil ($K_T = 4.5 \cdot 10^{-3}$ cm^2 s^{-1}) compared to a nonuniform soil with a wet layer in the top 5 cm ($K_T = 4.5 \cdot 10^{-3}$ cm^2 s^{-1}), underlain by a dry layer ($K_T = 2 \cdot 10^{-3}$ cm^2 s^{-1}). An analytical solution was used for the uniform soil and a numerical solution was used for the nonuniform soil.

flow by conduction; the second term represents heat flow by convection. The heat capacity of the soil or porous media [$C_p(\theta)$] and the soil apparent thermal conductivity [$\lambda(\theta)$] are functions of volumetric water content.

The soil volumetric heat capacity as a function of volumetric water content can be expressed as:

$$C_p(\theta) = C_s(1 - \varphi - \theta_o) + C_o\theta_o + C_w\theta + C_a a, \qquad (4.57)$$

where φ is porosity, a is the volumetric air content, θ_o is the volumetric fraction of organic matter, C_s is the heat capacity [ML^{-1}T^{-2}K^{-1}] for the solid phase, C_o is the heat capacity for the organic phase, C_w is the heat capacity for the water phase, and C_a is the heat capacity for the air phase. According to de Vries (1963), the heat capacities for organic matter, the solid phase, and water are approximately $2.51 \cdot 10^6$, $1.92 \cdot 10^6$, and $4.18 \cdot 10^6$ J m^{-3} °C^{-1}. The heat capacity for air is approximately zero.

The apparent thermal conductivity, $\lambda(\theta)$, combines the effective thermal conductivity, $\lambda_e(\theta)$, of the porous media (solid plus water) in the absence of flow and macrodispersivity, which is a linear function of the velocity (de Marsily 1986):

$$\lambda(\theta) = \lambda_e(\theta) + \beta_t C_w |J_w|, \tag{4.58}$$

where β_t is the thermal dispersivity [L]. Chung and Horton (1987) developed an equation for the soil effective thermal conductivity as a function of volumetric water content:

$$\lambda_e(\theta) = b_1 + b_2\theta + b_3\sqrt{\theta}, \tag{4.59}$$

where b_1, b_2, and b_3 are empirical parameters [MLT^{-3}K^{-1}] that vary with soil texture.

Two types of boundary conditions can be specified at the top and bottom boundaries of the soil profile. The type-1 (or Dirichlet) boundary condition specifies a temperature (that can vary with time):

$$T(z,t) = T_0(t) \quad \text{at } z = 0 \text{ or } z = L, \tag{4.60}$$

where L is the lower boundary. The *type-3* (or *Cauchy*) boundary condition is given in terms of the variable and its derivative (a *type-2 boundary condition* is given in terms of the derivative only). It specifies a heat flux (that can vary with time):

$$-\lambda\frac{\partial T}{\partial z} + TC_w J_w = T_0(t)C_w J_w\big|_z \quad \text{at } z = 0 \text{ or } z = L, \tag{4.61}$$

in which T_0 is the temperature of the incoming water at the boundary and J_w is the water flux at the boundary.

An option is available to specify an atmospheric boundary condition for daily fluctuations in soil temperature represented by a sine function (Kirkham and Powers 1972):

$$T_0 = \overline{T} + A\sin\left(\frac{2\pi t}{p_t} - \frac{7\pi}{12}\right), \tag{4.62}$$

where p_t [T] is the period (1 day), $\overline{T}$ is the average temperature at the soil surface [K], and A is the amplitude of the sine wave [K]. The second (phase shift) term in the argument of the sine function ensures that the day's temperature maximum occurs at 1:00 p.m. Note the similarity to Equation 4.28.

To demonstrate the use of HYDRUS-1D, temperature changes in response to daily fluctuations at the soil surface in a uniform profile consisting of a loam soil at a constant pressure head of 0 cm (saturated conditions) are predicted. The initial temperature in the soil profile is 25°C. The temperature at the surface fluctuates ± 5°C around an average of 25°C. This is HYDRUS Simulation 4.1 – Heat Flow Without Convection.

Open HYDRUS-1D and close any previous projects. Then select New Project (□) on the toolbar and give the project a name and description. The HYDRUS-1D overall display is shown in Figure 4.18. Select "Main Processes" at the top of the Preprocessing panel and check heat transport only. Move through the input windows in

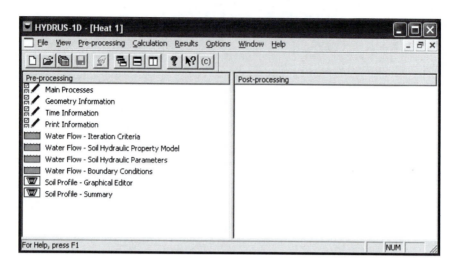

FIGURE 4.18 HYDRUS-1D overall display window.

sequence by selecting "Next". In the Geometry Information window, accept the default settings, including centimeters for length units and a depth of profile of 100 cm. In the Time Information window, choose hours for time units and a final time of 24 h. Set the maximum time step to 0.1 h to ensure a smooth curve in output graphs.

In the Print Information window, set the number of print times to three (once every 8 h). Select "Print Times" and use the "Default" button to get an even distribution of print times (or specify the print times desired).

Don't change any of the settings in the Iteration Criteria window. In the Soil Hydraulic Model window, choose the van Genuchten-Mualem form for the soil water retention and hydraulic conductivity curves. In the Water Flow Parameters window, select any of the cells with a numerical value, then in the lower left, use the drop-down Soil Catalog menu to select a loam.

In the Water Flow Boundary Conditions window and subsequent Constant Boundary Fluxes window, don't make any changes (if water flow is not selected in the Main Processes window, HYDRUS-1D simply takes the initial pressure heads and keeps them constant throughout the simulation).

In the Heat Transport Parameters window, note that the interval for temperature fluctuations at the soil surface is set to 24 h and the temperature amplitude is equal to 5°C. Select any of the numerical values in the material's table and then use the drop-down menu for the Chung and Horton thermal conductivity function to select a loam. In the Heat Transport Boundary Conditions window, choose "Temperature BC" and set a boundary value of 25°C for upper and lower boundaries.

In the HYDRUS-1D guide window, select "OK" to run the PROFILE application. This produces a display such as that shown in Figure 4.19. The green horizontal lines indicate the default setting of grid spacing that HYDRUS-1D has assigned. One can see this more clearly if the "Profile Discretization" icon (亘) on the toolbar is selected. There are 101 nodes and the grid is evenly spaced. One can change the

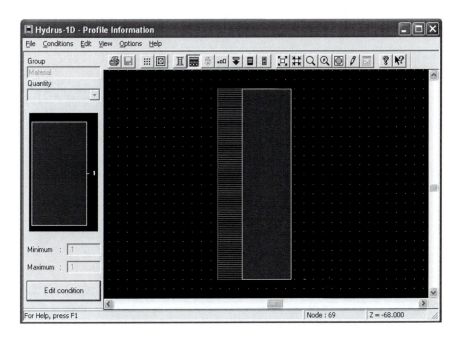

FIGURE 4.19   Profile Information window showing the material distribution for a uniform soil.

number of nodes in a panel on the left or change the density so that they are not evenly spaced. In this example, the default settings are used.

The next icon () to the right on the toolbar is the Material distribution display (shown in Figure 4.19). The display shows that the soil profile is uniform, consisting of Material 1, which in this case is a loam. Select the "Initial Conditions" icon ( ) farther to the right on the toolbar to see a display of initial soil water pressure heads. The profile is set to a uniform pressure head of −100 cm by default. Select the "Edit" button (from the Edit bar on the left) and move the hand cursor to select all the grid points by clicking on the top-most green line and then clicking on the bottom green line. The Condition Specification window will appear. Set the pressure at the top to zero and the bottom to 100 (remove the check mark to make the top and bottom pressures the same). This will ensure that the soil is saturated and at equilibrium so that there is no water flow. In the panel on the left, use the Quantity drop-down menu to select "Temperature" and use the same procedure to make all the initial temperatures 25°C.

Select the "Observation Points" icon ( ) on the toolbar and then use the "Insert" button to place observation points at depths of 0, −10, and −50 cm. To do this, move the hand cursor over the green horizontal line that represents each depth (depth and node number are displayed at the lower right corner of the screen) and click on the appropriate depth (see Figure 4.20).

Now that all the settings have been made, close the Profile Information window. The Soil Profile Summary window appears (if you don't see it, click on the icon at the bottom of the Pre-processing panel). It shows that the soil pressure head h

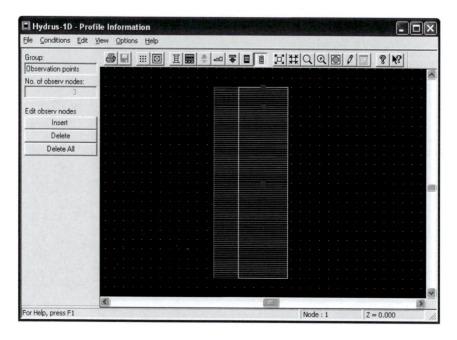

FIGURE 4.20   Profile Information window showing three observation points.

increases from 0 cm at the surface to 100 cm at the deepest depth. Scrolling to the right, one can see that all the temperatures are set to 25°C.

Run the program using the Execute HYDRUS icon ( ) on the toolbar. At this point, a disk operating system (DOS) window appears as HYDRUS-1D performs all the calculations to get a numerical solution (Figure 4.21). One can see the time steps

```
 Hydrus-1D Calculation: Heat 1
    22.2000  1    226   .52E+00  .12E+02   .00E+00   .00E+00     -.5    0.   100.
    22.3000  1    227   .52E+00  .12E+02   .00E+00   .00E+00     -.5    0.   100.
    22.4000  1    228   .52E+00  .12E+02   .00E+00   .00E+00     -.5    0.   100.
    22.5000  1    229   .52E+00  .12E+02   .00E+00   .00E+00     -.5    0.   100.
    22.6000  1    230   .52E+00  .12E+02   .00E+00   .00E+00     -.5    0.   100.
    22.7000  1    231   .52E+00  .12E+02   .00E+00   .00E+00     -.5    0.   100.
    22.8000  1    232   .52E+00  .12E+02   .00E+00   .00E+00     -.5    0.   100.
    22.9000  1    233   .52E+00  .12E+02   .00E+00   .00E+00     -.5    0.   100.
    23.0000  1    234   .52E+00  .12E+02   .00E+00   .00E+00     -.5    0.   100.
    23.1000  1    235   .52E+00  .12E+02   .00E+00   .00E+00     -.5    0.   100.
    23.2000  1    236   .52E+00  .12E+02   .00E+00   .00E+00     -.5    0.   100.
    23.3000  1    237   .52E+00  .12E+02   .00E+00   .00E+00     -.5    0.   100.
    23.4000  1    238   .52E+00  .12E+02   .00E+00   .00E+00     -.5    0.   100.
    23.5000  1    239   .52E+00  .12E+02   .00E+00   .00E+00     -.5    0.   100.
    23.6000  1    240   .52E+00  .12E+02   .00E+00   .00E+00     -.5    0.   100.

        Time  ItW   ItCum  vTop    SvTop    SvRoot    SvBot    hTop hRoot hBot

    23.7000  1    241   .52E+00  .12E+02   .00E+00   .00E+00     -.5    0.   100.
    23.8000  1    242   .52E+00  .12E+02   .00E+00   .00E+00     -.5    0.   100.
    23.9000  1    243   .52E+00  .12E+02   .00E+00   .00E+00     -.5    0.   100.
    24.0000  1    244   .52E+00  .12E+02   .00E+00   .00E+00     -.5    0.   100.
Real time [sec]   0.0000000000000000E+000
Press Enter to continue
```

FIGURE 4.21   DOS window showing that the execution of the numerical solution is complete.

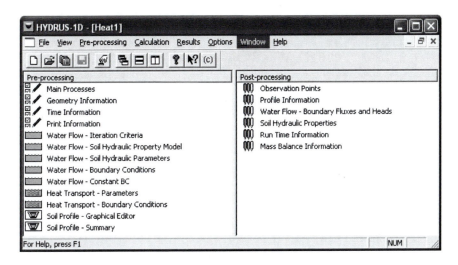

FIGURE 4.22 Overall display window showing results posted in the Post-processing panel.

that were chosen to maintain an accurate solution and the total time in seconds that completed the calculations. Click "Enter" to close the DOS window and to see the results displayed in the Post-processing panel of the HYDRUS-1D overall display (Figure 4.22).

Select "Observation Points" in the Post-processing panel. At the top of the Observation Nodes display, from the list box for the Vertical Variable choose "Temperature". This will produce a graph of the temperature at each of the three observation nodes as a function of time (see Figure 4.23). To see the legend for depths, right-click on the graph, from the pop-up menu select "Legend", and use the check box to make it visible.

Figure 4.23 shows the temperature fluctuations at the surface that vary by 5°C around the mean of 25°C. Temperatures at 10 cm below the surface are dampened and lag behind the surface temperatures. There is little variation in temperatures at 50 cm. The pattern is similar to the results seen for annual temperature fluctuations in the soil used in an analytical solution (Figure 4.12).

Next select "Profile Information" in the Post-processing panel. At the top of the Basic Profile Information display, choose "Temperature" for the horizontal variable. This will produce a graph of temperature as a function of depth at the initial time and three print times (8, 16, and 24 h) (Figure 4.24). Temperature fluctuations due to daily changes at the surface penetrate to the deepest depth where they are fixed by the lower boundary condition in this soil.

4.6 SUMMARY

In this chapter, it has been shown that part of the surface energy balance is a flux into and out of the soil. The steady soil heat flux term can be described by Fourier's law (Equation 4.6), which has a form very similar to Darcy's law. Nonsteady or

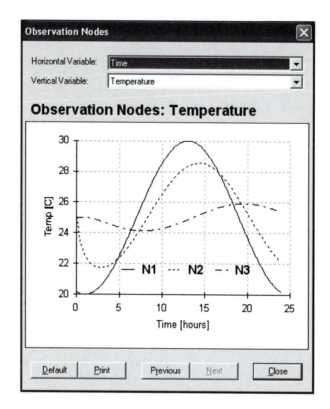

Class got a different output (handwritten note in margin)

FIGURE 4.23 Temperatures as a function of time at three observation nodes (N1 = 0, N2 = 10, and N3 = 50 cm).

transient flow is based on the heat conservation equation. This leads to a PDE describing temperature changes due to heat flow by conduction (Equation 4.21). An analytical solution to the heat flow equation for constant temperature at the soil surface and a uniform soil can be developed using the Laplace transform method (Equation 4.25). An analytical solution for sinusoidal fluctuations of soil temperature at the soil surface also exists (Equation 4.29). However, there are many problems in soil heat flow, such as layered soils, that cannot be addressed by analytical solution. Numerical solutions to the heat flow equation have been developed using finite difference and finite element approaches. HYDRUS-1D uses the finite element approach to solve a heat flow equation that accounts for conduction as well as convection (Equation 4.56).

4.7 DERIVATIONS

In this section, the properties of the Laplace transform are shown and an analytical solution to the heat flow equation is derived using the Laplace transform method. The properties of the complementary error function are also discussed.

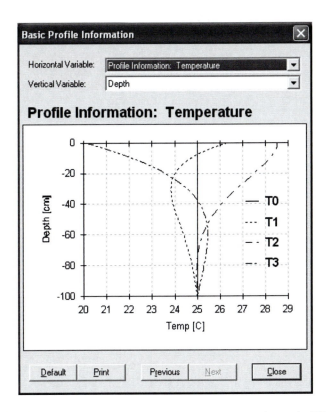

FIGURE 4.24 Temperatures as a function of depth at four time intervals (T0 = 0, T1 = 8, T2 = 16, and T3 = 24 h).

4.7.1 Properties of the Laplace Transform

The Laplace transform is used to transform a PDE to an ODE by eliminating time as a variable. The Laplace transform of a function of space and time, for example $f(z,t)$ is defined as:

$$\mathscr{L}\big[f(z,t)\big] \equiv \int_0^\infty f(z,t)\,e^{-st}\,dt = \hat{f}(z), \tag{4.63}$$

where the notation $\mathscr{L}\,[f(z,t)]$ indicates that the function inside the brackets is being transformed. Once the integral is evaluated, t will disappear from the equation and the transformed equation will be a new function that contains s, which is a new parameter that can be treated as a constant. The last term indicates the new transformed equation. It is still a function of z, but no longer a function of t. After the ODE is solved using one of the many methods for solving ODEs, taking the *inverse Laplace transform* will bring t back into the solution as a variable. The notation for the inverse Laplace transform is $\mathscr{L}^{-1}$.

Laplace transforms of a number of functions of t are shown in Table 4.1. The table can also be used to find inverse Laplace transforms. This table is compiled from entries in Jury and Roth (1990). Note that the exponential function can be written as e^x or as $\exp(x)$. The following abbreviations apply where a and b are constants:

$$\mathcal{A} = \frac{1}{\sqrt{\pi t}} \cdot \exp\left(\frac{-x^2}{4t}\right)$$

$$\mathcal{B} = \text{erfc}\left(\frac{x}{2\sqrt{t}}\right)$$

$$\mathcal{C} = \exp\left(a^2 t - ax\right) \cdot \text{erfc}\left(\frac{x}{2\sqrt{t}} - a\sqrt{t}\right)$$

$$\mathcal{D} = \exp\left(a^2 t + ax\right) \cdot \text{erfc}\left(\frac{x}{2\sqrt{t}} + a\sqrt{t}\right).$$

TABLE 4.1

Laplace transforms (Jury and Roth, 1990).

$f(x,t)$	$\mathcal{L}[f(x,t)]$	Transform #
$\delta(t)$	1	1*
1	$1/s$	2
t^N	$N!/s^{N+1}$	3**
$\exp(-at)$	$1/(s+a)$	4
$\sin(at)/a$	$1/(s^2 + a^2)$	5
$\cos(at)$	$s/(s^2 + a^2)$	6
$\mathcal{A}/(2t)$	$\exp(-xs^{1/2})$	7
$\mathcal{A}$	$[\exp(-xs^{1/2})]/s^{1/2}$	8
$\mathcal{B}$	$[\exp(-xs^{1/2})]/s$	9
$2t\mathcal{A} - x\mathcal{B}$	$[\exp(-xs^{1/2})]/(s^{3/2})$	10
$(\mathcal{C} + \mathcal{D})/2$	$[\exp(-xs^{1/2})]/(s - a^2)$	11
$(\mathcal{C} - \mathcal{D})/(2a)$	$[\exp(-xs^{1/2})]/[s^{1/2}(s - a^2)]$	12
$\mathcal{A} - a\mathcal{D}$	$[\exp(-xs^{1/2})]/(s^{1/2} + a)$	13
$\mathcal{D}$	$[\exp(-xs^{1/2})]/[s^{1/2}(s^{1/2} + a)]$	14
$(\mathcal{B} - \mathcal{D})/a$	$[\exp(-xs^{1/2})]/[s(s^{1/2} + a)]$	15
$t\mathcal{A} + \mathcal{C}/(4a) - (1 + 2ax + 4a^2 t)\mathcal{B}/(4a)$	$[\exp(-xs^{1/2})]/[(s - a^2)(s^{1/2} + a)]$	16

* $\delta(t)$ is the Dirac-delta function.
** $N!$ is "N factorial" $= N \cdot (N - 1) \cdot (N - 2) \cdot \ldots \cdot (1)$.

Some properties of the Laplace transform and inverse Laplace transform are

$$\mathcal{L}\big[\alpha f(t)+\beta g(t)\big]=\alpha\,\mathcal{L}\big[f(t)\big]+\beta\,\mathcal{L}\big[g(t)\big]$$

$$\mathcal{L}^{-1}\big[\alpha f(t)+\beta g(t)\big]=\alpha\,\mathcal{L}^{-1}\big[f(t)\big]+\beta\,\mathcal{L}^{-1}\big[g(t)\big],$$

(4.64)

where $f(t)$ and $g(t)$ are functions of t, and α and β are constants.

4.7.2 Laplace Transform Solution to the Heat Flow Equation

The Laplace transform will be applied to each of the terms in the heat flow equation (Equation 4.21) and to the boundary and initial conditions (Equations 4.23 and 4.24). For the term on the left side of Equation 4.21, the transform is

$$\mathcal{L}\left[\frac{\partial T(z,t)}{\partial t}\right]=\int_0^\infty \frac{\partial T(z,t)}{\partial t}e^{-st}dt.$$

The integral on the right is the product of two functions of t. *Integration by parts* can be used to change this integral to one that can be solved. According to this method, the integral of a product is

$$\int_a^b u(t)\cdot dv(t)=\big[u(t)\cdot v(t)\big]_{t=a}^{t=b}-\int_a^b v(t)\cdot du(t).$$

Let:

$$u(t)=e^{-st}$$

$$\frac{du(t)}{dt}=-se^{-st}$$

$$du(t)=-se^{-st}dt,$$

and:

$$dv(t)=\frac{\partial T(z,t)}{\partial t}dt=dT(z,t)$$

$$v(t)=T(z,t).$$

Making these substitutions:

$$\mathscr{L}\left[\frac{\partial T(z,t)}{\partial t}\right] = \left[e^{-st} \cdot T(z,t)\right]_{t=0}^{t=\infty} - \int_0^\infty T(z,t) \cdot (-se^{-st}dt)$$

$$= \left[e^{-s\cdot\infty} \cdot T(z,\infty) - e^{-s\cdot 0} \cdot T(z,0)\right] + s\int_0^\infty T(z,t)e^{-st}dt$$

$$= -T(z,0) + s\int_0^\infty T(z,t)e^{-st}dt$$

The last term above is the product of s and the Laplace transform of $T(z,t)$ (see the definition, Equation 4.63), so one can write the above equation as:

$$\mathscr{L}\left[\frac{\partial T(z,t)}{\partial t}\right] = -T(z,0) + s\mathscr{L}\left[T(z,t)\right]$$

$$= s\hat{T}(z) - T(z,0).$$

Note that $T(z,0)$ is the initial condition (Equation 4.24), which is zero (this is where this critical information enters the solution process). The transformed left side of the heat flow equation is

$$\mathscr{L}\left[\frac{\partial T(z,t)}{\partial t}\right] = s\hat{T}(z). \tag{4.65}$$

Now find the Laplace transform of the right side of the heat flow equation (Equation 4.21):

$$\mathscr{L}\left[K_T \frac{\partial^2 T(z,t)}{\partial z^2}\right] = \int_0^\infty K_T \frac{\partial^2 T(z,t)}{\partial z^2} e^{-st}dt$$

$$= K_T \int_0^\infty \frac{\partial^2 T(z,t)}{\partial z^2} e^{-st}dt.$$

On the right side, one must integrate with respect to t and take the second derivative of $T(z,t)$ with respect to z. The order of integration and differentiation can be reversed so that:

$$\mathscr{L}\left[K_T \frac{\partial^2 T(z,t)}{\partial z^2}\right] = K_T \frac{d^2}{dz^2} \int_0^\infty T(z,t)e^{-st}dt.$$

Once the functions inside the integral have been evaluated, it will no longer be a function of z and t, but simply a function of z, so the partial derivatives become ordinary derivatives above. The last integral term is simply the Laplace transform of $T(z,t)$ (again, see the definition, Equation 4.63):

$$\mathscr{L}\left[K_T \frac{\partial^2 T(z,t)}{\partial z^2}\right] = K_T \frac{d^2 \hat{T}(z)}{dz^2}. \tag{4.66}$$

Write the transformed heat flow equation using Equations 4.65 and 4.66:

$$s\hat{T}(z) = K_T \frac{d^2 \hat{T}(z)}{dz^2}. \tag{4.67}$$

The transformed Equation 4.67 is no longer a PDE because the time variable has been eliminated in the transform process. It is now an ODE with one independent variable (depth, z) and with a parameter s. Now a solution to the ODE using standard integration methods can be sought. But first one must find what the transformed boundary conditions are (the initial conditions have already been used).

The boundary condition at the surface ($z = 0$) is a constant temperature of T_0 (Equation 4.23) and the Laplace transform of a constant (Table 4.1) is simply the constant divided by s:

$$\mathscr{L}\left[T(0,t)\right] = \hat{T}(0) = \frac{T_0}{s}. \tag{4.68}$$

The bottom boundary condition in a semi-infinite soil ($z = \infty$) is also zero (Equation 4.23):

$$\mathscr{L}\left[T(\infty,t)\right] = \hat{T}(\infty) = 0. \tag{4.69}$$

An equation is now sought that satisfies the ODE, Equation 4.67. That is, when it is differentiated twice, the result is the same equation multiplied by some constants. A function that doesn't change when differentiated is e^z. To get the constant to appear, one must modify this slightly. Let:

$$\hat{T}(z) = e^{mz}, \tag{4.70}$$

where m is undefined at this point. If the second derivative of this function with respect to z is taken:

$$\frac{d^2 \hat{T}(z)}{dz^2} = \frac{d}{dz}\left[\frac{d(e^{mz})}{dz}\right]$$

$$= \frac{d}{dz}\left[me^{mz}\right]$$

$$= m^2 e^{mz}.$$

Substituting into Equation 4.67:

$$s \cdot e^{mz} = K_T m^2 e^{mz}$$

$$\left(K_T m^2 - s\right) \cdot e^{mz} = 0. \tag{4.71}$$

For this equation to be zero for any z from zero to infinity, the difference in parentheses must be zero:

$$K_T m^2 - s = 0$$

$$m^2 = \frac{s}{K_T}$$

$$m = \pm \sqrt{\frac{s}{K_T}}.$$

Note the ± sign in front of the radical. This indicates that Equation 4.71 will be satisfied if m is either:

$$m = \sqrt{\frac{s}{K_T}} \quad \text{or} \quad m = -\sqrt{\frac{s}{K_T}}.$$

Either value of m could be used in Equation 4.71 as a solution. This indicates that these are two different *particular* solutions to the ODE. There is a theorem that says that for a *linear* ODE, the most *general* solution is a linear combination of the particular solutions. The general solution is

$$\hat{T}(z) = A e^{\sqrt{(s/K_T)} \cdot z} + B e^{-\sqrt{(s/K_T)} \cdot z}, \tag{4.72}$$

where A and B are constants, but undefined at the moment because the linear combination that is the most general solution is not known. The boundary conditions can be used to define A and B.

According to Equation 4.69, the transformed boundary condition at $z = \infty$ is

$$\hat{T}(\infty) = 0$$

$$A e^{\sqrt{(s/K_T)} \cdot (\infty)} + B e^{-\sqrt{(s/K_T)} \cdot (\infty)} = 0$$

$$A e^{\infty} + B e^{-\infty} = 0$$

$$A e^{\infty} + B \cdot 0 = 0$$

$$A = 0.$$

To find the value of B, use the other boundary condition. According to Equation 4.69, the transformed boundary condition at $z = 0$ is

$$Be^{-\sqrt{(s/K_T)}\cdot 0} = \frac{T_0}{s}$$

$$B = \frac{T_0}{s}.$$

The general solution to the ODE (Equation 4.67) is

$$\hat{T}(z) = \frac{T_0}{s}e^{-\sqrt{(s/K_T)}\cdot z}. \tag{4.73}$$

Notice that this is a function of z, but not a function of t. To get the solution to the PDE, one must now use the inverse Laplace transform of Equation 4.73. Looking at the Laplace transforms in Table 4.1 in the middle column, transform #9 is

$$\hat{f}(x) = \frac{e^{-x\sqrt{s}}}{s}.$$

This is very close to Equation 4.73 and the inverse Laplace transform of this equation (see Table 4.1) is

$$f(x,t) = \text{erfc}\left(\frac{x}{2\sqrt{t}}\right),$$

where erfc is the complementary error function. Note that t reappears in the reverse transform. One can make Equation 4.73 the same as transform #9 if x is replaced with $z / K_T^{1/2}$ and multiplied by T_0. The inverse Laplace transform of Equation 4.73 is

$$T(z,t) = T_0\text{erfc}\left(\frac{z}{2\sqrt{K_T t}}\right), \tag{4.74}$$

where the property of the inverse Laplace transform that constant factors such as T_0 can be carried through the transformation (Equation 4.64) has been used. Equation 4.74 is the solution seen in Section 4.4 (see Equation 4.25).

4.7.3 PROPERTIES OF THE COMPLEMENTARY ERROR FUNCTION

What is the complementary error function that appears in the solution to the heat flow equation (Equation 4.25)? First examine the function:

$$f(x) = e^{-x^2}. \tag{4.75}$$

A plot of $f(x)$ vs. x for x near zero is shown in Figure 4.25. This function is similar to the *normal distribution*, a bell-shaped curve centered at $x = 0$. The function has a value of one at $x = 0$ and drops to almost zero for x less than about -2 or more than about $+2$.

The area under this curve is $\pi^{1/2}$ and since the curve is symmetric about $x = 0$, one can write for the positive half of the graph:

$$\frac{\sqrt{\pi}}{2} = \int_0^{\infty} e^{-x^2} dx$$

$$\tag{4.76}$$

$$1 = \frac{2}{\sqrt{\pi}} \int_0^{\infty} e^{-x^2} dx.$$

The same would be true for the negative half of the graph (limits of $x = -\infty$ to 0 on the integral). The *error function of x*, written as $\mathrm{erf}(x)$, is defined as $2/\pi^{1/2}$ times the area under the positive side of this curve; not from zero to infinity, but out to some finite value of x:

$$\mathrm{erf}(x) = \frac{2}{\sqrt{\pi}} \int_0^{x} e^{-x^2} dx. \tag{4.77}$$

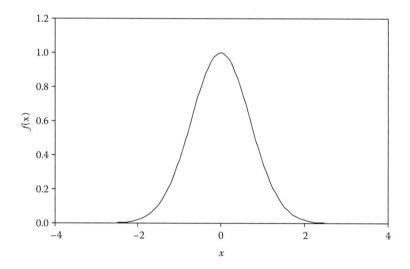

FIGURE 4.25   Plot of Equation 4.75 as a function of x.

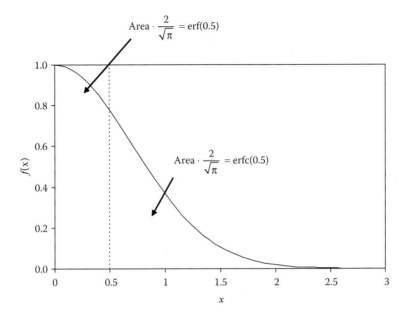

FIGURE 4.26 Error function and complementary error function of $x = 0.5$.

For example, erf(0.5) would represent the area under the curve to the left of $x = 0.5$ in Figure 4.26, multiplied by a factor of $2/\pi^{1/2}$. When $x = \infty$, erf(x) = 1 and when $x = 0$, erf(x) = 0.

The complementary error function *of x* is defined as:

$$\mathrm{erfc}(x) = \frac{2}{\sqrt{\pi}} \int_x^\infty e^{-x^2}\, dx. \tag{4.78}$$

It is the area under the curve *beyond x* in Figure 4.26, multiplied by $2/\pi^{1/2}$. For a given value of the *argument* of the complementary error function (x in Equation 4.78), the value of the function can be obtained from tables in standard math books or from a program such as Excel. The relationship between the error function and complementary error function, as one might expect from considering the areas in Figure 4.26, is

$$\mathrm{erf}(x) + \mathrm{erfc}(x) = 1. \tag{4.79}$$

4.8 PROBLEMS

1. Show that temperature is linearly distributed in a uniform soil under steady heat flow conditions. Follow the procedure used for steady water flow in Equations 3.6 through 3.8. Use a soil column of length L with a temperature

at the top of T_2 and a temperature at the bottom of T_1. Call the temperature at an intermediate depth $T(z)$ and solve for $T(z)$.

2. Use a calculator with Equation 4.41 to compute the temperatures at the first three time levels for 11 nodes with $\Delta z = 2$ cm and $\Delta t = 1$ s. The temperatures at the boundaries are 30°C at the surface and 0°C at the bottom. The initial temperatures are zero everywhere except the surface. Use $K_T = 4.5 \cdot 10^{-3}$ cm² s⁻¹. Check your answers against Figure 4.14 (show the calculations for the $j = 1$ through $j = 3$ columns in this figure).

3. What is the time lag for an annual cycle, compared to the surface temperatures, in temperatures at a depth of 300 cm in a soil with a thermal conductivity of $4.5 \cdot 10^{-3}$ cm² s⁻¹?

4. Use Excel and Equation 4.29 to make a plot of the temperature as a function of time at depths of 0, −25, −50, and −75 cm for a location where the mean annual temperature is 15°C, the maximum annual temperature at the soil surface is 35°C and the minimum annual temperature at the soil surface is −5°C. Use $K_T = 388.8$ cm² d⁻¹. Plot temperatures for $t = 1$ to 730 days. The phase shift is $-\pi/2$ so that when time is equal to zero and depth is equal to zero, $\sin(\varphi) = 0$. What is the shallowest depth that water lines could be buried at this site without freezing?

5. Use Excel and Equation 4.29 to find out how deep daily fluctuations in soil temperature will penetrate the soil (fluctuate by more than 1°C) if the average daily temperature is 25°C and the maximum and minimum surface temperatures are 35°C and 15°C, respectively. Use $K_T = 388.8$ cm² d⁻¹ (convert this to square centimeter per hour). The period is 24 h. The phase shift is $-\pi$ so that when time is equal to zero and depth is equal to zero, $\sin(\phi) = -1$.

6. Set up HYDRUS-1D to run the heat flow problem described in Section 4.5. Show graphs from the HYDRUS-1D output similar to Figures 4.23 and 4.24.

7. Set up HYDRUS-1D to run the heat flow problem described in Section 4.5, but allow water flow in order to see the effect of convection on soil temperatures. In the Water Flow Parameters window, increase K_s from 1.04 to 10.4 cm h⁻¹. In the Soil Profile Graphical Editor window, change the initial conditions to a uniform pressure head of zero (select the Edit button, move the hand cursor to select all the grid points, and set the pressure at the top and bottom to 0). This will create a saturated, unit-gradient condition (see Equation 3.29) and cause steady downward flow at a rate equal to K_s (10.4 cm h⁻¹ in this soil). Show graphs from the HYDRUS-1D output similar to Figures 4.23 and 4.24.

8. Show that Equation 4.73 is a solution to the ordinary differential Equation 4.67. That is, show that if you multiply Equation 4.73 by s, or if you take the second derivative of Equation 4.73 with respect to z and multiply by K_T, you get the same equation.

9. Find the Laplace transform of the function $f(z,t) = c$ using Equation 4.63 (not Table 4.1), where c is a constant.

10. Show that Equation 4.25:

$$T(z,t) = T_0 \cdot \mathrm{erfc}\left(\frac{z}{2\sqrt{K_T t}} \right),$$

is a solution to the heat flow Equation 4.21:

$$\frac{\partial T}{\partial t} = K_T \frac{\partial^2 T}{\partial z^2}.$$

To do this, follow the steps below:

Step 1: Take the derivative of Equation 4.25 with respect to t (that's the left side of the heat flow equation).

Step 2: Take the derivative of Equation 4.25 with respect to z.

Step 3: Take the derivative of the resulting equation from Step 2 with respect to z again and multiply by K_T (this is the right side of the heat flow equation).

Step 4: Compare the equations from Step 1 and 3 and show that they are the same.

The derivative of the complementary error function is

$$\frac{\partial}{\partial x}\mathrm{erfc}(u) = -\frac{2}{\sqrt{\pi}} e^{-u^2} \frac{\partial u}{\partial x}.$$

11. The essence of the numerical approach to solving differential equations, such as the heat flow equation, is that given a set of temperatures at one time level (the j time level), the temperatures at the next time level (the $j + 1$ time level) can always be calculated. Computer codes that use a numerical solution usually do not save all the temperatures calculated at each time level, since this would take up a lot of memory. Instead, once the temperatures have been calculated at the $j + 1$ time level, they are used to write over the temperatures at the j time level. A similar problem arises with Excel spreadsheets. For example, using a numerical solution such as that shown in Figure 4.14 to go to large times (larger values of j) will require a very large spreadsheet with many columns. You can avoid doing this by using a *macro* that takes the temperatures that have been calculated at the $j + 1$ level and copies them over into the j time level and then repeating the process as many times as required to reach the desired time level. To create a macro in Excel, look for the Developer tab at the top of the tool bar. If it is not present, click on the Microsoft Office Button at the top left. Then click Excel Options and in the Popular category, check the box to show the Developer tab. Then select the tab and click on Record Macro at the left end of the tool bar. You will see the Record Macro window. You can give the macro a name and assign a letter that will activate the macro whenever you press the control key plus that letter. Once this is done, click OK and use your mouse

to go through the steps that you want to be part of the macro. When you are finished with the steps, click on Stop Recording at the left of the tool bar. Do the same simulation as that shown in Figure 4.14, but use a spreadsheet with a macro. That is, find the temperature distribution after 1000 s in a soil with $K_T = 4.5 \cdot 10^{-3}$ cm^2 s^{-1}. The initial temperatures are zero and the boundary conditions are a temperature of 30°C at the surface and zero at a depth of 30 cm. Use $\Delta z = 2$ cm and $\Delta t = 100$ s (instead of 1 s as in Figure 4.14). The depths of the nodes are shown in cells C9–C39 of Figure 4.27. The initial values for j, t, and temperatures in the soil profile are shown in cells D7–D39. Create a macro to copy these over into the known temperature column, cells F7–F39, to initialize the program. The new value for j is calculated in cell G7 by using a formula to add one to the value in cell F7. The new value for t is calculated in cell G8 by using a formula to add Δt to the value in cell F8. The temperatures at the next time level are calculated in cells G10–G39 using the equation shown in the formula bar (Equation 4.41). The temperature at the boundary in the next time level is simply the boundary condition (cell G9 is equal to the value in cell F9). Use a macro to copy the calculated values for $j + 1$, t, and temperatures from cells G7–G39 back into the known time level, cells F7–F39. Do not copy the formulas from the cells, just the values. To do this, use "copy," then "paste special,"

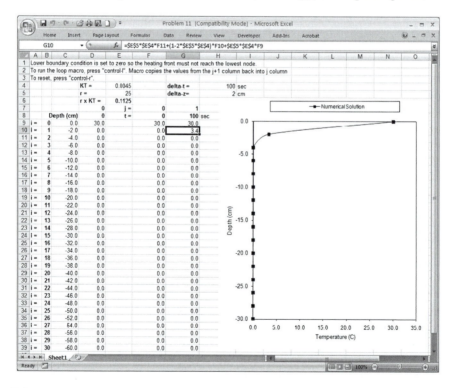

FIGURE 4.27 Excel spreadsheet for explicit finite difference numerical solution to heat flow equation.

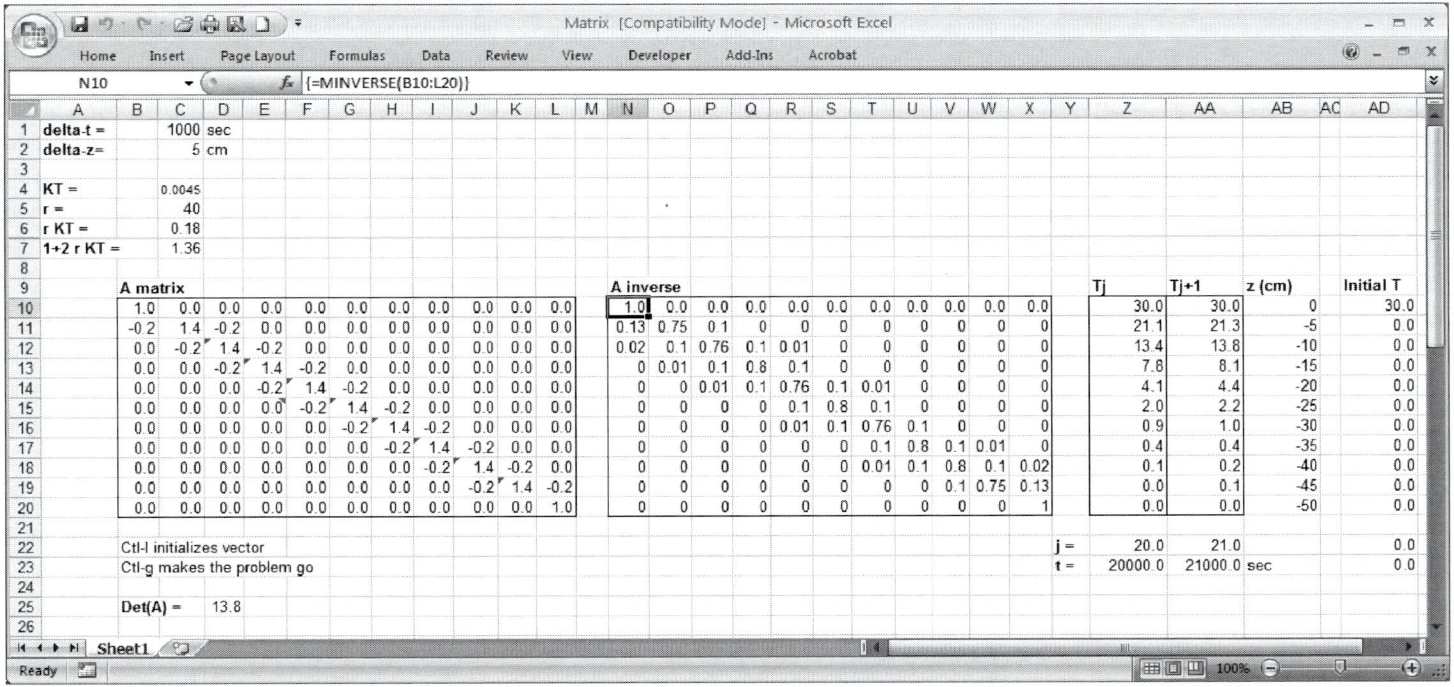

Matrix [Compatibility Mode] - Microsoft Excel
Home Insert Page Layout Formulas Data Review View Developer Add-Ins Acrobat
N10
{=MINVERSE(B10:L20)}
A B C D E F G H I J K L M N O P Q R S T U V W X Y Z AA AB AC AD
1 delta-t = 1000 sec
2 delta-z= 5 cm
3
4 KT = 0.0045
5 r = 40
6 r KT = 0.18
7 1+2 r KT = 1.36
8
9 A matrix A inverse Tj Tj+1 z (cm) Initial T
10 1.0 0.0 0.0 0.0 0.0 0.0 0.0 0.0 0.0 0.0 0.0 1.0 0.0 0.0 0.0 0.0 0.0 0.0 0.0 0.0 0.0 0.0 30.0 30.0 0 30.0
11 -0.2 1.4 -0.2 0.0 0.0 0.0 0.0 0.0 0.0 0.0 0.0 0.13 0.75 0.1 0 0 0 0 0 0 0 0 21.1 21.3 -5 0.0
12 0.0 -0.2 1.4 -0.2 0.0 0.0 0.0 0.0 0.0 0.0 0.0 0.02 0.1 0.76 0.1 0.01 0 0 0 0 0 0 13.4 13.8 -10 0.0
13 0.0 0.0 -0.2 1.4 -0.2 0.0 0.0 0.0 0.0 0.0 0.0 0 0.01 0.1 0.8 0.1 0 0 0 0 0 0 7.8 8.1 -15 0.0
14 0.0 0.0 0.0 -0.2 1.4 -0.2 0.0 0.0 0.0 0.0 0.0 0 0 0.01 0.1 0.76 0.1 0.01 0 0 0 0 4.1 4.4 -20 0.0
15 0.0 0.0 0.0 0.0 -0.2 1.4 -0.2 0.0 0.0 0.0 0.0 0 0 0 0 0.1 0.8 0.1 0 0 0 0 2.0 2.2 -25 0.0
16 0.0 0.0 0.0 0.0 0.0 -0.2 1.4 -0.2 0.0 0.0 0.0 0 0 0 0 0.01 0.1 0.76 0.1 0 0 0 0.9 1.0 -30 0.0
17 0.0 0.0 0.0 0.0 0.0 0.0 -0.2 1.4 -0.2 0.0 0.0 0 0 0 0 0 0 0.1 0.8 0.1 0.01 0 0.4 0.4 -35 0.0
18 0.0 0.0 0.0 0.0 0.0 0.0 0.0 -0.2 1.4 -0.2 0.0 0 0 0 0 0 0 0.01 0.1 0.8 0.1 0.02 0.1 0.2 -40 0.0
19 0.0 0.0 0.0 0.0 0.0 0.0 0.0 0.0 -0.2 1.4 -0.2 0 0 0 0 0 0 0 0 0.1 0.75 0.13 0.0 0.1 -45 0.0
20 0.0 0.0 0.0 0.0 0.0 0.0 0.0 0.0 0.0 0.0 1.0 0 0 0 0 0 0 0 0 0 0 1 0.0 0.0 -50 0.0
21
22 Ctl-I initializes vector j = 20.0 21.0 0.0
23 Ctl-g makes the problem go t = 20000.0 21000.0 sec 0.0
24
25 Det(A) = 13.8
26
Sheet1
Ready
100%

FIGURE 4.28 Excel spreadsheet for solving heat flow equation using matrix algebra in Problem 12. The coefficient matrix A is on the left in cells B10–L20. The formula bar shows the equation for finding the inverse of the coefficient matrix A. The temperatures at the known time level T^j are in cells Z10–Z20.

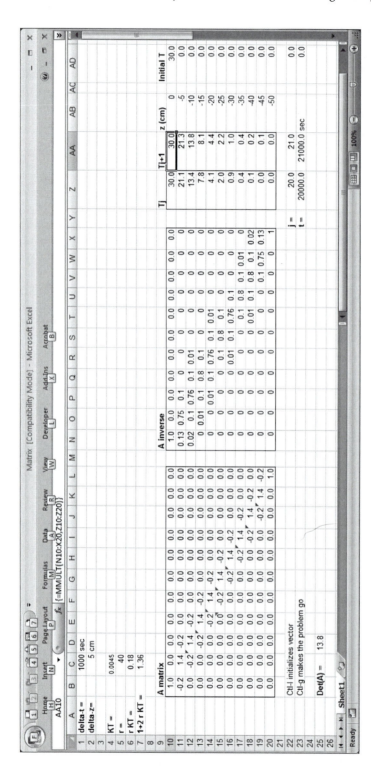

FIGURE 4.29 Excel spreadsheet for solving the heat flow equation using matrix algebra in Problem 12. The formula bar shows the equation for finding the product of the inverse of the coefficient matrix A^{-1} and the vector of temperatures at the known time level, T.

and select the radio button for "values." Make a graph that uses the values of depth and temperature in cells C9–C39 and G9–G39.

12. Excel can be used to solve the heat flow Equation 4.21 using matrix algebra and a fully implicit numerical scheme such as that shown in Equations 4.47 through 4.55 (Gottfried 2007). Do this for a simulation similar to that in Problem 11. That is, find the temperature distribution after 20,000 s in a soil with $K_T = 4.5 \cdot 10^{-3}$ cm^2 s^{-1}. The initial temperatures are zero and the boundary conditions are a temperature of 30°C at the surface and zero at a depth of 30 cm. Use $\Delta z = 5$ cm and $\Delta t = 1000$ s. See Figure 4.28. The coefficient matrix A is on the left in cells B10–L20. The formula bar shows the equation for finding the inverse of the A matrix. To do this, select the cells that will become the A-inverse matrix, cells N10–X20. Then type into the formula bar what is shown in Figure 4.28, "=MINVERSE(B10:L20)," but do not include the brackets at the beginning and end of the formula. Then press Ctrl-Shift-Enter and the brackets should appear. This identifies these cells as the inverse of the matrix in cells B10–L20. The temperatures at the known time level T^j are in cells Z10–Z20. The temperatures at the next time level are calculated in the T^{j+1} vector in cells AA10–AA20. The formula bar in Figure 4.29 shows the equation for finding the product of the inverse of the coefficient matrix A^{-1} and the vector of temperatures at the known time level T^j. To do this, select the cells that will become the T^{j+1} vector, cells Z10–Z20. Then type into the formula bar what is shown in Figure 4.29, "=MMULT(N10:X20,Z10:Z20)" (again, do not include the brackets at the beginning and end of the formula). Then press Ctrl-Shift-Enter and the brackets should appear. This identifies these cells as a product of the A-inverse matrix and the vector of temperatures at the known time level. These new temperatures in the T^{j+1} vector are copied back over into the known time level T^j vector using a macro. A macro is also used to initialize the problem by copying the initial temperature vector, cells AD10–AD20, into the known time level of vector of temperatures T^j, cells Z10–Z20. The temperature distribution after 20,000 s is shown in the T^j vector in Figure 4.29. Make a graph of these temperatures (on the x-axis) vs. depths, cells AB10–AB20 (on the y-axis). The determinant of the A matrix can be calculated using the formula "=MDETERM(B10:L20)". It is not near zero so the numerical scheme should be stable. See cell D25 in Figure 4.29.

13. Solve Equation 4.26 for ϕ. Show the intermediary steps.

14. In Equation 4.45, it was shown that the explicit finite difference algorithm uses a weighted average of the temperatures at the same node and the two nearest nodes at the previous time level. What sort of weighted average does the fully implicit finite difference scheme use? Hint: solve Equation 4.47 for T_i^{j+1}. Do the weights sum to one? Does this look like a better algorithm than Equation 4.45?

5 Transient Water Flow in Soils

[handwritten annotation: Type 1 - constant concentration, ponded infiltration]

[handwritten annotation: Type 2 - flux boundary, unit gradient - free draining]

[handwritten annotation: Type 3 - flux concentration specified]

5.1 INTRODUCTION

Water flow can be either steady or transient with regard to time. In Chapter 3, steady water movement was described. In this chapter, transient water movement is addressed. Water typically enters the soil surface in the form of precipitation or irrigation, or by means of industrial and municipal spills (Figure 5.1). Some of the rainfall or irrigation water may be intercepted by the plant canopy. If the rainfall or irrigation intensity is larger than the infiltration capacity of the soil, water will be removed by surface runoff, or will accumulate at the soil surface until it evaporates back to the atmosphere or infiltrates into the soil. Part of the water that infiltrates into the soil profile is returned back to the atmosphere by evaporation. Another part may be taken up by plant roots and eventually returned to the atmosphere by plant transpiration. The processes of evaporation and transpiration are often combined into the single process of evapotranspiration. Only water that is not returned to the atmosphere by evapotranspiration may percolate to the deeper soil zone and eventually reach groundwater. If the water table is close enough to the soil surface, the process of capillary rise may move water from the water table through the capillary fringe toward the root zone and the soil surface.

Governing equations for water flow can be developed for either steady or transient conditions, but analytical solutions usually exist only for steady conditions. Numerical methods can be used to solve the transient flow problems. In the following sections, the governing equations and initial and boundary conditions for transient flow are described along with a brief introduction to numerical methods for transient water flow. Then the important transient flow processes are described: infiltration, redistribution, evaporation, transpiration, preferential flow, and groundwater recharge/discharge. The HYDRUS numerical models are used to illustrate these processes as well as the inverse method for determining flow parameters.

5.2 TRANSIENT WATER FLOW

In this section, the Richards (1931) equation that governs transient water flow is discussed along with the initial conditions and boundary conditions that apply.

5.2.1 THE RICHARDS EQUATION

A conservation equation for nonsteady vertical water flow can be developed by considering a representative elementary volume as was done in Chapter 4 for heat flow (Section 4.4.1). The water flux into this volume during a small time interval, Δt, must

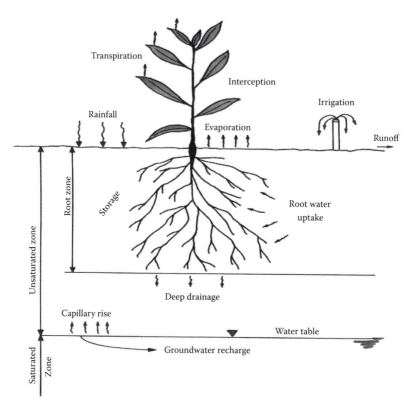

FIGURE 5.1   Transient water flow processes in soil. (From Šimůnek et al., 2008.)

be equal to the water flux out of the volume, plus the change in water stored within the volume and any sources or sinks. The resulting <u>conservation equation</u> is:

$$\frac{\partial \theta}{\partial t} = -\frac{\partial J_w}{\partial z} - S(h),$$

(5.1)

where θ is the volumetric water content [L³L⁻³], t is time [T], z is the spatial coordinate [L], J_w is the volumetric flux density [LT⁻¹], and $S(h)$ is a sink function [L³L⁻³T⁻¹] (a negative $S(h)$ is a source), usually accounting for the root water uptake (transpiration). Equation 5.1 is often referred to as the *mass conservation equation* or the *continuity equation*. The conservation equation must be combined with one or several equations describing the volumetric flux density, J_w, to produce the governing equation for variably saturated water flow. Formulations of governing equations for the different types of flow (uniform and preferential flow) are all based on this continuity equation. One can substitute the Buckingham-Darcy Equation 3.15 for J_w in this equation to obtain:

Richard's Equation

$$\frac{\partial \theta(h)}{\partial t} = \frac{\partial}{\partial z}\left(K(h)\frac{\partial h}{\partial z}\right) + \frac{\partial K(h)}{\partial z} - S(h).$$

(5.2)

capillarity   gravity   root water uptake sink

This equation was first developed by Richards (1931) and is known as the *Richards equation*. It can be easily expanded to two and three dimensions. This partial differential equation is the equation governing variably saturated flow in the unsaturated zone.

Equation 5.2 is the *mixed form* of the Richards equation in that there are two dependent variables: θ and h. The first term on the right side accounts for the effect of capillarity on water movement and the second term on the right side accounts for the effect of gravity on water movement. Various other formulations of the Richards equation are possible. For example, when the time derivative of the water content is expanded using the chain rule as follows:

$$\frac{\partial \theta(h)}{\partial t} = \frac{d\theta(h)}{dh}\frac{\partial h}{\partial t} = C_w(h)\frac{\partial h}{\partial t}, \tag{5.3}$$

one obtains the pressure head formulation:

$$C_w(h)\frac{\partial h}{\partial t} = \frac{\partial}{\partial z}\left(K(h)\frac{\partial h}{\partial z}\right) + \frac{\partial K(h)}{\partial z} - S(h), \tag{5.4}$$

where $C_w(h)$ is the soil water (or hydraulic) capacity function $[L^{-1}]$ that characterizes the slope of the retention curve (see Equation 2.51). The only dependent variable is h.

The Richards equation (in any form) is a nonlinear partial differential equation. It is similar to the heat flow equation that includes convection and conduction (Equation 4.56), but differs in one very important way. The coefficient $K(h)$ that multiplies the gradient term, is a function of the dependent variable (h or θ). This makes it a *nonlinear* partial differential equation. Because of its strongly nonlinear nature, only a relatively few simplified analytical solutions can be derived. Most practical applications of the Richards equation require a numerical solution (see Section 5.3). Alternatively, the equation is simplified in some manner as happens with infiltration equations (Section 5.4.3).

5.2.2 Initial Conditions

The Richards equation can be solved analytically or numerically provided that applicable initial and boundary conditions are specified. Initial conditions characterizing the initial state of the system can be specified either in terms of the water content:

$$\theta(z,t) = \theta_i(z,0), \tag{5.5}$$

or the pressure head:

$$h(z,t) = h_i(z,0), \tag{5.6}$$

where θ_i $[L^3L^{-3}]$ and h_i $[L]$ are initial values of water content or pressure head, respectively. For numerical models, it is generally recommended to specify initial conditions in terms of the pressure head, since this variable is the driving force for water flow. Specifying initial conditions in terms of the water content often leads to unrealistically large pressure head gradients and, consequently, water fluxes across textural boundaries, after water contents are converted into pressure heads using the soil water retention curves.

5.2.3 BOUNDARY CONDITIONS

Two types of conditions can be specified on boundaries of the transport domain: system-dependent and system-independent boundary conditions. System-independent boundary conditions are boundary conditions for which the specified boundary value (i.e., pressure head, water content, water flux, or gradient) does not depend on the status of the soil system. System-dependent boundary conditions are boundary conditions for which the actual boundary condition (pressure head, water content, water flux, or gradient) depends on the status of the system and is calculated by the model itself.

5.2.3.1 System-Independent Boundary Conditions

Several system-independent boundary conditions may be applied to the transport domain boundaries. When the pressure head at the boundary is known, one can use the *Dirichlet* or *type-1 boundary condition*:

$$h(z, t) = h_0(z, t), \tag{5.7}$$

where h_0 is a prescribed pressure head [L]. This boundary condition is often referred to as a *pressure head boundary condition* or a *type-1 boundary condition*. This boundary condition must be used when simulating ponded infiltration; for describing the hydrostatic pressure at the boundary between the soil and standing or flowing water in a furrow, lake or river; to specify the water level in a well; or define the position of the water table. The water flux across a Dirichlet boundary is not known *a priori*, but must be calculated from the mathematical solution (either analytical or numerical) of the governing flow problem.

When the water flux across the boundary is known, one can use the *Neumann* or *type-2 boundary condition*:

$$-K(h)\left(\frac{\partial h}{\partial z} + 1\right) = J_0(z, t), \tag{5.8}$$

where J_0 is a prescribed water flux [LT^{-1}]. This boundary condition is often referred to as a *flux boundary condition*, or a *type-2 boundary condition*. A Neumann boundary condition must only be used along boundaries where the flux is known, provided the flux does not depend on the soil system. This boundary condition hence cannot be used to model precipitation or irrigation, since the precipitation or irrigation rate may exceed the infiltration capacity of the soil, in which case ponding will occur and the actual boundary flux will decrease.

A particular form of the type-2 boundary condition is the *gradient-type boundary condition* of the form:

$$\frac{\partial h}{\partial z} + 1 = g_0(z,t), \qquad (5.9)$$

where g_0 is a prescribed total gradient [LL^{-1}]. Equation 5.9 is commonly used to specify a unit vertical hydraulic gradient simulating free drainage from the bottom of a soil profile when the water table is situated far below the domain of interest. The condition is consistent with unit gradient conditions (see Equation 3.29) often observed in field studies of water flow, especially during redistribution (Sisson 1987; McCord 1991).

5.2.3.2 System-Dependent Boundary Conditions

In many applications, neither the flux across nor the pressure head or gradient along a boundary is known *a priori*, but follows from interactions between the soil and its surroundings (e.g., the atmosphere or deeper subsurface). The boundary representing the soil-air interface, which is exposed to atmospheric conditions, is one example of the system-dependent boundary. The potential fluid flux across this interface is controlled exclusively by external conditions (precipitation, evaporation). However, the actual flux also depends on the (transient) moisture conditions in the soil. Soil surface boundary conditions may change from prescribed flux to prescribed head conditions (and vice-versa). This occurs, for example, when the precipitation rate exceeds the infiltration capacity of the soil, resulting in either surface runoff or accumulation of excess water at the soil surface, depending on the soil conditions. The infiltration rate in that case is no longer controlled by the precipitation rate, but instead by the infiltration capacity of the soil. A system-dependent boundary condition also occurs when the potential evaporation rate, as calculated from meteorological conditions (the evaporative demand of the atmosphere), exceeds the capability of the soil to deliver enough water to the soil surface. In this case, the potential evaporation rate can be significantly reduced to an actual evaporation rate that is controlled by the soil.

System-dependent atmospheric boundary conditions can be implemented mathematically using an approach of Neuman et al. (1974), which limits the absolute value of the flux such that the following two conditions are satisfied:

$$\left| K(h) \left(\frac{\partial h}{\partial z} + 1 \right) \right| \leq E, \qquad (5.10)$$

and

$$h_A \leq h \leq h_S, \qquad (5.11)$$

where E is the maximum potential rate of infiltration or evaporation under the current atmospheric conditions [LT^{-1}], h is the pressure head at the soil surface [L], and h_A and h_S are, respectively, minimum and maximum pressure heads allowed under the prevailing soil conditions [L]. The value for h_A is determined from the equilibrium conditions between soil water and atmospheric water vapor, whereas h_S is usually set equal to zero (which would initiate instantaneous surface runoff) or results from the accumulation of excess water in the surface ponding layer, in which case its value must be calculated from the difference between the infiltration and precipitation (or irrigation) rates. When one of the limits of Equation 5.11 is reached, a prescribed head boundary condition will be used to calculate the actual surface flux. Methods for calculating E and h_A on the basis of atmospheric data have been discussed by Feddes et al. (1974).

Another example of a system-dependent boundary condition is a seepage face through which water leaves the saturated part of the flow domain. The boundary condition in this case can be formulated mathematically as follows:

$$J_0(z,t) = 0 \quad \text{for } h(z,t) < 0$$
$$h_0(z,t) = 0 \quad \text{for } h(z,t) \geq 0. \tag{5.12}$$

This boundary condition states that there is no flux across the boundary as long as the boundary is unsaturated, and that the pressure head changes to zero once saturation is reached. The flux across the boundary is then calculated from the flow field by solving the governing flow equations. This boundary condition can be used at the bottom of certain types of (finite) lysimeters, along tile drains, or at seepage faces in a landscape.

A special system-dependent boundary condition is sometimes implemented in one-dimensional numerical models (e.g., van Dam et al. 1997; Šimůnek et al. 2006) to account for flow to horizontal subsurface tile drains. An equivalent drainage flux from the bottom of the simulated soil profile is then calculated using an appropriate analytical solution for the tile drainage system (e.g., Hooghoudt 1940; Ernst 1962; van Hoorn 1997) (see Section 3.2.6). The drainage equations involved generally hold for steady-state flow into the drain and depend on the geometry of the system (e.g., depth of tile drain, depth to impermeable layer, location of water table midway between two drains, and possibly information about soil layering).

5.3 NUMERICAL SOLUTIONS TO THE RICHARDS EQUATION

Numerical methods can be used to solve the Richards Equation 5.2 under conditions where analytical solutions do not exist. These methods were first applied to soil water movement in the early 1960s (Ashcroft et al. 1962; Hanks and Bowers 1962). The two most common methods are finite differences and finite elements (Huyakorn and Pinder 1983; Smith 1985). The HYDRUS models use a finite element numerical approach in space and a finite difference approach in time. In this section, we describe briefly a one-dimensional finite difference numerical approach because it is much simpler and results in a similar set of simultaneous equations that must be solved using matrix algebra.

The finite difference approach is used to develop a set of algebraic equations from one of the forms of the Richards equation. The mixed form of the Richards equation (Equation 5.2) has been used most often to solve problems of transient flow since the work by Celia et al. (1990). For the sake of simplicity, we use the h form of the Richards equation (Equation 5.4). The development of the set of equations for one-dimensional vertical water movement with no source/sink term and known pressure heads at the boundaries is demonstrated.

Just as shown in Chapter 4 for transient heat flow (see Section 4.4.3), the first step is to discretize the soil profile into N depth intervals, Δz, such that $z = i \Delta z$ and $i = 0...N$ (Smith 1985). Depth is positive downward. The point at which intervals join is called a *node*, so there are $N + 1$ nodes in the profile. Time is also discretized by intervals of Δt such that $t = j \Delta t$. The depth and time intervals can vary but will be kept constant in this case. Then the derivatives in Equation 5.4 are written as discrete differences divided by the appropriate interval:

node number i *(handwritten annotation)*

j is time step (handwritten annotation)

$$C_{\mathrm{w}}\left(h_i^{j+1/2}\right)\cdot\left(\frac{h_i^{j+1}-h_i^{j}}{\Delta t}\right) = \frac{\left[K(h)\cdot\left(\frac{\partial h}{\partial z}+1\right)\right]\Big|_{i+1/2} - \left[K(h)\cdot\left(\frac{\partial h}{\partial z}+1\right)\right]\Big|_{i-1/2}}{\Delta z}$$

$$= \frac{K\left(\frac{h_i+h_{i+1}}{2}\right)\cdot\left(\frac{h_{i+1}-h_i}{\Delta z}+1\right) - K\left(\frac{h_{i-1}+h_i}{2}\right)\cdot\left(\frac{h_i-h_{i-1}}{\Delta z}+1\right)}{\Delta z}$$

$$= \frac{K\left(\frac{h_i+h_{i+1}}{2}\right)}{(\Delta z)^2}\left(h_{i+1}-h_i+\Delta z\right) - \frac{K\left(\frac{h_{i-1}+h_i}{2}\right)}{(\Delta z)^2}\left(h_i-h_{i-1}+\Delta z\right).$$

$$(5.13)$$

Subscripts denote depth nodes and superscripts denote time steps. Note that in Equation 5.13 the midpoint hydraulic conductivities are evaluated for an arithmetic mean of pressure heads in neighboring nodes. Other averaging approaches, such as geometric or harmonic means, can also be used. At this point, the time step for evaluating the terms on the right side of Equation 5.13 is unspecified. If all the terms on the right side are evaluated at the known (j) time step and $C_{\mathrm{w}}(h_i^{j+1/2})$ is approximated by $C_{\mathrm{w}}(h_i^{j})$, then there is only one unknown in the equation (h_i^{j+1}) from the term on the left side. The equation can be solved explicitly for h_i^{j+1} at each node ($i = 1...N-1$, with h_0^{j+1} and h_N^{j+1} known from the boundary conditions). This is known as the *explicit* finite difference method and has the disadvantage of being unstable for all but very small time steps.

If all the terms on the right side of Equation 5.13 are evaluated at the unknown ($j + 1$) time step, then there are three unknowns: h_{i-1}^{j+1}, h_i^{j+1}, and h_{i+1}^{j+1}. Collecting coefficients of these terms on the left side gives:

$$a_i h_{i-1}^{j+1} + b_i h_i^{j+1} + c_i h_{i+1}^{j+1} = d_i, \tag{5.14}$$

where:

$$a_i = -r \frac{K_{i-1/2}}{C_{wi}}$$

$$c_i = -r \frac{K_{i+1/2}}{C_{wi}} \tag{5.15}$$

$$b_i = 1 - a_i - c_i$$

$$d_i = h_i^j + \Delta z (a_i - c_i),$$

and $r = \Delta t / \Delta z^2$, $K_{i \pm 1/2} = K(h_{i \pm 1/2}^{j+1})$, $C_{wi} = C_w(h_i^{j+1/2})$, $h_{i \pm 1/2}^{j+1}$ is the arithmetic average of h_i^{j+1} and $h_{i \pm 1}^{j+1}$, and $h_i^{j+1/2}$ is the arithmetic average of h_i^j and h_i^{j+1}. Equation 5.14 can be written for each node so there are $N + 1$ equations with $N - 1$ unknowns (h_i^{j+1}) to be solved simultaneously (the pressure heads at the boundaries are known). This system of equations can be written as a combination of a matrix and two vectors:

$$
\begin{bmatrix}
1 & 0 & 0 & & & & 0 \\
a_1 & b_1 & c_1 & & & & 0 \\
0 & a_2 & b_2 & c_2 & & & 0 \\
 & . & & . & & . & \\
0 & & & a_{N-2} & b_{N-2} & c_{N-2} & 0 \\
0 & & & & a_{N-1} & b_{N-1} & c_{N-1} \\
0 & & & & 0 & 0 & 1
\end{bmatrix}
\begin{bmatrix}
h_0^{j+1} \\
h_1^{j+1} \\
h_2^{j+1} \\
. \\
h_{N-2}^{j+1} \\
h_{N-1}^{j+1} \\
h_N^{j+1}
\end{bmatrix}
=
\begin{bmatrix}
h_0^j \\
h_1^j + \Delta z (a_1 - c_1) \\
h_2^j + \Delta z (a_2 - c_2) \\
. \\
h_{N-2}^j + \Delta z (a_{N-2} - c_{N-2}) \\
h_{N-1}^j + \Delta z (a_{N-1} - c_{N-1}) \\
h_N^j
\end{bmatrix}.
$$

$$\tag{5.16}$$

This can also be written in matrix notation as:

$$[A] \cdot \{h\}^{j+1} = \{B\}^j, \tag{5.17}$$

where $[A]$ is the square tridiagonal matrix in Equation 5.16 with the coefficients a_i, b_i, and c_i along the subdiagonal, diagonal, and superdiagonal, $\{h\}^{j+1}$ is the vector of the $N - 1$ unknown values of h and the two known values at the boundaries, and $\{B\}^j$ is a vector containing the values of h at the known time step. Computer algorithms are used to invert $[A]$ and solve Equation 5.17 for $\{h\}^{j+1}$ at a given time step. The only remaining difficulty is that the coefficients a_i, b_i, and c_i contain $K_{i \pm 1/2}$ and C_{wi}, which require pressure heads at the unknown time level in order to be evaluated. An iterative procedure is usually used at each time step wherein h_i^{j+1} is estimated, $K_{i \pm 1/2}$ and

C_{wi} are evaluated, and the set of equations is solved for an improved estimate of h_i^{j+1}. This is repeated until the difference between the two estimates is acceptably small, and then the algorithm goes to the next time step (e.g., Paniconi et al. 1991). If it's not possible to reduce the difference to an acceptable level, the numerical solution is said not to *converge*.

As noted in Chapter 4 (see Section 4.4.3), inverting large matrices (e.g., $N = 100$) involves many computations, but computer algorithms have been developed to do this efficiently. If the *determinant* of the A matrix is near zero it is not possible to invert the matrix. In this case, the numerical approach will fail. The determinant depends on the values of a_i, b_i, and c_i in Equation 5.15, which depend in turn on the choice of values for Δt and Δz and the values of K_i and C_{wi}.

Specifying all the terms on the right side of Equation 5.13 at the $j + 1$ time step is known as the *fully implicit* finite difference method. Another approach is used where the terms on the right side of Equation 5.13 are evaluated at the $j + 1/2$ time step (by taking an average on the right side at the j and $j + 1$ time step) and this is known as the Crank-Nicolson finite difference method (Smith 1985). The fully implicit and Crank-Nicolson methods are stable for much larger time steps than the

---

TABLE 5.1

Widely used numerical models for simulating variably saturated water flow in soils

Model name	Internet address	Reference
COUP	http://www.lwr.kth.se/vara%20 datorprogram/CoupModel	Jansson and Karlberg (2001)
DAISY	http://code.google.com/p/daisy-model/	Hansen et al. (1990)
HYDRUS-1D	http://pc-progress.com/en/Default .aspx?hydrus-1d	Šimůnek et al. (2008)
HYDRUS-2D	http://pc-progress.com/en/Default .aspx?hydrus-2d	Šimůnek et al. (1999a)
HYDRUS (2D/3D)	http://pc-progress.com/en/Default .aspx?hydrus-3d	Šimůnek et al. (2006)
MACRO	http://www.mv.slu.se/BGF/Macrohtm/ macro.htm	Jarvis (1994)
RZWQM	http://www.ars.usda.gov/Main/docs .htm?docid = 17740	Ahuja and Hebson (1992)
SHAW	http://www.ars.usda.gov/Services/docs .htm?docid = 16931	Flerchinger et al. (1996)
SWAP	http://www.swap.alterra.nl/	van Dam et al. (1997)
SWIM	http://www.clw.csiro.au/products/swim/	Verburg et al. (1996)
TOUGH2	http://www-esd.lbl.gov/TOUGH2/	Pruess (1991)
UNSAT-H	http://hydrology.pnl.gov/resources/ unsath/unsath.asp	Fayer (2000)
VS2DI	http://wwwbrr.cr.usgs.gov/projects/ GW_Unsat/vs2di1.2/	Healy (1990)

explicit method, but water flow problems that involve large gradients in h such as infiltration, still require very small time steps (fast computers).

Commonly used numerical models that solve the Richards equation to simulate water flow in variably saturated soils are shown in Table 5.1. These models also solve the equations that describe solute transport (Chapter 6). In the following sections, HYDRUS-1D and HYDRUS (2D/3D) are used to illustrate important transient flow processes and parameter optimization methods.

5.4   INFILTRATION

Infiltration is a key process because it determines how much water from rainfall, irrigation, or a contaminant spill enters the soil and how much becomes runoff (or *overland flow* in hydrology terminology). It is also a key process in erosion in that there can be no erosion without runoff to transport and scour sediment.

5.4.1   Infiltration into a Uniform Soil

In this example (HYDRUS Simulation 5.1 – Ponded Infiltration), we show how to set up HYDRUS-1D to simulate infiltration into a uniform soil with the hydraulic properties of the Cecil Ap horizon (see Table 2.5 for the water retention parameters and Table 3.2 for K_s). HYDRUS-1D is opened and in the New Project window (not shown) a name and description are entered. Only water flow is selected in the Main Processes window (Figure 5.2)

In the Geometry Information window (not shown), length units of centimeters are selected and the depth of the profile is set to 250 cm. Hours are selected for time units in the Time Information window, the final time is set to 2 h, and the time discretization settings are as shown in Figure 5.3.

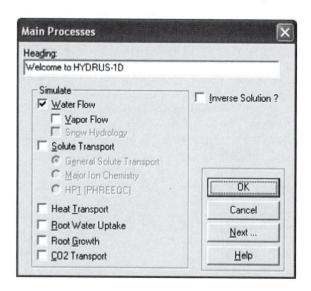

FIGURE 5.2   Main Processes window in the infiltration example for a uniform soil.

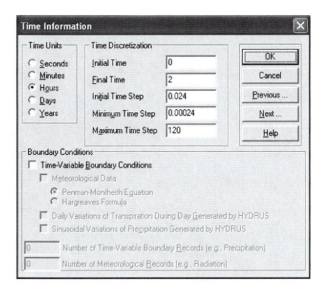

FIGURE 5.3 Time Information window in the infiltration example for a uniform soil.

Three print times are selected in the Print Information window (not shown) and the print times of 0.1, 1.0, and 2.0 h are entered (not shown). In the Iteration Criteria window (not shown), the default values for water simulation are used. The Soil Hydraulic Model window settings are the same as the default except the "air-entry value of −2 cm" option is chosen (not shown). This option improves numerical stability in soils with a $K(h)$ function that is highly nonlinear near saturation. The van Genuchten (1980a) hydraulic properties from Tables 2.5 and 3.2 are entered in the Water Flow Parameters window (Figure 5.4).

In the Water Flow Boundary Conditions window (not shown), a constant pressure head condition is selected for the upper boundary condition and a free drainage condition for the lower boundary condition. The initial conditions are in terms of pressure heads (the default setting). The Profile Information window is shown in Figure 5.5. The default number of nodes (101) is used, but the distribution is changed

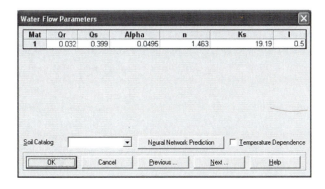

FIGURE 5.4 Water Flow Parameters window in the infiltration example for a uniform soil.

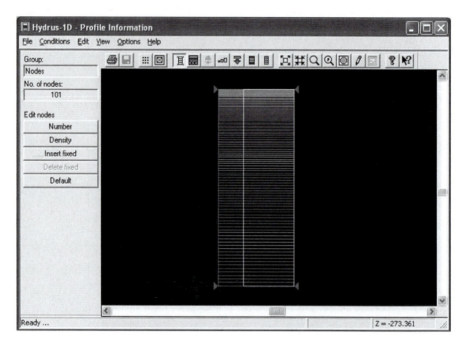

FIGURE 5.5 Profile Information window showing grid spacing in the infiltration example for a uniform soil.

from uniform spacing to a high-density distribution near the surface (where pressure heads and water contents are changing most rapidly). This is done by clicking the "Density" button on the left panel and then selecting the top node and entering a density value of 0.1. This produces a distance between nodes at the top of the model space that is 1/10 that of the distance between nodes at the bottom.

The initial conditions for the pressure head are set to −100 (the default) at all depths in the Profile Information window:

$$h(z,0) = -100. \tag{5.18}$$

In the Soil Profile Summary window (Figure 5.6), the pressure head at the surface ($z = 0$) is changed from −100 to 0 cm to simulate the boundary condition that water is ponding with no surface storage (Dirichlet boundary condition):

$$h(0,t) = 0. \tag{5.19}$$

After the program has run, the Profile Information window from the Post-processing panel shows the distributions of the pressure head for the selected print times (Figure 5.7). During infiltration, a wetting front of higher pressure head moves down through the soil over time. For a given soil, the abruptness of the wetting front will depend on the shape of the $\theta(h)$ and $K(h)$ functions. Both of these depend on

Soil Profile Summary

	z	h	Root	Axz	Bxz	Dxz	Mat
1	0	0	0	1	1	1	1
2	0.454545	-100	0	1	1	1	1
3	0.950413	-100	0	1	1	1	1
4	1.4876	-100	0	1	1	1	1
5	2.06612	-100	0	1	1	1	1
6	2.68595	-100	0	1	1	1	1
7	3.34711	-100	0	1	1	1	1
8	4.04959	-100	0	1	1	1	1
9	4.79339	-100	0	1	1	1	1
10	5.57851	-100	0	1	1	1	1
11	6.40496	-100	0	1	1	1	1
12	7.27273	-100	0	1	1	1	1
13	8.18182	-100	0	1	1	1	1
14	9.13223	-100	0	1	1	1	1
15	10.124	-100	0	1	1	1	1

OK Cancel Previous Next Help

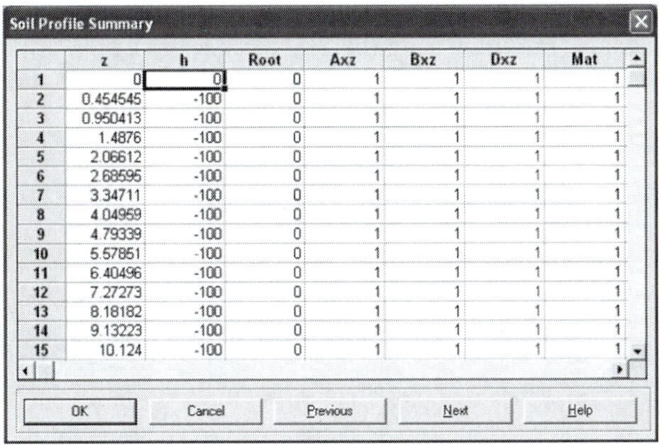

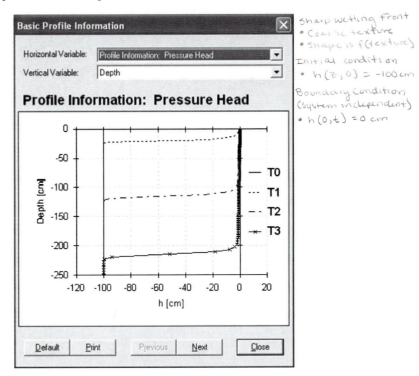

FIGURE 5.6 Soil Profile Summary window in the infiltration example for a uniform soil.

pore-size distribution. For coarse-textured soils with a narrow pore-size distribution, the wetting front will be more abrupt; in a fine-textured soil, the wetting front will be more diffuse. In this case, the wetting front is very sharp, as one might expect in a loamy sand. The wetting front is a combination of new water added by the rain and old water displaced to lower depths.

Sharp wetting front
• Coarse texture
• Shape is f(texture)
Initial condition
• h(z,0) = -100 cm
Boundary condition
(system independent)
• h(0,t) = 0 cm

FIGURE 5.7 Pressure heads as a function of depth for various times in the infiltration example for a uniform soil. T0 = 0.0, T1 = 0.1, T2 = 1.0, and T3 = 2.0 h.

A numerical solution can also be used to estimate the infiltration rate $i(t)$ [LT^{-1}] as a function of time. The infiltration rate is the water flux at the soil surface so it is described by the Buckingham-Darcy equation:

$$i(t) = -J_w(t)\Big|_{z=0} = \left[K(h)\frac{\partial h(z,t)}{\partial z} + K(h) \right]_{z=0}, \qquad (5.20)$$

where the vertical bar and subscript indicates that the flux is evaluated at the soil surface.

The infiltration rate can be seen by selecting Water Flow – Boundary Fluxes and Heads from the post-processing panel (Figure 5.8). It is shown as a negative flux when Actual Surface Flux is chosen using the pull-down menu for the vertical variable (top of Figure 5.8). Excel can be used to plot this data as a positive flux in the more conventional manner. In Excel, the T.LEVEL.OUT file from the HYDRUS-1D folder with the project name is opened (using All Files for file type). The columns labelled "Time" and "vTop" (the infiltration rate) are copied to a new worksheet where a new column is created with the negative of the vTop values.

Surface Flux
• Initially high
• saturated conditions
• nearly constant over time after 0.8 hr

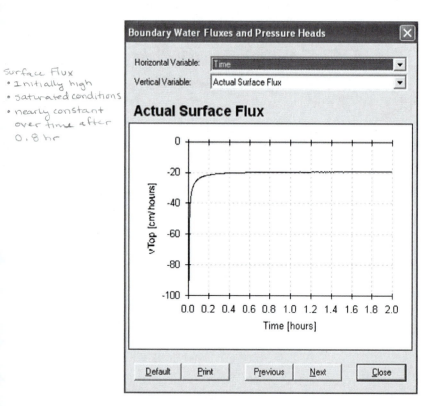

FIGURE 5.8 Boundary water fluxes and pressure heads window showing the actual surface flux in the infiltration example for a uniform soil.

The infiltration rate plotted in Excel is shown in Figure 5.9. The rate decreases with time and approaches an asymptotic minimum infiltration rate. Experimental observations confirm that $i(t)$ follows the pattern in this figure. This is for a uniform soil with no layers or crusting at the surface and relatively wet antecedent water contents. Equation 5.20 can be used to understand this pattern. When infiltration first starts, the wetting front is steep and very close to the surface (Figure 5.7). As a result, $\partial h/\partial z$ at the surface is large and the first term in Equation 5.20 produces a large value for $i(t)$. Under these conditions, the gradient in pressure head (due mainly to capillarity) is responsible for the rapid movement of water into a moist soil. The second term in Equation 5.20 represents the effect of gravity and may have little effect on water movement during the initial stages of infiltration. Later in the infiltration event, the wetting front has moved deeper into the soil (Figure 5.7). As a result, $\partial h/\partial z$ at the surface is much smaller (maybe even zero) and the first term in Equation 5.20 (and capillarity) has little effect. When the first term approaches zero, $i(t) \approx K_s$ and only gravity causes flow (unit gradient flow, see Equation 3.29). Under these circumstances, a minimum infiltration rate is reached and it is approximately the saturated hydraulic conductivity of the Cecil Ap horizon, 19.19 cm h^{-1} (see Table 3.2), as demonstrated in Figure 5.9.

Because the initial high infiltration rate in a dry soil is due to the pressure head gradient term, if the soil is wetter before the rainfall event, the initial infiltration rate will be lower. However, the final infiltration rate in both a wet and a dry soil will be the same, approximately K_s.

Cumulative infiltration $I(t)$ [L] is the integral of the infiltration rate:

$$I(t) = \int_0^t i(t)dt. \tag{5.21}$$

It is the area under the rate curve in Figure 5.9 (see Example 5.2 and Figure 5.21). Once the infiltration rate becomes constant, cumulative infiltration increases linearly.

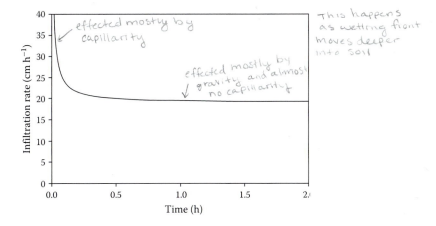

FIGURE 5.9   Infiltration rate as a function of time in the infiltration example for a uniform soil.

To get a more realistic simulation, one can specify slightly different initial and boundary conditions. For the surface boundary condition, a system-dependent boundary condition (see Equations 5.10 and 5.11) is used that starts with a very high flux equal to a rainfall rate of 25 cm h^{-1}. When the soil near the surface wets up sufficiently that the pressure head at the first node reaches zero (h_S in Equation 5.11), the boundary condition changes to a constant pressure head boundary condition equal to zero (ponding with no surface storage). For the bottom boundary condition, a constant pressure head boundary condition of zero is used to represent a water table at a depth of 250 cm. For the initial conditions, a pressure head distribution starting with −250 cm at the soil surface and increasing linearly to a pressure head of 0 cm at the bottom of the model space is specified. This represents a soil profile where the pressure heads are in equilibrium with a water table at a depth of 250 cm.

To do this in HYDRUS-1D (HYDRUS Simulation 5.2 – Flux Infiltration), the previous example for infiltration into a uniform soil is used with the following changes. In the Time Information window (shown in Figure 5.3), the Time Variable Boundary Conditions box is checked and a value of 1 is placed in the box for Number of Time Variable Boundary Records (e.g., Precipitation). In the Water Flow Boundary Conditions window, the upper boundary condition is changed to Atmospheric BC with Surface Run Off and the bottom boundary condition is changed to Constant Pressure Head (Figure 5.10).

After the Water Flow Boundary Conditions window, a new window will appear labeled Time Variable Boundary Conditions (Figure 5.11). A value of 2 h is entered for time and a rainfall rate of 25 cm h^{-1} is entered for precipitation. This indicates that

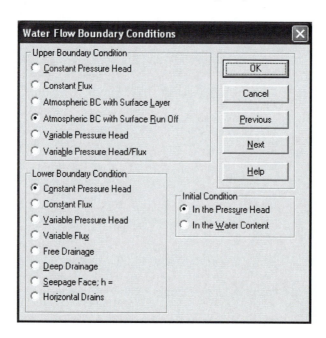

FIGURE 5.10 Water Flow Boundary Conditions window in the infiltration example with a system-dependent boundary condition.

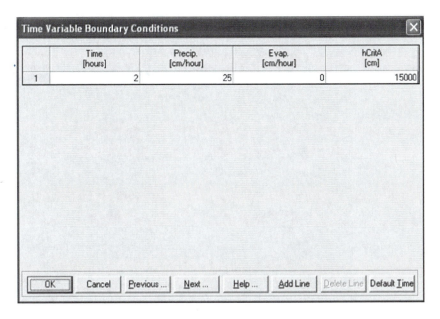

FIGURE 5.11 Time Variable Boundary Conditions window in the infiltration example with a system-dependent boundary condition.

from time zero up until 2 h, the rainfall rate will be 25 cm h^{-1}. In this window, hCritA is the absolute value of the minimum allowed pressure head at the soil surface when the atmospheric boundary condition is used (see the Help file for this window). This is h_A in Equation 5.11. It is set to indicate a soil at wilting point: 15,000 cm.

In the Profile Information window, the initial conditions are changed using the Edit Condition button on the left (Figure 5.12). The hand cursor that appears is used to select all the nodes. Then, when the Condition Specification window appears, a pressure head of −250 cm is entered for the top value and a pressure head of 0 cm is entered for the bottom value.

After running HYDRUS-1D, the simulated distribution of pressure heads as a function of depth for various times can be seen in the Profile Information window (Figure 5.13). The initial equilibrium distribution of pressure heads is apparent. The movement of a wetting front down the profile is also apparent with the front reaching a depth of about 80 cm at the end of the simulation. Compare this figure with the earlier example of infiltration into the same soil with an initial distribution of pressure heads equal to a constant −100 cm (Figure 5.7). The results are similar except for the initial distribution of pressure heads.

The simulated infiltration rate as a function of time is shown in Figure 5.14. Infiltration is a constant flux equal to 25 cm h^{-1} until about 0.2 h and then decreases to a steady rate of 19.19 cm h^{-1}, which is equal to K_s for the soil. The effect of the system-dependent boundary condition at the surface is apparent when this figure is compared to the earlier infiltration example (Figure 5.9). With the rainfall rate of 25 cm h^{-1}, once the infiltration rate for the soil drops below this rate, water starts to run off (assuming that surface storage is negligible). In the earlier example, it was

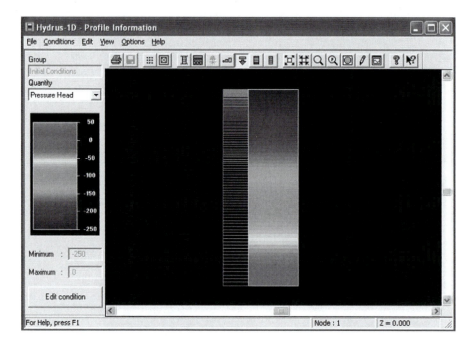

FIGURE 5.12 Profile Information window showing initial conditions in the infiltration example with a system-dependent boundary condition.

assumed that water was ponded at the surface from the beginning of the infiltration event. It would take a very high-intensity rainfall event to cause ponding early in the event (more than 40 cm h^{-1}, see Figure 5.9). The system-dependent boundary condition used in this example is more realistic.

HYDRUS-1D computes the absolute error in the mass balance of water within the model space at each time step. This is computed as the change in the volume of water within the flow domain minus the net flow across domain boundaries including root uptake of water during the entire simulation. In a perfect numerical solution, there would be no difference between these two quantities. The error is reported as a percentage of the maximum of two quantities: (1) the sum of the absolute change in water content over all elements and (2) the sum of the absolute values of all fluxes in and out of the flow domain. The Mass Balance Information is contained in the last item in the Post-processing panel. In Figure 5.15, the mass balance in the infiltration example with a system-dependent boundary condition is shown. Results are shown for each print time (0.1, 1.0, and 2.0 h) and in all cases the error (WatBalR) is less than 0.00%. In general, water mass balance errors should be less than 1%.

5.4.2 SOIL CRUSTS AND SUBSURFACE LAYERS

Another factor that can cause the infiltration rate to decrease is the formation over time of a *surface seal* or *crust* at the soil surface. A surface seal is a very thin layer (1–5 mm) at or just below the soil surface that forms due to the breakdown of soil aggregates and chemical dispersion of clay particles under raindrop impact. The clay particles fill the

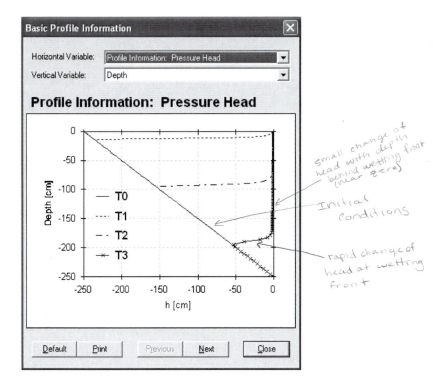

Basic Profile Information ✕

Horizontal Variable: Profile Information: Pressure Head ▾

Vertical Variable: Depth ▾

Profile Information: Pressure Head

(handwritten annotations:)
small change of head with depth behind wetting front (near zero)

Initial conditions

rapid change of head at wetting front

Default | Print | Previous | Next | Close

FIGURE 5.13 Resulting pressure head distributions in the infiltration example with a system-dependent boundary condition. T0 = 0.0, T1 = 0.1, T2 = 1.0, and T3 = 2.0 h.

soil pores and create a layer with a saturated hydraulic conductivity several orders of magnitude less than the undisturbed soil (Miller and Radcliffe 1992). This low conductivity layer can prevent saturation of the soil just beneath the seal due to the suction that occurs at the interface (see Section 3.2.4), further reducing the infiltration rate.

Subsurface clay layers near the surface can also reduce infiltration rates. An unstructured clay layer will usually have a lower K_s than an overlying sand layer and reduce K_{eff} (Chapter 3, Section 3.2.4) and $i(t)$ once the wetting front enters the clay layer. Again, HYDRUS-1D can be used to plot distributions of pressure heads and water contents in a soil profile that consists of a Cecil soil profile (Bruce et al. 1983). This is HYDRUS Simulation 5.3. – Infiltration into Layered System. The boundary and initial conditions are the same as those in the first uniform soil infiltration example (Figures 5.2 to 5.9). The final time is extended to 4 h. The soil hydraulic properties and depths of horizons are taken from Table 2.5 (water retention parameters) and Table 3.2 (K_s). The van Genuchten soil hydraulic model is chosen with the "air-entry value of −2 cm" option. This is especially important in this simulation where many of the horizons are clays with a $K(h)$ function that is highly nonlinear near saturation.

The pressure heads at various times are shown in Figure 5.16. The curves are remarkably different from Figure 5.7 for a uniform soil due primarily to the presence of positive values that peak at a depth of ≈100 cm after 1 h. This depth corresponds to the top of the Bt2 horizon. This horizon and the horizons below it have sharply lower

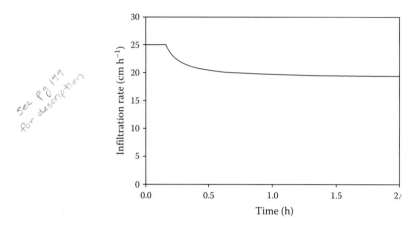

See pg 179 for description

FIGURE 5.14 Infiltration rate as a function of time in the infiltration example with a system-dependent boundary condition.

K_s than the upper horizons (Table 3.2). When the wetting front reaches this layer, water flow is slowed causing pressure to build above the layer. This produces a larger gradient in pressure head at the entrance to the layer and compensates to a degree for the low conductivity. At a depth of ≈160 cm, pressure heads have become more negative than the initial conditions. This is because of the increase in K_s that occurs in the C horizon, which starts at 160 cm (Tables 2.5 and 3.2).

As noted in Section 3.2.4, under steady flow, pressure heads are distributed linearly with depth within a horizon. In Figure 5.16, the distribution is linear above the

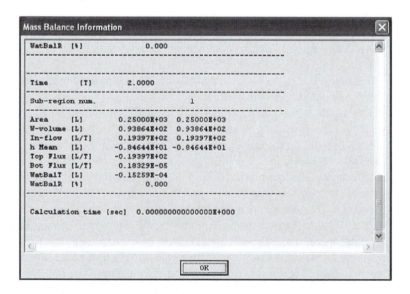

FIGURE 5.15 Mass Balance window in the infiltration example with a system-dependent boundary condition.

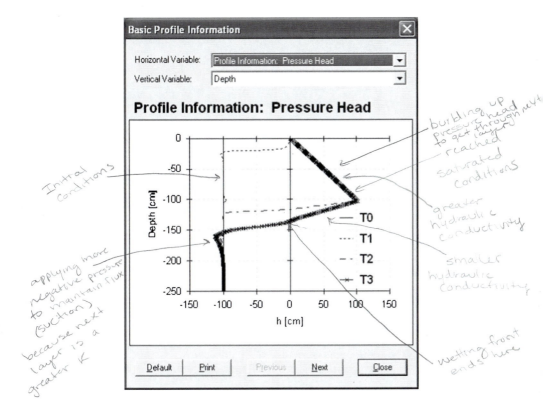

(handwritten annotations:) Initial conditions — applying more negative pressure (suction) to maintain flux (because next layer is a greater K) — bubbling up pressure head next to get (caught?) reached saturated conditions — greater hydraulic conductivity — smaller hydraulic conductivity — wetting front ends here

FIGURE 5.16 Pressure heads as a function of depth for various times during infiltration into a layered soil. T0 = 0.0, T1 = 0.1, T2 = 1.0, and T3 = 4.0 h.

wetting front, indicating that θ is constant (saturated) in this region. At the wetting front, the lines curve, indicating that θ is changing over time in this region (transient flow). The distribution of pressure heads is similar to what was predicted using the K_{eff} approach to predict interlayer pressure heads in Chapter 3 (see Problem 7 and Figure 3.26), except the curves are not linear in Figure 5.16 below a depth of about 150 cm and they are not as negative as predicted in the steady-flow analysis in Problem 7.

The effect of soil layers on the infiltration rate is shown in Figure 5.17. The infiltration rate for a uniform soil with the same properties as the first horizon in the layered soil (the curve in Figure 5.9) is included for comparison purposes. Initially, the infiltration rates are the same, but once the wetting front reaches the second (BA) horizon after about 0.2 h, the infiltration rate in the layered soil decreases. This is due to the lower K_s in the BA horizon (see Table 3.2). A more severe drop in infiltration rate occurs after about 0.4 h when the wetting front reaches the Bt2 horizon. After entering this horizon, the infiltration rate in the layered soil approaches a constant value. Thus, subsurface layers have an important effect on the infiltration rate at the surface through their effect on movement of the wetting front.

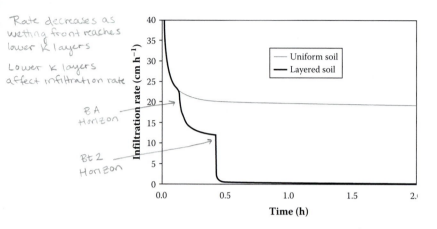

Rate decreases as
wetting front reaches
lower K layers

Lower K layers
affect infiltration rate

B A
Horizon

Bt 2
Horizon

FIGURE 5.17   Infiltration rate as a function of time in a uniform soil and a layered soil.

A dry coarse-textured layer under a fine-textured layer will also impede deeper movement of the wetting front and reduce the infiltration rate, but through a different mechanism. To illustrate this problem, a two-dimensional water flow model, HYDRUS (2D/3D), is used for a soil profile consisting of a clay loam layer to a depth of 20 cm, underlain by a sand layer that extends to a depth of 50 cm. This is HYDRUS Simulation 5.4 – 2D Infiltration into Layered System. The van Genuchten (1980a) hydraulic parameters are taken from the textural class averages of Carsel and Parrish (1988) (clay loam: $\theta_r = 0.095$, $\theta_s = 0.41$, $\alpha = 0.019$ cm^{-1}, $n = 1.31$, $K_s = 0.26$ cm h^{-1}, and $l = 0.5$; sand: $\theta_r = 0.045$, $\theta_s = 0.43$, $\alpha = 0.145$ cm^{-1}, $n = 2.68$, $K_s = 29.7$ cm h^{-1}, and $l = 0.5$). The model region is 100 cm in width and a line source of water (as might occur with furrow irrigation) is placed at a distance 50 cm from the left side (boundary pressure head = 0). The initial pressure heads are −500 cm everywhere.

The distribution of water contents 10 h after the point source of water is started is shown in Figure 5.18. The location of the sand layer is apparent in that it has a lower water content ($\theta = 0.05$ cm^3 cm^{-3}) under the initial pressure head than the clay loam layer ($\theta = 0.25$ cm^3 cm^{-3}). At this early time, the wetting front has not reached the sand layer and the wetting front is semicircular in shape.

The distribution of water contents after 24 h is shown in Figure 5.19. The wetting front has become distorted because water cannot enter the dry sand layer. The reason for this is that pressure heads at the leading edge of the wetting front are negative and water cannot enter the smallest air-filled pores in the sand layer until pressure heads increase to the point where capillarity will draw water into these pores (the water-entry pressure head for the sand). This stalls the wetting front until pressure heads rise to the critical level for entry. Since the pore-size distribution is narrower in the sand, it is not long after water first enters the sand that the pressure is high enough at the wetting front to fill the largest capillary pores in the sand. Once the sand is saturated, it no longer impedes flow because K_s is high in the sand compared to the clay layer above. Baver et al. (1972) referred to the action of a buried dry sand layer in temporarily impeding water flow and infiltration as a *check valve*. This principle is used in the design of golf greens, which have a sand

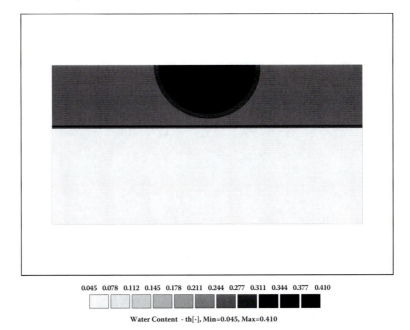

0.045 0.078 0.112 0.145 0.178 0.211 0.244 0.277 0.311 0.344 0.377 0.410

Water Content - th[-], Min=0.045, Max=0.410

FIGURE 5.18 Distribution of water contents after 10 h in a soil profile consisting of a clay loam layer from 0 to 20 cm and a sand layer from 20 to 50 cm. The model space is 100 cm in width and a source is placed at the center of the soil surface.

surface layer over a coarse gravel layer at about 50 cm. The gravel layer prevents drainage from the root zone under frequent light irrigations, but if there is a heavy rain, the gravel layer will fill and drain the root zone so that the green does not become waterlogged.

5.4.3 Infiltration Equations

Because of the importance of the infiltration process, simplified solutions to the Richards equation have been developed to predict infiltration. Most of the infiltration equations have been developed for conditions when rainfall does not limit infiltration. In this case, the infiltration rate is less than the rainfall or irrigation rate and runoff occurs. Soil hydraulic properties control the infiltration rate so it is *profile controlled* (Hillel 2004), also called *ponded* conditions, although the depth of ponding may be negligible if surface storage is small. Many infiltration equations have been developed, including those by Kostiakov (1932), Horton (1940), Philip (1957), and Parlange et al. (1985).

5.4.3.1 Green–Ampt Equations

Green and Ampt (1911) developed a simplified mechanistic equation for infiltration by assuming that the wetting front in a soil was a square wave or sharp front.

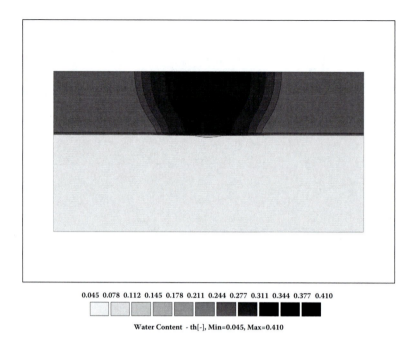

0.045 0.078 0.112 0.145 0.178 0.211 0.244 0.277 0.311 0.344 0.377 0.410

Water Content - th[-], Min=0.045, Max=0.410

FIGURE 5.19 Distribution of water contents after 24 h in a soil profile consisting of a clay loam layer from 0 to 20 cm and a sand layer from 20 to 50 cm.

Although this is approximately true only in coarse-textured soils, there is no error in predicting the infiltration rate as long the amount of water behind the predicted square front is equal to the amount of new water behind the true wetting front.

The Green-Ampt equation for cumulative infiltration in a uniform soil, including the effect of gravity, is (see Section 5.12 for derivation):

$$I(t) = K(h_0)t + \Delta h \Delta \theta \ln\left(1 + \frac{I(t)}{\Delta h \Delta \theta}\right),$$ (5.22)

where

[handwritten: hydraulic conductivity as function of initial head]

$$\Delta \theta = \theta_0 - \theta_i$$ [handwritten: initial condition or antecedent condition]

$$\Delta h = h_0 - h_f,$$ (5.23)

where θ_i is the initial or antecedent water content and θ_0 is the water content at the surface and extends to the wetting front. The pressure head, h_0, corresponds to the water content, θ_0. The pressure head, h_f, is the pressure head at the wetting front. When the rainfall rate is greater than or equal to the infiltration rate, ponding occurs. In this case,

θ_0 and $K(h_0)$ can be approximated by θ_s and K_s, respectively, and h_0 can be assumed to be zero (for negligible ponding) or the depth of ponded water. The unknowns in Equations 5.22 and 5.23 are then $I(t)$ and h_f. White and Sully (1987) showed that the wetting front pressure head can be approximated using the macroscopic capillary length, λ_c (Tables 3.1 and 3.3):

$$h_f = -\frac{\lambda_c}{2b}, \quad \text{about } 0.55$$

(5.24)

where b is a dimensionless factor that has a theoretical range of 1/2 to $\pi/4$, but can be assumed to be equal to 0.55 in most cases (Warrick and Broadbridge 1992). Hence, the only unknown in Equation 5.22 is $I(t)$, but the equation cannot be solved directly for this variable because it appears both inside and outside the natural log function. Therefore, it must be solved iteratively. Alternatively, Equation 5.22 can be solved for t and the times corresponding to a range of values for cumulative infiltration can be calculated. In contrast to White and Sully (1987), Haverkamp et al. (1985) found that it was best to consider h_f as a fitting parameter.

Once the cumulative infiltration curve as a function of time is known, the infiltration rate can be calculated as the instantaneous slope of this curve:

θ_i = initial water content
θ_0 = water content at surface and extending to the wetting front
h_0 = pressure head corresponds θ_0
h_f = wetting front pressure head
$I(t)$ = cumulative infiltration

$$i(t) = \frac{dI(t)}{dt}.$$

(5.25)

Example 5.1

Use the Green and Ampt (1911) equations to determine how long it will take for 0.3 cm of rain to infiltrate into a compacted, structureless clay soil with $K_s = 0.01$ cm h^{-1}. The field saturated water content is 0.39 cm^3 cm^{-3} and the water content before the event is 0.18 cm^3 cm^{-3}. Assume that there is negligible ponding at the soil surface throughout the rainfall event.

Calculate h_f using Equation 5.24 and a value of $\lambda_c = 100$ cm from Table 3.3:

$$h_f = -\frac{\lambda_c}{2b} = -\frac{100\,\text{cm}}{2(0.55)}$$

$$h_f = -90.9\,\text{cm}.$$

Calculate Δh and $\Delta \theta$:

$$\Delta h = h_0 - h_f = 0\,\text{cm} - (-90.9\,\text{cm}) = 90.9\,\text{cm}$$

$$\Delta \theta = 0.39\frac{\text{cm}^3}{\text{cm}^3} - 0.18\frac{\text{cm}^3}{\text{cm}^3} = 0.21\frac{\text{cm}^3}{\text{cm}^3}.$$

Solve Equation 5.22 for t and substitute values:

$$t = \frac{I - \Delta h \Delta\theta \ln\left[1 + \dfrac{I}{\Delta h \Delta\theta}\right]}{K_s}$$

$$= \frac{0.3\,\text{cm} - (90.9\,\text{cm})(0.21)\ln\left[1 + \dfrac{0.3\,\text{cm}}{(90.9\,\text{cm})(0.21)}\right]}{0.01\dfrac{\text{cm}}{\text{h}}}$$

$$= 0.23\,\text{h}.$$

Example 5.2

Use an Excel spreadsheet to make a graph of cumulative infiltration $I(t)$ vs. time t and infiltration rate $i(t)$ vs. time for $I(t)$ up to 0.3 cm for the soil and situation in Example 5.1.

Enter the parameter values and calculate h_f, $\Delta\theta$, and Δh (Figure 5.20). In the first column, enter values for $I(t)$ starting at zero and increasing in increments of 0.01 h. Solve the Green-Ampt equation for t as in the example above and enter that as a formula in the next column. In the third column, calculate the infiltration rate as the difference in cumulative infiltration divided by the difference in time over the last time step (see the formula bar in Figure 5.20).

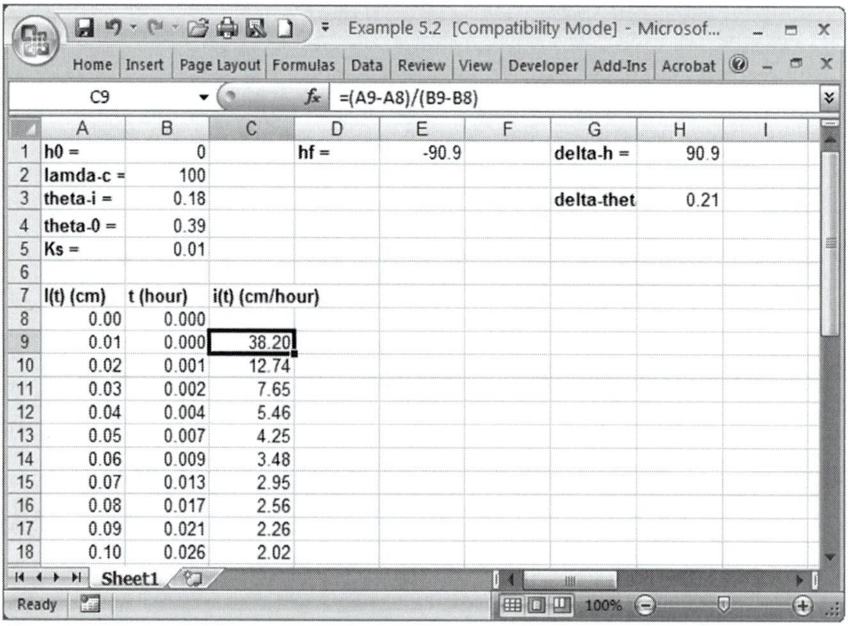

	A	B	C	D	E	F	G	H	I
1	h0 =	0		hf =	-90.9		delta-h =	90.9	
2	lamda-c =	100							
3	theta-i =	0.18					delta-thet	0.21	
4	theta-0 =	0.39							
5	Ks =	0.01							
6									
7	I(t) (cm)	t (hour)	i(t) (cm/hour)						
8	0.00	0.000							
9	0.01	0.000	38.20						
10	0.02	0.001	12.74						
11	0.03	0.002	7.65						
12	0.04	0.004	5.46						
13	0.05	0.007	4.25						
14	0.06	0.009	3.48						
15	0.07	0.013	2.95						
16	0.08	0.017	2.56						
17	0.09	0.021	2.26						
18	0.10	0.026	2.02						

FIGURE 5.20 Excel spreadsheet for the Green-Ampt infiltration equation.

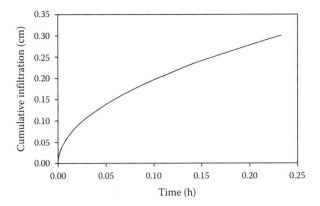

FIGURE 5.21   Cumulative infiltration as a function of time predicted using the Green-Ampt equation.

The cumulative infiltration curve and the infiltration rate curve are shown in Figures 5.21 and 5.22, respectively. Cumulative infiltration increases rapidly during the initial stage and then reaches a nearly linear rate of increase. The infiltration rate starts very high and declines sharply to a steady residual rate equal to K_s.

5.4.3.2   Curve Number Approach   *Based on observations not physics*

The *curve number* method is an empirically determined rainfall-runoff relationship that provides an indirect estimate of the total depth of water that infiltrates during a storm (McCuen 1982; NRCS 2004). It does not give the infiltration rate $i(t)$, but simply the cumulative runoff, R, at the end of a storm. The method is based on numerous measurements of runoff for many soil types and considers soil texture, soil drainage

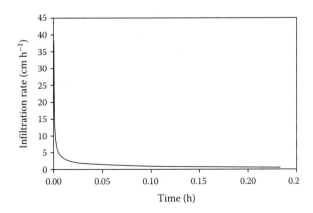

FIGURE 5.22   Infiltration rate as a function of time calculated as the derivative with respect to time of the Green-Ampt cumulative infiltration curve.

class, antecedent runoff conditions (ARC), and vegetative cover. The total runoff (R in centimeters) during an event is:

$$R = \frac{(P - 0.2S)^2}{P + 0.8S} \qquad P > I_a$$

Potential maximum soil moisture retention after runoff begins

$$R = 0 \qquad P \leq I_a,$$

(5.26)

where P is the total precipitation in centimeters and S is the water storage capacity of the soil in centimeters. The water storage capacity is calculated using the curve number (CN):

$$S = \frac{2540}{CN} - 25.4.$$

(5.27)

Curve numbers are tabulated in Chapter 9 of *The National Engineering Handbook* (NRCS 2004) and vary with land use, ranging from near zero for a dry, fully vegetated, highly permeable surface to near 100 for an impervious surface. The curve number method assumes that an initial amount of rain must occur before runoff starts and this is called the *initial abstraction, $I_a = 0.2\ S$*. Runoff is zero until precipitation exceeds I_a (Equation 5.26). *The amount of storage available in your soil*

The curve number also varies with soil hydrologic groups, which range from Group A (soils with low runoff potential when thoroughly wet) to Group D (soils with high runoff potential when thoroughly wet). Hydrologic groups are described in Chapter 7 of *The National Engineering Handbook* (NRCS 2004) and are also given for all U.S. soil series in the National Soil Information System (NASIS) maintained by the U.S. National Resource Conservation Service (NRCS 2009). This information is also available through the Web Soil Survey at http://websoilsurvey.nrcs.usda.gov/app. The curve number can be adjusted for antecedent runoff conditions I through III, representing dry, normal, and wet conditions, but some studies have indicated that this does not improve the prediction (see Chapters 9 and 10, NRCS 2004). With the curve number approach, there is an emphasis on management and cover. This is a recognition of the importance of crusting (and impervious surface percentage) on infiltration. The use of soil hydrologic groups is also a recognition of the importance of the soil profile (e.g., subsurface impeding horizons), not just the hydraulic properties of the surface layer.

A distinction can be made between *infiltration excess* runoff and *saturation excess* runoff (Beven 2001). Infiltration excess runoff occurs when downward movement of water into the soil is less than the rainfall rate. This is also called Hortonian flow, after Horton (1933). This type of process is described by the Green-Ampt (1911) equations. Saturation excess runoff occurs when water moves laterally within the soil during a storm and causes lower hillslope positions to be saturated, thereby decreasing the infiltration rate to less than the infiltration capacity. This is also called the *variable source area concept* (Hewlett and Troendle 1975) because the area where runoff occurs expands with the size of the storm. The curve number approach can be considered a saturation excess approach (Steenhuis et al. 1995). Another way to look at the difference between these processes is to consider the gradients and conductivities (components of Darcy's Law, see Equation 5.20). In infiltration excess runoff, there is a large gradient favoring infiltration, but the soil conductivity is not sufficiently large to prevent runoff. In saturation excess runoff, the conductivity may be very large, but the gradient is small or even favors exfiltration (a seep). A final distinction is that the infiltration excess approach is essentially a one-dimensional (vertical) view of the infiltration process. The saturation excess approach requires a two-dimensional or three-dimensional view.

5.4.4 BOREHOLE INFILTRATION

Infiltration from a borehole is a three-dimensional flow problem (Chapter 3, Section 3.3.4.3). It can be modeled in two dimensions, however, using cylindrical coordinates available in HYDRUS (2D/3D).

In this example (HYDRUS Simulation 5.5 – Borehole Infiltration), infiltration is simulated from an unlined borehole with a constant head of water into a loamy sand ($\theta_r = 0.0485$, $\theta_s = 0.3904$, $\alpha = 0.0347$ cm^{-1}, $n = 1.7466$, and $K_s = 105.12$ cm d^{-1}). Since infiltration from a borehole will be symmetrical around the vertical centerline of the borehole (in the absence of any slope or heterogeneities in the soil surrounding the borehole), a two-dimensional slice through the soil from the centerline of the borehole on the left (axis of symmetry) radiating out into the soil to the right is simulated. The horizontal dimension of the model space is 100 cm and the vertical dimension is also 100 cm, with the soil surface at the top (Figure 5.23). A borehole with a radius of 5 cm penetrates to a depth 30 cm below the soil surface. The boundary condition at the bottom of the model space is a free drainage boundary condition simulating a deep water table (see Section 5.2.3.1). To simulate water ponded in the borehole to a height 10 cm above the bottom of the borehole, a constant-head boundary condition is used. Along the boundary representing the bottom of the borehole and up the side wall to a height of 10 cm, the boundary condition is a constant pressure head of 10 cm in equilibrium with the lowest nodal point on this boundary. This ensures that the pressure at the bottom of the borehole is 10 cm and pressure decreases linearly up the side wall to a value of zero at 10 cm above the borehole bottom. All other points along the model-space boundary are a no-flux boundary condition. The distribution of pressure heads after 3 h of simulation is shown in Figure 5.23. Capillarity has drawn the wetting front into the soil in all directions including upward toward the soil surface. The influence of gravity is most apparent in the saturated region immediately surrounding the ponded portion of the borehole where a teardrop shape to the contour is apparent.

The infiltration rate from the borehole as a function of time is shown in Figure 5.24. The volumetric infiltration rate (Q in cubic centimeters per hour) has been converted to the infiltration flux (i in centimeters per hour) by dividing Q by the bottom area of the borehole (πr^2). The infiltration rate declines to a near steady value within one half hour, much like one-dimensional infiltration into a uniform soil with water ponded at the surface (see Figure 5.9). Unlike one-dimensional infiltration, however, the steady value in the borehole infiltration example is much greater than the soil K_s

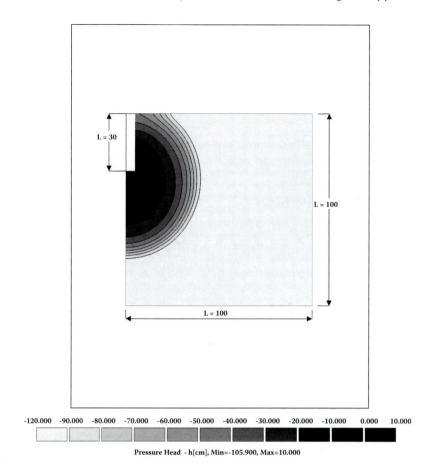

FIGURE 5.23 Distribution of pressure heads after 3 h in the borehole infiltration example.

(4.3 cm h^{-1}). This is because flow is occurring in three dimensions (Darcy's Law for one-dimensional flow does not apply) and because the gradient is much more than a unit-gradient. This is why the borehole equation (Chapter 3, Equation 3.34) must be used to calculate K_s.

5.4.5 SUBSURFACE IRRIGATION

Irrigation using a buried irrigation line results in flow of water that can be modeled as two-dimensional infiltration (HYDRUS Simulation 5.6 – Subsurface Drip Irrigation). Skaggs et al. (2004) compared measured water contents in a Hanford sandy loam soil with those simulated using HYDRUS-2D. Soil properties were estimated using the Rosetta database for van Genuchten (1980a) parameters ($\theta_r = 0.021$, $\theta_s = 0.34$, $\alpha = 0.023$ cm^{-1}, $n = 1.4$, $K_s = 1.6$ cm h^{-1}, and $l = -0.92$). The irrigation line was buried at a depth of 5 cm. The model space consisted of a two-dimensional soil block, 50 cm wide and 60 cm deep with the soil surface at the top (Figure 5.25). It

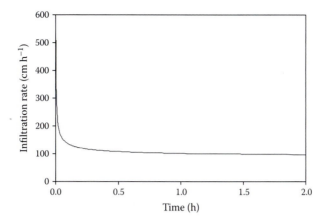

FIGURE 5.24 Infiltration rate as a function of time in the borehole infiltration example.

was assumed that water flow would be symmetrical about the vertical centerline of the irrigation pipe, so only half the area around the pipe was simulated. All the boundaries were considered to be zero flux except for a 2-cm diameter semicircle on the left boundary, which represented the irrigation line. The boundary condition on this semicircle was specified as a flux of 6.37 cm h^{-1} (the irrigation rate in liters per

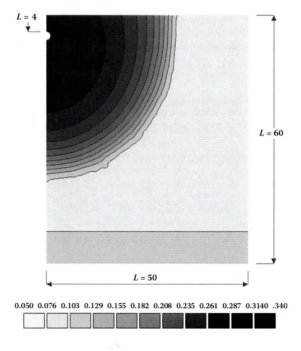

FIGURE 5.25 Distribution of water contents around a subsurface irrigation line simulated with HYDRUS (2D/3D) after 10 h. (Adapted from Skaggs, T.H., et al., *J. Irrig. Drain. Eng.*, 130, 304, 2004.)

hour for a given segment of irrigation line divided by the surface area of the irrigation line segment). Water contents are shown after 10 h of simulated irrigation in Figure 5.25. The wetting front has moved out in a nearly cylindrical manner due to capillarity, but some distortion of the wettest contour due to gravity is apparent. The HYDRUS simulations agreed quite well with the measured distributions of water contents (Skaggs et al. 2004).

5.5 REDISTRIBUTION

Once a rainfall event is over, water movement does not cease. For example, the final distribution of pressure heads in Figure 5.13 will continue to change after 2 h even if there's no further precipitation. This process is known as *redistribution* or drainage (Jury and Horton 2004). In theory, with a water table at a depth of 250 cm, pressure heads will return to the initial distribution of an equilibrium profile with pressure heads increasing linearly from −250 cm at the surface to 0 cm at the water table (Figure 5.13). In reality, it takes so long for redistribution to occur that an equilibrium profile rarely occurs unless the water table is at a very shallow depth. To see this, the previous infiltration simulation shown in Figure 5.13 is extended to 1 week (168 h) with no precipitation (or evapotranspiration) beyond 2 h using HYDRUS-1D. This is HYDRUS Simulation 5.7 – Redistribution. The earlier run is saved as a new project and the following changes are made. In the Time Information window, the final time is changed to 168 h, the Time Variable Boundary Conditions box is checked, and two time-variable records are specified (Figure 5.26).

Print times are selected at 2 (end of rainfall), 84, 126, and 168 h. In the Time Variable Boundary Conditions window (Figure 5.27), two precipitation records are entered to produce a 2 h rainfall period followed by a period with no rainfall that

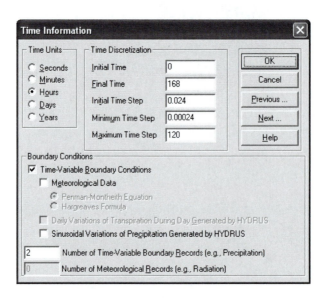

FIGURE 5.26 Time Information window in the redistribution example.

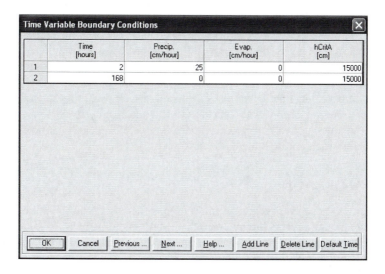

FIGURE 5.27   Time Variable Boundary Conditions window in the redistribution example.

lasts to the final time. The evaporation rate is set to zero so that only redistribution will occur after 2 h.

The predicted pressure head distribution after running HYDRUS-1D is shown in Figure 5.28. The pressure head at time T1 = 2 h shows the depth to which the wetting front moved (about 200 cm) during the rainfall event and is the last time in

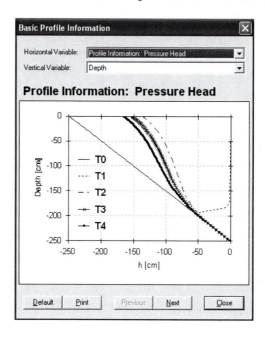

FIGURE 5.28   Simulated pressure head distributions in the redistribution example. T0 = 0, T1 = 2, T2 = 84, T3 = 126, and T4 = 168 h.

the earlier infiltration example (Figure 5.13). It's clear that pressure heads continued to change beyond 2 h with pressure heads decreasing at more shallow depths. The curves are moving in the direction of the equilibrium distribution (the initial conditions in Figure 5.13), but even after a week of redistribution they are not at a true equilibrium distribution. This is because as pressure heads decrease at shallow depths, unsaturated hydraulic conductivity decreases and water movement is slower. The average pressure head in the profile after about a week is close to −100 cm, which is the pressure head often used to estimate field capacity in coarse-textured soils (Romano and Santini 2002).

5.6 EVAPORATION   *Know this

Evaporation is the reverse process of infiltration: a flux *out* of the soil at the surface. In this example (HYDRUS Simulation 5.8 – Evaporation), HYDRUS-1D is used to illustrate the evaporation process with a uniform loamy sand ($\theta_r = 0.05$, $\theta_s = 0.39$, $\alpha = 0.0347$ cm^{-1}, $n = 1.75$, $K_s = 4.38$ cm h^{-1}, and $l = 0.5$). The surface boundary condition is set to a system-dependent boundary condition for evaporation and the simulation is run for 72 h with print times at 3, 6, 12, 24, and 48 h. The surface boundary condition is an evaporative flux of 1 cm d^{-1} (0.042 cm h^{-1}) until the surface dries to a pressure head of −15,000 cm ($E = 0.042$ cm h^{-1}, $h_A = -15,000$ cm, see Equations 5.10 and 5.11). At that time the boundary condition switches to a constant pressure head of −15,000 cm. The resulting water content profiles for different times after evaporation started are shown in Figure 5.29.

The numerical solution shows that water contents decrease with depth as time progresses because of the loss of water through evaporation at the surface. In this case, instead of a wetting front seen in the infiltration example, a drying front progresses deeper into the soil. The evaporation rate for the same soil is shown in Figure 5.30. Evaporation is high initially and constant. During this period, evaporation from the soil is able to keep up with the potential evaporation rate of 0.042 cm h^{-1}. The reason for the high initial evaporation rate is the steep gradient in water potential between the wet soil and the dry air above the soil and the high hydraulic conductivity of the wet soil surface layer (Figure 5.29). During this period, the decrease in the hydraulic conductivity of the surface layer is fully compensated by the increase in the hydraulic gradient. This period is referred to as *stage one* evaporation (Hillel 2004). After about 10 h, the evaporation rate starts to drop. At this point, the soil surface has dried completely (reached the minimum pressure head of −15,000 cm) and the actual evaporation rate becomes less than the potential rate. Although the hydraulic conductivity of the surface remains constant (the surface pressure head is constant), the evaporation rate decreases with time because the hydraulic gradient at the surface gradually decreases. This period of decreasing evaporation rate is referred to as *stage two* evaporation. Eventually the evaporation rate approaches a constant residual rate where most of the flux at the surface is in the vapor phase. This final period is referred to as *stage three* evaporation.

The heat and radiant energy balance at the soil surface determines the potential evaporation rate, ET_0 (Chapter 4, Equation 4.3). This varies from day to day and even minute to minute as wind speed, radiation, and relative humidity change. The actual

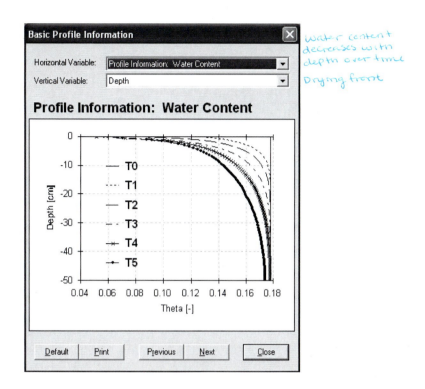

(handwritten annotations) water content decreases with depth over time / Drying front

FIGURE 5.29 Water contents as a function of depth for various times during evaporation from a uniform loamy sand. T0 = 0, T1 = 3, T2 = 6, T3 = 12, T4 = 24, and T5 = 48 h.

evapotranspiration rate, ET, is usually less than ET_0. Part of the evapotranspiration demand is satisfied by soil evaporation. As can be seen in Figure 5.30, evaporation can be high when the soil is wet (e.g., shortly after a rainfall event), but rapidly drops to a very low rate. Thus, evaporation alone usually will not meet the potential evapotranspiration demand, ET_0.

5.7 TRANSPIRATION

Where nondormant plants are present, most of the evaporative demand is satisfied by transpiration. Photosynthesis uses only about 1% of the water transpired by plants (Hillel 2004). For photosynthesis to occur, however, CO_2 must enter the plant leaves and reach chloroplasts. As a result, plants must maintain open stomata that allow CO_2 to enter the interior of leaves (Figure 5.1). Plant cells must be moist to function, so the open stomata result in a continuous loss of water from the interior of the leaf through the stomata due to evaporation. This water must be replenished by plant uptake of water through the root system or the plant leaf cells will desiccate and the plant will wilt.

Water moves through the plant to the atmosphere along a continuous path. This is called the soil-plant-atmosphere continuum (Philip 1966). Water moves from the soil across the plant root to the xylem vessels. It then moves up the xylem vessels to the

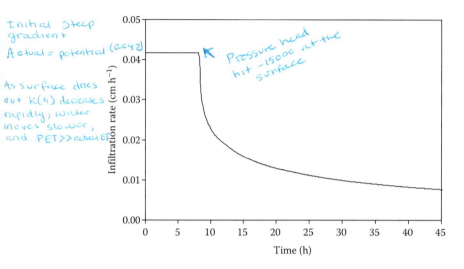

Handwritten annotations (left margin and on figure):

Initial steep gradient

Actual = potential (0.042)

As surface dries out K(h) decreases rapidly, water moves slower, and PET >> actual ET

Pressure head hit -15000 at the surface

FIGURE 5.30   Evaporation rate as a function of time in a uniform loamy sand.

plant leaves. In the leaves, water exits the xylem vessels and moves through leaf cells to the stomatal cavities where it evaporates. A steady gradient of decreasing water potential pulls water from the soil through the plant to the atmosphere. Water is drawn first from the surface horizons where roots are more numerous. As this water is depleted, the plant draws water from deeper in the soil, but this requires more of an energy gradient due to the effect of gravity. Also root densities (length of root per volume of soil) decrease with depth.

There is a limiting leaf water potential below which stomatal guard cells begin to lose turgor and close. As soon as this happens, the actual evapotranspiration rate drops below the potential rate: $ET < ET_0$. Stomata can be partially closed, which reduces transpiration but doesn't completely stop it. If stomata close completely then $ET = 0$.

Root water uptake can be considered a sink term in the Richards Equation 5.2. Various terms have been developed, including those by Molz (1981), Jarvis (1989), and Vrugt et al. (2001). Feddes et al. (1978) modeled root water uptake as a function of soil water pressure head, h:

$$S(h) = \alpha(h)S_p, \qquad (5.28)$$

where $S(h)$ is the root water uptake or volume of water removed from a unit volume of soil per unit time [T^{-1}], S_p is the potential water uptake rate [T^{-1}], and $\alpha(h)$ is a dimensionless stress response function of the pressure head ($0 \le \alpha \le 1$). The stress response function is shown in Figure 5.31a. Water uptake is assumed to be zero close to saturation due to a lack of oxygen in the root zone (pressure heads greater than h_1). For pressure heads less than h_4 (the wilting point pressure head), water uptake is also assumed to be zero. Water uptake is optimal between pressure heads of h_2 and h_3.

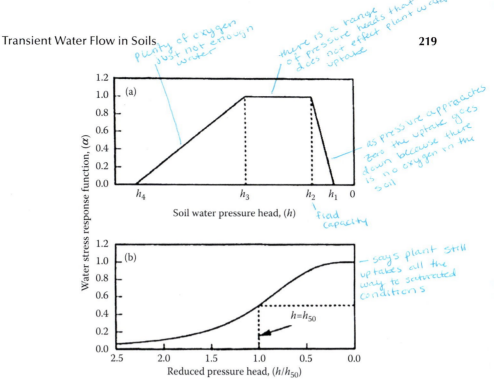

[handwritten annotations:] Plenty of oxygen just not enough water

there is a range of pressure heads that does not effect uptake

as pressure approaches zero the uptake goes down because there is no oxygen in the soil

field capacity

— says plant still uptakes all the way to saturated conditions

FIGURE 5.31 Plant water stress response function $\alpha(h)$ as used by (a) Feddes et al. (1978) and (b) van Genuchten (1987). (Feddes, R.A., et al., *Simulation of Field Water Use and Crop Yield*, John Wiley, New York, 1978; van Genuchten, *Research Report No. 121*, U.S. Salinity Laboratory, USDA, ARS, Riverside, CA, 1987.)

The pressure head, h_3, may be adjusted depending on the transpiration rate so that it is more negative when transpiration rates are low (optimal root water uptake occurs over a wider range in pressure heads at lower transpiration rates).

Van Genuchten (1987) suggested an S-shaped root water stress response function (Figure 5.31b). The parameter h_{50} represents the pressure head at which root water uptake is reduced by 50%. Note that in contrast to the response function of Feddes et al. (1978), this response function does not consider the reduction near saturation due to low oxygen levels. Van Genuchten (1987) also expanded the root water stress response function to include osmotic stress. *[handwritten:] ✱ know this*

HYDRUS-1D can be used to simulate evapotranspiration (HYDRUS Simulation 5.9 – Transpiration). In the Main Processes window, water flow and root water uptake are checked. The simulation is run for 365 days (time-variable boundary condition with 365 records specified in the Time Information window). A profile depth of 250 cm is used. The same loamy sand as in the previous example ($\theta_r = 0.0485$, $\theta_s = 0.3904$, $\alpha = 0.0347$ cm^{-1}, $n = 1.7466$, and $K_s = 105.12$ cm d^{-1}) is used to simulate plant transpiration. A system-dependent atmospheric boundary condition with surface runoff is used for the upper boundary condition to simulate infiltration and evaporation. The free drainage boundary condition is used to simulate a deep water table for the lower boundary condition. The Feddes et al. (1978) water uptake reduction model (Equation 5.28) is used with threshold pressure heads selected for grass from the database pull-down menu: $h_1 = -10$ cm, $h_2 = -25$ cm, $h_3 = -300$ cm for high transpiration rates and $-1,000$ cm for low transpiration rates, and $h_4 = -15,000$ cm (Figures 5.31 and 5.32).

In the Time Variable Boundary Conditions window (shown for previous problems in Figure 5.11 and Figure 5.27), 365 records for daily precipitation and potential evapotranspiration from Athens, Georgia, for 1995 are copied from a spreadsheet (see Problem 7 at the end of this chapter). In this year, total precipitation was 141 cm, slightly greater than the long-term average of 127 cm. A full canopy is assumed and consequently there is no evaporation. It is also assumed that the grass does not go dormant in the winter. A root distribution that decreases linearly from the soil surface to a depth of 100 cm is specified by selecting the root distribution icon from the tool bar in the Profile Information window using the Edit Condition button (Figure 5.33). Observation points are placed at depths of 5 and 50 cm.

The potential surface flux as a function of time is shown in Figure 5.34. Since there is no evaporation, the only potential surface flux is precipitation, so this is a plot of daily rainfall (as negative fluxes) for 1995 in Athens, Georgia. There is a dry period in early spring and large rainfall events in the early fall (tropical storms).

Potential and actual evapotranspiration (root water uptake in this case) are shown in Figure 5.35. It's clear that potential evapotranspiration is low during the winter months and reaches peak values of nearly 1 cm d^{-1} in the summer. Actual evapotranspiration is identical to potential evapotranspiration until about day 80 when the first dry spell starts. At this point, actual transpiration drops below the potential rate due to plant water stress and remains below the potential rate for much of the summer. In the fall when potential evapotranspiration rates decrease and the soil profile water contents start to increase, actual and potential evapotranspiration rates again coincide.

The pressure head at a depth of 5 cm is shown in Figure 5.36. When the first dry period occurs in the early spring, the pressure head at this depth drops quickly to the minimum value $-15,000$ cm (wilting point) due to root uptake. The pressure head at

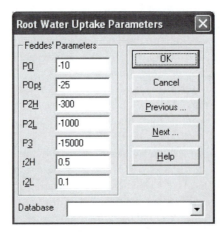

FIGURE 5.32 Root Water Uptake Parameters window for the transpiration example.

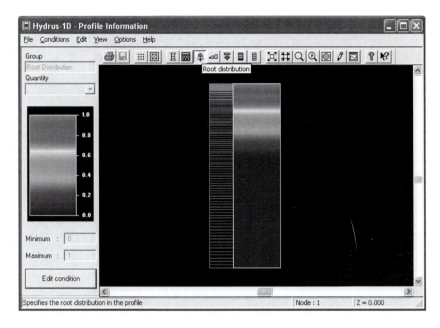

FIGURE 5.33   Profile Information window showing the root distribution for the transpiration example.

a depth of 50 cm (Figure 5.37) also drops to wilting point, but the drop occurs later. With the rainfall that starts shortly after 100 days, the pressure head near the surface recovers, but the pressure head at 50 cm remains at wilting point because this rainfall does not penetrate to deeper depths in the soil profile. At about 120 days, there's sufficient rainfall that pressure heads at both depths return to near zero. However, the pressure head at the deeper depth soon returns to wilting point because water from small rainfall events that keep the shallow depths relatively wet do not penetrate to deeper depths. A higher rainfall period from about 140 to 180 days results in pressure heads near zero at both depths. After about 180 days, the pressure head at the deeper depth returns to wilting point and stays there until the large rainfall event on about day 230. By contrast, the pressure head near the surface responds to the small rainfall events during this interval. Late in the season, a large rainfall event on about day 280 and lower potential evapotranspiration rates result in pressure heads at both depths returning to near zero.

Overall, these results show that even in a slightly wetter than normal year in a region of the United States with relatively high rainfall, this lawn experienced considerable water stress during the summer. This is due in large part to the low plant-available water in such a coarse-textured soil.

5.8   PREFERENTIAL FLOW

Increasing evidence exists that variably saturated flow in many field soils is not consistent with the uniform flow pattern typically predicted with the Richards equation

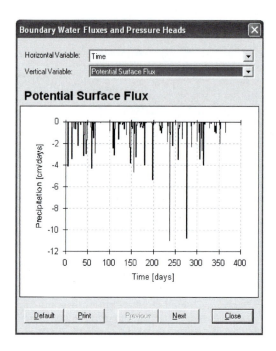

FIGURE 5.34 Precipitation (a negative surface flux) as a function of time in the transpiration example.

(Quisenberry and Phillips 1976; Luxmoore 1981; Flury et al. 1994; Hendrickx and Flury 2001). This is due to the presence of macropores, fractures, or other structural voids or biological channels through which water and solutes may move preferentially, while bypassing a large part of the matrix pore space. Preferential flow and transport processes are probably the most frustrating in terms of hampering accurate predictions of contaminant transport in soils and fractured rocks. Contrary to uniform flow, preferential flow results in irregular wetting of the soil profile as a direct consequence of water moving faster in certain parts of the soil profile than in others. Hendrickx and Flury (2001) defined preferential flow as constituting all phenomena where water and solutes move along certain pathways, while bypassing a fraction of the porous matrix. Water and solutes for these reasons can propagate to far greater depths, and much faster, than would be predicted with the Richards equation describing uniform flow.

The most important causes of preferential flow are the presence of macropores and other structural features, development of flow instabilities (i.e., fingering) caused by profile heterogeneities or water repellency (Hendrickx et al. 1993), and funneling of flow due to the presence of sloping soil layers that redirect downward water flow (Kung 1990a, 1990b). While the latter two processes (i.e., flow instability and funneling) are usually caused by textural differences and other factors at scales significantly larger than the pore scale, macropore flow and transport are usually generated at the pore or slightly larger scales, including scales where soil structure first manifests itself (i.e., the pedon scale) (Šimůnek et al. 2003).

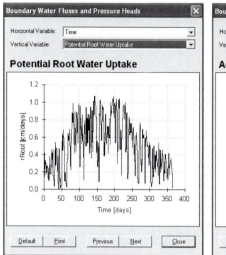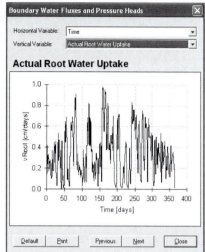

FIGURE 5.35 Potential root uptake (left) and actual root uptake (right) of water as a function of time in the transpiration example.

5.8.1 Macropores

Uniform flow in granular soils and preferential flow in structured media (both macroporous soils and fractured rocks) can be described using a variety of single-porosity, dual-porosity, dual-permeability, multiporosity, and/or multipermeability models. While single-porosity models assume that a single pore system exists that is fully accessible to both water and solute, dual-porosity and dual-permeability models both assume that the porous medium consists of two interacting pore regions, one associated with the inter-aggregate, macropore, or fracture system, and one comprising the micropores (or intra-aggregate pores) inside soil aggregates or the rock/soil matrix. While dual-porosity models assume that water in the matrix is stagnant, dual-permeability models allow for water flow within the soil or rock matrix.

Figure 5.38 illustrates a hierarchy of conceptual formulations that can be used to model variably saturated water flow and solute transport in soils. The simplest formulation (Figure 5.38a) is a *single-porosity* (equivalent porous medium) model applicable to uniform flow in soils. The other models apply in some form or another to preferential flow or transport. Of these, the *dual-porosity* model in Figure 5.38c assumes the presence of two pore regions, with water in one region being immobile and water in the other region being mobile. This model allows the exchange of both water and solute between the two regions (Šimůnek et al. 2003). Conceptually, this model views the soil as consisting of a soil matrix containing grains/aggregates with a certain internal microporosity (intra-aggregate porosity) and a macropore or fracture domain containing the larger pores (inter-aggregate porosity). While water and solutes are allowed to move through the larger pores and fractures, they can also flow in and out of aggregates. By comparison, the intra-aggregate pores represent immobile pockets that can exchange, retain, and store water and solutes, but do not

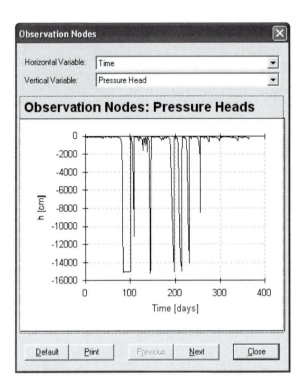

FIGURE 5.36 Pressure head as a function of time at a depth of 5 cm in the transpiration example.

contribute to advective (or convective) flow. Models that assume *mobile-immobile* flow regions (Figure 5.38b) are conceptually somewhere in between the single- and dual-porosity models. While these models assume that water will move similarly as in the uniform flow models, the liquid phase for purposes of modeling solute transport is divided in terms of mobile and immobile fractions, with solutes allowed to move by advection and dispersion only in the mobile fraction and by diffusion between the two pore regions. This model has long been applied to solute transport studies (e.g., van Genuchten and Wierenga 1976).

Finally, *dual-permeability* models (Figure 5.38d) are those in which water can move in both the inter- and intra-aggregate pore regions (or matrix and fracture domains). These models in various forms are now also becoming increasingly popular (Pruess and Wang 1987; Gerke and van Genuchten 1993). Available dual-permeability models differ mainly in how they implement water flow in and between the two pore regions. Approaches to calculating water flow in macropores or interaggregate pores include those invoking Poiseuille's equation (Ahuja and Hebson 1992), the Green and Ampt or Philip infiltration models (Chen and Wagenet 1992), the kinematic wave equation (Germann 1985), and the Richards equation (Gerke and van Genuchten 1993). *Multiporosity* and/or *multipermeability* models (not shown in Figure 5.38) are based on the same concept as dual-porosity and dual-permeability models, but include additional interacting pore regions. These models can be readily simplified to the

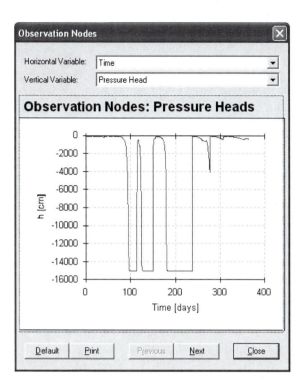

FIGURE 5.37 Pressure head as a function of time at a depth of 50 cm in the transpiration example.

dual-porosity/permeability approaches. Recent reviews of preferential flow processes and available mathematical models are provided by Hendrickx and Flury (2001), Šimůnek et al. (2003), and Jarvis (2007).

Another way to simulate the effect of macropores on water flow is through the use of *scaling*. This is a procedure designed to simplify the description of the spatial

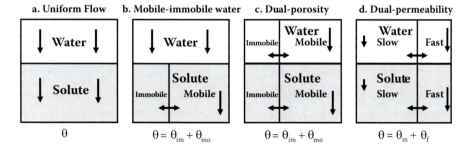

FIGURE 5.38 Conceptual models of water flow and solute transport. θ is the water content, θ_{mo} and θ_{im} in (b) and (c) are water contents in the mobile and immobile flow regions, respectively, and θ_m and θ_f in (d) are water contents in the matrix and macropore (fracture) regions, respectively. (Šimůnek and van Genuchten, 2006.)

variability in the soil hydraulic properties. Here we describe scaling as implemented in the HYDRUS models. It is assumed that variability in the hydraulic properties of a given soil profile can be approximated by means of a set of linear scaling transformations, which relate the soil hydraulic characteristics $\theta(h)$ and $K(h)$ at a given point in the soil to reference characteristics $\theta^*(h^*)$ and $K^*(h^*)$. The technique is based on the similar media concept introduced by Miller and Miller (1956) for porous media that differ only in the scale of their internal geometry. The concept was extended by Simmons et al. (1980) to materials that differ in morphological properties, but which exhibit *scale-similar* soil hydraulic functions. Three independent scaling factors are embodied in HYDRUS. These three scaling parameters may be used to define a linear model of the actual spatial variability in the soil hydraulic properties as follows (Vogel et al. 1991):

$$K(h) = \alpha_K K^*(h^*)$$

$$\theta(h) = \theta_r + \alpha_\theta \left[\theta^*(h^*) - \theta_r^* \right] \tag{5.29}$$

$$h = \alpha_h h^*,$$

in which, for the most general case, α_K, α_θ, and α_h are mutually independent, dimensionless scaling factors for the hydraulic conductivity, water content, and the pressure head, respectively. Less general scaling methods arise by invoking certain relationships between α_K, α_θ, and α_h. In HYDRUS, scaling factors for each node can be generated stochastically by specifying the mean and standard deviation of the factors. Spatial correlation of the scaling factors can also be included.

As an example of scaling (HYDRUS Simulation 5.10 – Scaling), HYDRUS (2D/3D) is used to simulate infiltration under conditions of incipient ponding of water at the surface into a two-dimensional block of soil 300 cm wide and 100 cm deep. The soil is a loam with hydraulic properties taken from the Rosetta-Lite database ($\theta_r = 0.061$, $\theta_s = 0.399$, $\alpha = 0.011$ cm^{-1}, $n = 1.474$, and $K_s = 0.50$ cm h^{-1}). A standard deviation of 1.0 is specified for log α_K with a correlation length ten times larger in the vertical direction than the horizontal direction. The spatial distribution of K_s for a given realization is shown in Figure 5.39. In this case, the stochastic assignment of scaling factors has produced two zones of high K_s and the larger correlation length in the vertical direction has stretched these zones along the z axis. The distribution of water contents in this soil profile 0.3 h after infiltration began is shown in Figure 5.40. Preferential flow in the two zones causes the wetting front to advance further in these areas.

5.8.2 FINGERING AND FUNNEL FLOW

Fingering and funnel flow occur in layered soil profiles where a fine-textured layer overlies a coarse-textured layer. Fingering can occur when there is no slope to the coarse-textured layer. In an experiment to examine fingering, Baker and Hillel (1990) used a plexiglass chamber 75 cm wide, 52 cm high, and 1.3 cm thick. The top layer was about 6 cm in depth and consisted of a fine sand with particle sizes between 45 and 106 µm. The sublayer consisted of a coarse-textured sand with a particle-size range that varied from experiment to experiment.

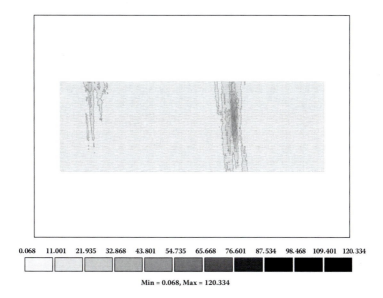

0.068   11.001   21.935   32.868   43.801   54.735   65.668   76.601   87.534   98.468   109.401   120.334

Min = 0.068, Max = 120.334

FIGURE 5.39   Spatial distribution of K_s (cm h^{-1}) generated by HYDRUS (2D/3D) using scaling to simulate preferential flow in a loam soil block that is 300 cm in width and 100 cm in depth.

When the wetting front reached the coarse layer, water could not enter because the wetting front pressure head was less than the water-entry value for the coarse sublayer (as shown in the HYDRUS simulation in Figure 5.19). The wetting front would pause at the interface and over time the wetting front pressure head would increase. When the wetting front pressure head increased to the value required to enter the smallest pore in the coarse sublayer, water started to infiltrate. Eventually, the wetting front pressure head increased to a value where the flux through the coarse sublayer was equal to or greater than the flux through the top layer. Baker and Hillel (1990) called this the effective water-entry value, h_e. If the flux was greater through the coarse-textured sublayer than through the fine-textured top layer, then the only way for flow to be equal was for water to flow through a fraction of the sublayer in discreet *fingers* (Figure 5.41). The effective water-entry value corresponded with the inflection point on the water release curve. This represented the dominant pore size. The fraction of wetted soil in the coarse sublayer at steady state was equal to the ratio of the flux through the top layer and $K(h_e)$ in the sublayer. Since fingering results in water (and solutes) penetrating only a portion of a soil profile, it is a type of preferential flow.

When the coarse sublayer has a slope, *funneling* occurs. In another chamber experiment, Walter et al. (2000) packed fine sand into a plexiglass chamber 1 cm in thickness with a discontinuous, embedded layer of coarse sand (Figure 5.42). They investigated the effect of slope and infiltration rate. Different color dyes were applied at intervals to the infiltration surface once steady flow was achieved, to show streamlines.

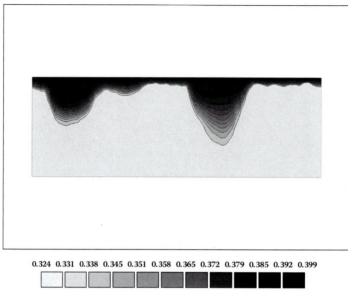

0.324 0.331 0.338 0.345 0.351 0.358 0.365 0.372 0.379 0.385 0.392 0.399

Water Content - th[-], Min = 0.324, Max = 0.399

FIGURE 5.40 The effect of preferential flow on water contents 0.3 h after infiltration began is simulated by HYDRUS (2D/3D) in a loam soil block that is 300 cm in width and 100 cm in depth.

During infiltration when the wetting front reached the coarse layer, it could not enter this layer because the wetting front pressure heads were less than the water-entry value. Since there was a slope to the layer, water funneled laterally to the end of the layer before penetrating deeper into the chamber (Figure 5.43). Above the coarse layer, the pressure head increased at the interface in the downslope direction because more and more water was being diverted. If the infiltration rate was high enough, the pressure head reached the water-entry value and infiltration occurred before the end of the embedded layer. At the tip of the embedded layer was a toe area where water never infiltrated. This was because the pressure was relieved by flow around the toe. Thus, funneling above sloping coarse layers results in water (and solutes) penetrating only a portion of a soil profile and is therefore another type of preferential flow (Kung 1990a, 1990b).

5.8.3 CAPILLARY BARRIERS

The funnel flow principle is used in designing capillary barriers or caps for diverting infiltrating water around hazardous waste storage facilities. Mallants et al. (1999) used HYDRUS-2D to simulate the effect of a capillary barrier with a 5% slope under different rainfall rates. The proposed cap consisted of a 100 cm-thick layer of loam for plant growth and water storage. Beneath this was the capillary barrier consisting of a 30 cm-thick layer of sand over a 100 cm-thick layer of gravel. Beneath the gravel was a 30 cm-thick layer of loam and a 200 cm-thick layer of clay to further reduce

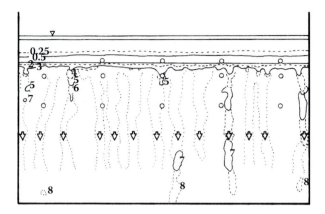

FIGURE 5.41 Tracings (alternating solid and dashed lines) of wetting front position with time during infiltration in the experiment by Baker and Hillel (1990). The numbers in the figure show the time (min) of the wetting fronts. The sublayer in this case consisted of a coarse sand with particle sizes between 500 and 710 μm. (From Baker, R.S., and D. Hillel. 1990. Laboratory tests of a theory of fingering during infiltration into layered soils. *Soil Sci. Soc. Am. J.* 54:20–30. With permission.)

infiltration. As the rainfall rate increased from 0.25 to 1.41 mm d^{-1}, the amount of percolating water diverted by the barrier at steady state increased from 45% to 87%. The effect of a capillary barrier consisting of a single layer of coarse material ($\theta_r = 0.012$, $\theta_s = 0.348$, $\alpha = 0.151$ cm^{-1}, $n = 7.35$, and $K_s = 1811$ cm d^{-1}) embedded in fine material ($\theta_r = 0.020$, $\theta_s = 0.348$, $\alpha = 0.045$ cm^{-1}, $n = 12.18$, and $K_s = 258$ cm d^{-1}) is shown in a simulation developed by Heiberger (1996). This is HYDRUS Simulation 5.11 – Capillary Barrier. The model space consists of a soil block 402 cm long and 250 cm deep with a 5% slope (Figure 5.44). The coarse material begins

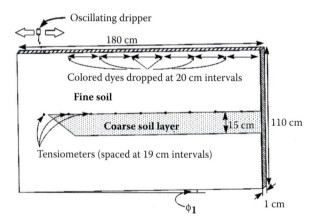

FIGURE 5.42 Schematic of the experimental setup of a plexiglass chamber 180 cm wide, 110 cm tall, and 1 cm thick, filled with fine soil and a discontinuous 15 cm-thick coarse layer used by Walter et al. (2000). The slope angle is φ_1. (From Walter, M.T., et al., *Water Resour. Res.*, 36, 841, 2000. With permission.)

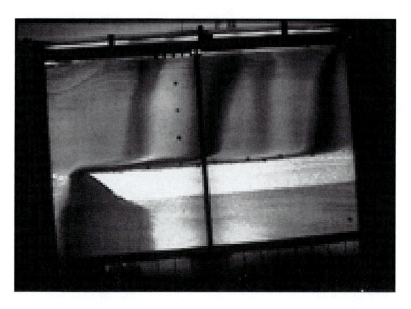

FIGURE 5.43   Photograph of an experimental run with an angle of 7.1° and an infiltration rate of 120 mm d^{-1}. (From Walter, M.T., et al., *Water Resour. Res.*, 36, 841, 2000. With permission.)

at a depth of 65 cm below the surface and is 80 cm thick, extending about 3/4 of the distance across the model space. The steady infiltration rate at the soil surface is 4 mm h^{-1}. The finite element mesh is apparent with a dense grid at the interface between the material layers. Flow vectors show the path and direction of the fastest flow after 15 h. Water is diverted downslope along the capillary barrier and around the end of the barrier.

5.9   GROUNDWATER RECHARGE AND DISCHARGE

The processes of groundwater recharge, groundwater discharge, the position of the water table, and streamflow are all interrelated in a humid environment (Fetter 1988). This can be seen by simulating the wetting of an initially dry hillslope using HYDRUS (2D/3D) (HYDRUS Simulation 5.12 Hillslope A). In this case, the two-dimensional model space is much larger than any of the earlier simulations. The hillslope is 100 m long in the horizontal direction and extends from a ridgetop on the left to a stream channel on the right (Figure 5.45). At the ridgetop, the soil surface is 50 m above the bottom of the model space. At the stream channel, the soil surface is 20 m above the bottom of the model space. The stream channel is 5 m deep (indicated by the nickpoint in Figure 5.45). The soil surface is given a convex shape. The boundary condition at the soil surface is a constant flux of 2 cm d^{-1}. The boundary condition on the left is a no-flux condition appropriate for a ridgetop where no lateral flow (only recharge) is expected. The bottom boundary condition is also a no-flux condition indicating impermeable rock. The boundary condition beneath the stream channel where no lateral flow (only discharge) is expected is

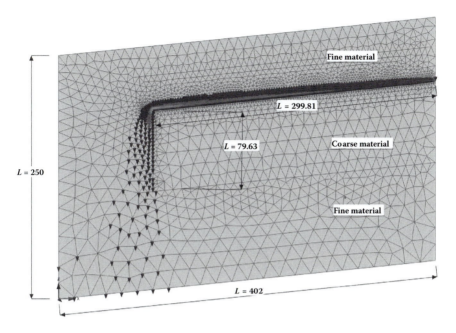

FIGURE 5.44 Water velocity vectors showing the path and direction of the fastest water flow around a capillary barrier after 15 h of simulation (From Heiberger, T.S., PhD Thesis, Oregon State University, 1996.)

a no-flux boundary condition. The boundary condition at the stream channel is a seepage condition, a type of system-dependent condition discussed previously in Section 5.2.3.2. The initial pressure heads are −100 m everywhere in the hillslope. The hydraulic properties of the hillslope are that of a loam soil textural class in the Carsel and Parish (1988) database ($\theta_r = 0.078$, $\theta_s = 0.43$, $\alpha = 3.6$ cm^{-1}, $n = 1.56$, and $K_s = 24.96$ cm d^{-1}). The simulation is run for 1,000 days.

To do this simulation accurately, so that the short-term response of the hillslope to a rainfall event would produce timely streamflow, would require many more nodes than we have used here (we used 3,921). Fiori and Russo (2007) found that a three-dimensional numerical model that included preferential flow was required to simulate the short-term response. Our example used a low constant flux at the soil surface instead of daily rainfall. Neither evaporation nor root water uptake driven by daily potential evapotranspiration were included. However, our simulation will be adequate to show the long-term response to wetting.

The distributions of hillslope pressure heads over time are shown in Figures 5.45 through 5.47. After 100 days, a wetting front of unsaturated pressure heads has moved uniformly downward in the hillslope to a depth of about 5 m (Figure 5.45). After 570 days, the wetting front has reached the bottom of the model space on the right (where the distance from the soil surface is the shortest) and a zone of saturated soil has expanded from the right corner to cover nearly the entire lower region of the hillslope (Figure 5.46). The top of the 0-cm pressure head contour represents the water table. The last portion of the hillslope to wet up is the lower

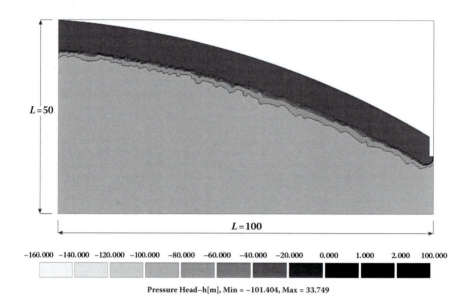

FIGURE 5.45   Pressure heads after 100 days in the hillslope example.

left corner because this region is farthest from the soil surface and the extending wetting zone coming in from the right. On the right boundary, the water table has reached the bottom of the channel, so water will start exiting the hillslope through the seep into the stream channel. Further rise in the water table next to the stream channel is limited due to the loss of water to the stream. After 800 days, the water

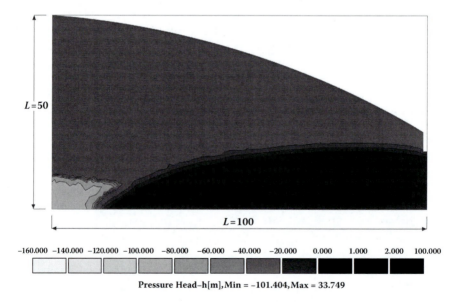

FIGURE 5.46   Pressure heads after 570 days in the hillslope example.

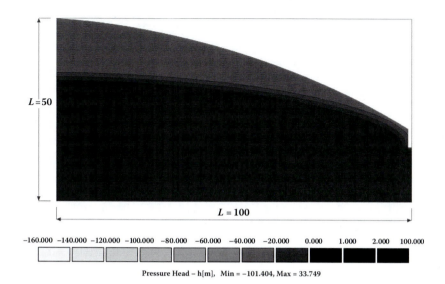

L = 50

L = 100

−160.000 −140.000 −120.000 −100.000 −80.000 −60.000 −40.000 −20.000 0.000 1.000 2.000 100.000

Pressure Head – h[m], Min = −101.404, Max = 33.749

FIGURE 5.47   Pressure heads after 800 days in the hillslope example.

table beneath the ridgetop has risen to a height of about 35 m above the bottom of the model space (Figure 5.47). On the right, the water table comes very close to the soil surface at about 5 m upslope from the stream channel. The water table at the stream channel has risen to the point that it is about 3 m above the bottom of the channel, indicating that the height of water in the stream should be about 3 m. There was no further change in the water table from the positions seen for day 800 through the remainder of the simulation (up to 1,000 days), indicating that the system has reached a steady state of dynamic equilibrium. The shape of the water table is a subdued replica of the hillslope topography, just as shown in the diagram in Chapter 2, Figure 2.15.

In Figure 5.48, the flow through the seepage boundary is shown as a function of time. There is no seepage to the channel until about day 500 when the water table intersects the stream channel. The seepage flux increases until about 800 days when it attains a steady value of 2 $m^2 d^{-1}$, which is equivalent to the steady flux through the soil surface (input equals output). At that point, the slope of the water table is such that a sufficient gradient has been generated, given the hydraulic conductivity of the loam, and the hillslope can transmit the incoming water at the rate at which it enters the hillslope. Hence, the water table can never be completely flat or there would be no gradient and no flow through the hillslope (unless there is no input). The steepness of the water table will depend on the rainfall rate, the soil hydraulic conductivity, and the length of the hillslope.

The effect of hydraulic conductivity on the slope of the water table can be seen by rerunning the simulation with a saturated hydraulic conductivity that is 10 times the value for the loam in the example so far (249.6 instead of 24.96 cm d^{-1}) (HYDRUS Simulation 5.13 Hillslope B). The distribution of pressure heads after 1000 days is shown in Figure 5.49. The water table is nearly flat because only a very small

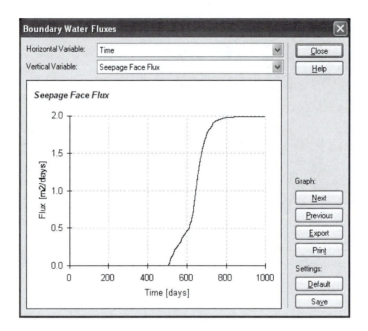

FIGURE 5.48 Seepage flow in the hillslope example.

gradient is required to produce the equilibrium flow through the hillslope, given the high K_s in this example. In this case, the shape of the water table is not as reflective of the hillslope topography.

Overall, these results are similar to what was shown in Chapter 3, Section 3.2.6, in regard to flow to drains. In that simplified analysis of steady flow, it was found that the water table had an elliptical shape and the maximum height of the water table was a function of the infiltration rate at the soil surface, the saturated hydraulic conductivity of the soil, and the distance between the drains (which would correspond to the hillslope length in these examples).

These simulations imply that during wet periods (such as winter months in eastern United States) the water table will rise to an average height characteristic of the climate and hillslope hydraulic properties. After a long drought, it may take several winter seasons to reestablish the water table at its equilibrium height, even though water contents at the soil surface are near normal. Stream baseflow would not return to normal levels until the equilibrium winter water table was fully established.

The shape of the hillslope can also have important effects. If the surface is changed from a convex to a concave hillslope surface near the stream, the water table may rise to the surface at the bottom of the footslope in the concave area. This area can be expected to produce saturation-excess runoff as described in Section 5.4.3. In contrast to the previous examples, the water table will be deeper higher up the hillslope at the *shoulder* position. Daniels and Gamble (1967) noted that the shoulder position in hillslopes with broad ridges and narrow valleys in the Coastal Plain region of southeastern United States were drier than other areas in the landscape

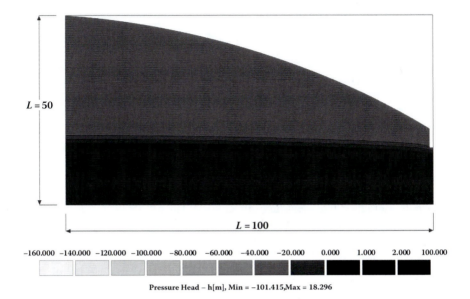

$L = 50$

$L = 100$

−160.000 −140.000 −120.000 −100.000 −80.000 −60.000 −40.000 −20.000 0.000 1.000 2.000 100.000

Pressure Head − h[m], Min = −101.415,Max = 18.296

FIGURE 5.49 Pressure heads after 1000 days in the hillslope example using a saturated hydraulic conductivity 10 times the value for a loam soil.

and produced red soils indicative of oxidized iron (aerobic conditions). Gray soils indicative of reduced iron formed at the base of the hillslope and near the center of the broad ridge. They attributed these soil formation processes, not to differences in parent material, but rather to the position of the water table.

5.10   INVERSE SOLUTIONS AND PARAMETER OPTIMIZATION

Modeling transient water flow requires an accurate estimation of a number of hydraulic parameters, including the soil water retention curve, saturated hydraulic conductivity, and the unsaturated hydraulic conductivity function. As an alternative to direct measurement of these parameters (Sections 3.4.1 and 3.4.2), they can be estimated indirectly using *inverse modeling*. This is commonly done in solute transport where analytical solutions to the solute transport equation (Chapter 6, Section 6.4.2) are fit to observed data to find the optimum value of transport parameters. Recently, however, numerical solutions to transient water flow problems (as well as solute transport problems) have been used to determine hydraulic parameters. This differs from the direct methods for estimating hydraulic parameters where steady flow conditions are required by the analytical solutions that are used. Inverse methods with numerical solutions (that do not require steady flow conditions) have become possible because advances in computer science have made obtaining numerical solutions much quicker so that programs can be run repeatedly to find the optimum parameter values.

This is much like the RETC program (described in Section 2.10), which fits a particular water retention curve equation to the observed water contents in order to find optimal values for parameters in the water retention curve equation. Inverse modeling is a form of model calibration, which is frequently used in hydrology (Gupta et al. 2003). It requires a set of observed data, such as measured water contents or pressure heads. In model calibration, the objective is usually to obtain better model predictions. Parameter optimization differs in that the objective is to determine the best estimate of the parameters as an alternative method to directly measuring them.

The HYDRUS models have an inverse modeling capability. An objective function, which includes deviations between measured and predicted variables such as pressure heads or water contents at different depths and times, is minimized. Predicted and measured fluxes or cumulative fluxes across boundaries can also be used as variables. The objective function can also include a penalty function for deviations between prior knowledge of the soil hydraulic parameters and their final estimates. Minimization of the objective function is accomplished using the Levenberg-Marquardt nonlinear minimization (Marquardt 1963).

Transient flow parameter optimization methods are described in detail by Hopmans et al. (2002). With these methods, both the soil retention and unsaturated hydraulic conductivity functions can be estimated from a single transient experiment. Whereas steady-state methods invert Darcy's equation, transient methods invert the Richards equation. Inverse methods can be used for laboratory experiments or for field experiments. Common transient flow experiments that can be used for parameter optimization include one- and multistep outflow methods, evaporation experiments, use of a tension infiltrometer, and field drainage experiments (Hopmans et al. 2002).

A common laboratory experiment is the multistep outflow method. A pressure cell similar to that used for the determination of a retention curve is employed (see Figure 2.29). Often the pressure cell is fitted with a small tensiometer so the pressure head in the soil can be monitored over time. There's also an apparatus for collecting water outflow from the cell and measuring this over time. The air pressure on the cell is increased once (the one-step method) or a number of times to successively higher pressures (the multistep method), much like what is done when measuring the water retention curve. However, instead of waiting until the cell reaches equilibrium at a given pressure to measure the water content, pressure head in the cell and water flow out of the cell are measured continuously during the experiment.

Example 5.3

Use the inverse mode in HYDRUS-1D to find optimal values of θ_r, θ_s, α, n, and K_s for a multistep method experiment (Hopmans, personal communication). This is HYDRUS Simulation 5.14 – Parameter Estimation (and MULTSTEP in the HYDRUS-1D inverse directory). The data for the experiment are shown in a spreadsheet in Figure 5.50 and available on the HYDRUS website (HYDRUS Simulation 5.14 Data).

In the experiment, four pressure settings were used and the pressure head at the bottom of the core was recorded over time. The pressure head was approximately −100 cm until 12 h, −200 cm from 12 to 48 h, −400 cm from 48 to 102 h, and −700 cm to the end of the experiment (190 h) (Figure 5.50). Tensiometer measurements at the center of the soil core were recorded along with the cumulative outflow from the core. In addition, the water content of the core was measured at the beginning of the experiment (at a pressure head of −22.1896 cm) and found

Time (hr)	hbot (cm)	h (cm)	Cbf (cm)		Time (hr)	hbot (cm)	h (cm)	Cbf (cm)		Time (hr)	hbot (cm)	h (cm)	Cbf (cm)
0.01	-100.822				32.1338	-198.592	-199.427	-0.44144		108.8	-706.489	-401.225	-0.67329
0.0671	-100.205	-45.8032	-0.0638		35.2171	-198.594	-198.507	-0.44111		109.8	-706.457	-409.263	-0.6743
0.0838	-100.078	-50.0953	-0.0769		48.1171	-198.398				111.217	-705.391	-418.02	-0.6779
0.1005	-99.9819	-54.1931	-0.0868		48.2671	-402.48	-213.538	-0.46058		112.634	-706.329	-425.327	-0.68109
0.1338	-99.8256	-60.6825	-0.103		48.5505	-402.379	-219.766	-0.47078		114.3	-706.262	-432.487	-0.68423
0.1671	-99.7143	-65.6414	-0.1145		48.9671	-402.308	-228.816	-0.47796		116.384	-706.181	-440.978	-0.68793
0.2005	-99.6235	-69.481	-0.1239		49.3838	-402.239	-236.349	-0.48486		118.3	-706.11	-448.218	-0.69086
0.2505	-99.497	-71.6596	-0.1369		49.8838	-402.16	-243.682	-0.49273		121.05	-706.344	-454.154	-0.69271
0.3505	-99.3603	-77.3143	-0.1511		50.4671	-402.086	-250.935	-0.50005		123.05	-706.274	-462.059	-0.69611
0.4505	-99.2538	-83.6799	-0.1621		51.2171	-402.011	-258.641	-0.50742		125.134	-706.2	-468.141	-0.69967
0.5671	-99.1572	-88.806	-0.1721		52.0505	-401.931	-266.088	-0.51515		127.55	-706.12	-474.916	-0.70328
0.7338	-99.061	-92.2685	-0.182		52.9671	-401.853	-273.37	-0.52279		129.717	-706.058	-480.226	-0.70552
1.0171	-98.9595	-94.673	-0.1925		53.9671	-401.787	-280.561	-0.529		131.884	-705.994	-485.416	-0.70796
1.4005	-98.8646	-97.107	-0.2023		55.0505	-401.713	-287.745	-0.53609		133.967	-705.927	-490.82	-0.71087
1.9338	-98.7733	-98.4074	-0.2117		56.3005	-401.633	-294.917	-0.54369		136.051	-705.874	-495.904	-0.71231
2.8671	-98.6825	-99.3448	-0.2211		57.6338	-401.56	-302.171	-0.55041		138.134	-705.812	-500.988	-0.71468
4.2005	-98.5912	-100.086	-0.2306		59.3838	-401.487	-310.639	-0.55704		140.217	-705.756	-505.832	-0.7164
6.2505	-98.5258	-100.464	-0.2373		61.0505	-401.425	-318.204	-0.56252		142.301	-705.701	-510.477	-0.71802
8.7171	-98.4946	-100.893	-0.2405		62.8838	-401.355	-325.158	-0.56874		144.384	-705.65	-514.949	-0.71933
10.7838	-98.4643	-100.862	-0.2437		64.8838	-401.281	-332.261	-0.57532		147.217	-706.086	-518.822	-0.72122
12.4171	-98.2314				67.0505	-401.221	-339.099	-0.58029		149.717	-706.032	-523.879	-0.7225
12.4505	-200.273	-121.349	-0.2677		69.1338	-401.173	-344.755	-0.58412		152.134	-705.969	-528.657	-0.72481
12.5005	-200.161	-126.84	-0.2792		71.2171	-401.257	-346.427	-0.58759		154.301	-705.921	-532.596	-0.72599
12.6338	-200.046	-134.748	-0.2912		73.6338	-401.214	-350.68	-0.59118		156.384	-705.874	-536.243	-0.7273
12.8005	-199.911	-142.358	-0.3051		75.7171	-401.164	-353.535	-0.59567		158.801	-705.816	-540.622	-0.7291
12.9671	-199.808	-147.519	-0.3157		77.8838	-401.125	-356.799	-0.59893		161.717	-705.752	-545.466	-0.73068
13.2171	-199.685	-153.15	-0.3284		80.8005	-401.094	-360.62	-0.60116		164.384	-705.69	-549.765	-0.73246
13.4671	-199.587	-157.442	-0.3386		82.9671	-401.062	-363.559	-0.60377		166.884	-705.631	-553.917	-0.73421
13.8005	-199.481	-162.136	-0.3495		86.1338	-401.028	-366.869	-0.60619		168.967	-705.582	-557.218	-0.73562
14.2171	-199.377	-166.699	-0.3603		88.4671	-401.001	-369.127	-0.60819		171.301	-705.541	-560.385	-0.73587
14.7171	-199.271	-171.198	-0.3713		90.6338	-400.972	-371.059	-0.6104		173.551	-705.935	-562.381	-0.73634
15.3005	-199.174	-175.376	-0.3812		92.8838	-400.945	-373.224	-0.61244		175.634	-705.902	-566.174	-0.73679
16.1338	-199.076	-179.916	-0.3914		95.7171	-401.011	-373.952	-0.61325		177.717	-705.866	-568.929	-0.73762
17.1338	-198.984	-184.146	-0.4008		97.8005	-400.986	-375.996	-0.61588		180.217	-705.814	-572.097	-0.73947
18.5505	-198.889	-188.286	-0.4107		99.9671	-400.965	-377.93	-0.61901		182.301	-705.77	-574.51	-0.74117
20.6338	-198.803	-192.495	-0.4196		102.384	-400.935	-379.265	-0.62112		184.634	-705.722	-577.285	-0.74284
23.2171	-198.713	-195.506	-0.4289		104.467	-400.91	-380.695	-0.62365		187.384	-705.66	-580.862	-0.7454
25.3005	-198.681	-197.243	-0.4322		105.917	-400.89				190.384	-705.601	-584.501	-0.74738
27.5505	-198.652	-198.274	-0.4352		107.217	-706.577	-382.406	-0.66777					
29.9671	-198.615	-198.998	-0.439		107.717	-706.546	-389.579	-0.66976					

FIGURE 5.50 Spreadsheet with the data for the MULSTEP example: time in hours, pressure head at the bottom of the core (hbot) in centimeters, pressure head in the core (h) in centimeters, and cumulative bottom flow (Cbf) in centimeters.

to be 0.429895 cm³ cm⁻³. The total number of data points was 229: 114 pressure heads, 114 cumulative bottom flows, and one measurement of $h(\theta)$.

In the Main Processes window (not shown), water flow and the inverse solution option are selected. In the Inverse Solution window (Figure 5.51), soil hydraulic properties are selected for optimization, weighting is by standard deviation, the maximum number of iterations is set to 50, and the total number of data points (229) is entered.

Centimeters are chosen for length units and the depth of the profile is set to 6 cm (the length of the core). In the Time Information window (not shown), a final time of 190.384 h is entered, the Time-Variable Boundary Conditions box is checked, and a value of 118 is placed in the box for Number of Time-Variable Boundary Records. The van Genuchten-Mualem hydraulic model is chosen in the Soil Hydraulic Model window (not shown). The settings for the Water Flow Parameters window are shown in Figure 5.52. Check marks are placed below the parameters to be fitted (θ_r, θ_s, α, n, and K_s) and initial estimates (from the Soil Catalog for a loam), minimum values, and maximum values are specified. When

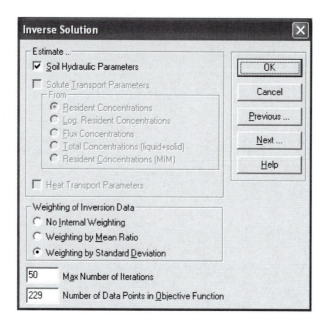

FIGURE 5.51   Inverse Solution window in the parameter optimization example.

no minimum and maximum values are entered (or they are both zero), there are no constraints in the optimization.

A constant flux condition is set for the surface boundary and a variable pressure head condition is set for the bottom boundary in the Water Flow Boundary Conditions window (not shown). The flux at the upper boundary is set to zero. The Time Variable Boundary Conditions window is shown in Figure 5.53. From the spreadsheet, the times and pressure heads at the bottom of the core are entered in the 118 lines within this window (only the first 14 entries are shown in this figure). The column labeled GWL is for groundwater level measurements or any other time-variable pressure heads at a bottom boundary (see the Help file).

The measured data to be used in the objective function are entered in the Data for Inverse Solution window. The first 15 entries are shown in Figure 5.54.

Water Flow Parameters - Inverse Solution - Material 1

	Qr	Qs	Alpha	n	Ks	I
Initial Estimate	0.078	0.43	0.036	1.56	1.04	0.5
Minimum Value	0	0.31	0	1.1	0	0
Maximum Value	0.3	0	0	0	0	0
Fitted ?	☑	☑	☑	☑	☑	☐

Soil Catalog for Initial Estimate: Loam ▼   Neural Network Prediction

OK   Cancel   Previous ...   Next ...   Help

FIGURE 5.52   Water Flow Parameters window in the parameter optimization example.

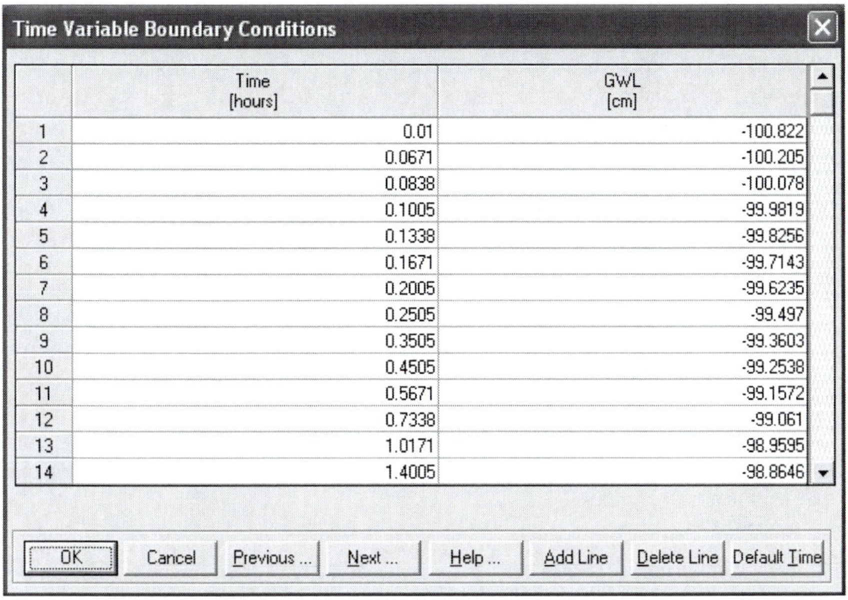

	Time [hours]	GWL [cm]
1	0.01	-100.822
2	0.0671	-100.205
3	0.0838	-100.078
4	0.1005	-99.9819
5	0.1338	-99.8256
6	0.1671	-99.7143
7	0.2005	-99.6235
8	0.2505	-99.497
9	0.3505	-99.3603
10	0.4505	-99.2538
11	0.5671	-99.1572
12	0.7338	-99.061
13	1.0171	-98.9595
14	1.4005	-98.8646

Time Variable Boundary Conditions

OK Cancel Previous ... Next ... Help ... Add Line Delete Line Default Time

FIGURE 5.53 Time Variable Boundary Conditions window in the parameter optimization example.

The Help file can be used to determine the meaning of the column labels: X, Y, Type, Position, and Weight. The values to be entered in the X and Y columns depend on the value for Type. Type is set to a value of 1 for all pressure head measurements. In this case, X is time, Y is the observed data (pressure heads in this case), and Position is the observation node number corresponding to where

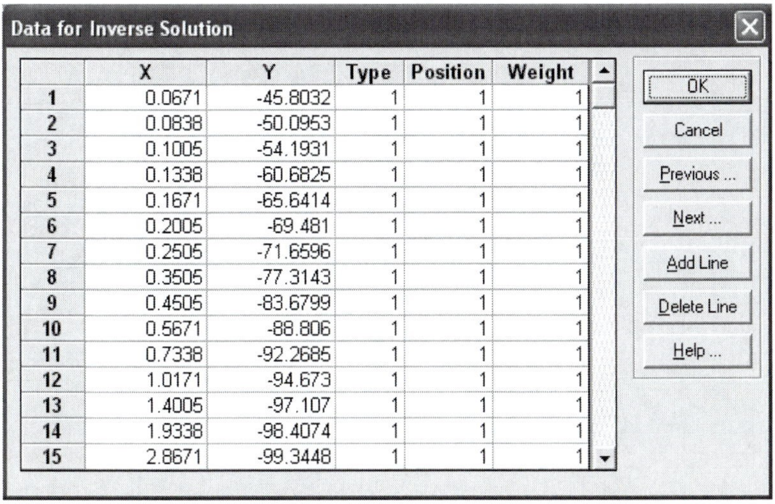

Data for Inverse Solution

	X	Y	Type	Position	Weight
1	0.0671	-45.8032	1	1	1
2	0.0838	-50.0953	1	1	1
3	0.1005	-54.1931	1	1	1
4	0.1338	-60.6825	1	1	1
5	0.1671	-65.6414	1	1	1
6	0.2005	-69.481	1	1	1
7	0.2505	-71.6596	1	1	1
8	0.3505	-77.3143	1	1	1
9	0.4505	-83.6799	1	1	1
10	0.5671	-88.806	1	1	1
11	0.7338	-92.2685	1	1	1
12	1.0171	-94.673	1	1	1
13	1.4005	-97.107	1	1	1
14	1.9338	-98.4074	1	1	1
15	2.8671	-99.3448	1	1	1

OK Cancel Previous ... Next ... Add Line Delete Line Help ...

FIGURE 5.54 Top of the file in the Data for Inverse Solution window in the parameter optimization example.

the pressure head is measured. Hence, the first entry in Figure 5.54 is the first recorded pressure head within the core (see cells A3 and C3 in the spreadsheet, Figure 5.50). The pressure heads continue to line 114 in the file. All these pressure head observations are assigned an equal weight (default value of 1), since there is no reason to believe that any observations are more or less reliable than the others.

Starting with line 115 in the Data for Inverse Solution window, the 114 measurements of cumulative flux are entered. For these entries, Type is set to 0 (see Figure 5.55 for the bottom of the file in the Data for Inverse Solution window). When Type = 0, X is time, Y is the cumulative flow across a specified boundary, and Position is a value that indicates the boundary. In this case, a value of 2 indicates it is the bottom boundary (see the linked table in the Help file). These entries continue to line 228 in the file and they all have the same weight of 1. The last entry in the file is the $h(\theta)$ data point. Type is set to 5, which specifies that X is the pressure head, Y is the volumetric water content, and Position is the material number (in this case there is only one material). The $h(\theta)$ data point is given a weight of 10 so that it will have a significant effect on the optimization process (and not be overshadowed by the 228 values of pressure heads and fluxes).

In the Profile Information window, 50 nodes are specified with a density value of 3 at the top and 1 at the bottom (Figure 5.56). An observation point is placed at the center of the core (a depth of 3 cm). The initial conditions are set to −25 cm at the top and −19 cm at the bottom of the core. In the Soil Profile Summary window (not shown), the initial pressure head at the bottom of the core is changed to −100.82, corresponding to the first measured pressure head at the bottom boundary (see Figure 5.53).

The observed and predicted pressure heads and cumulative outflows after running the example are shown in Figure 5.57. The final model simulation was a good fit to the observed pressure heads and cumulative bottom flow. Figure 5.58 shows the predicted water retention curve. The retention curve indeed passes

Data for Inverse Solution

	X	Y	Type	Position	Weight
218	166.884	-0.73421	0	2	1
219	168.967	-0.73562	0	2	1
220	171.301	-0.73587	0	2	1
221	173.551	-0.73634	0	2	1
222	175.634	-0.73679	0	2	1
223	177.717	-0.73762	0	2	1
224	180.217	-0.73947	0	2	1
225	182.301	-0.74117	0	2	1
226	184.634	-0.74284	0	2	1
227	187.384	-0.7454	0	2	1
228	190.384	-0.74738	0	2	1
229	-22.1896	0.429895	5	1	10

OK
Cancel
Previous ...
Next ...
Add Line
Delete Line
Help ...

FIGURE 5.55 Bottom of the file in the Data for Inverse Solution window in the parameter optimization example.

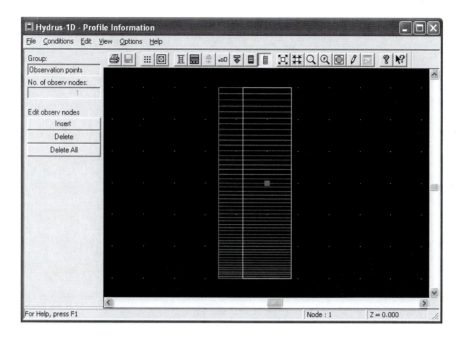

FIGURE 5.56 Profile Information window showing the grid and location of the observation point in the parameter optimization example.

through the single observed data point (circle) for $h(\theta)$ because of its large weight in the objective function.

The initial estimates of θ_r (WCR), θ_s (WCS), α, n, and K_s (CONDS) are shown in Figure 5.59 in the row for iteration 0. The initial sum of squares (SSQ) for the objective function is also shown. After eight iterations, HYDRUS-1D reduced the objective function sufficiently for the final parameters to be considered optimized. The final values for the parameters are $\theta_r = 0.20$ cm^3 cm^{-3}, $\theta_s = 0.44$ cm^3 cm^{-3}, $\alpha = 0.0101$ cm^{-1}, $n = 1.434$, and $K_s = 0.52$ cm h^{-1}.

5.11 SUMMARY

Soil water movement is a key process that affects both water quantity and quality in our environment. Water movement in soils occurs primarily under unsaturated, transient conditions. The governing equation for transient flow is the Richards equation and the most important hydraulic parameters are the soil saturated hydraulic conductivity and unsaturated hydraulic conductivity function.

Analytical solutions of the Richards equation for soil water flow under unsaturated, transient conditions are very few. Numerical methods such as those implemented in the HYDRUS models are commonly used to solve the equation. Numerical methods can also be used to determine soil hydraulic properties under transient flow conditions in an inverse mode. Simplifications of the unsaturated flow equations have been developed to describe infiltration because of the importance of this transient flow process. However, most of the infiltration equations assume one-dimensional

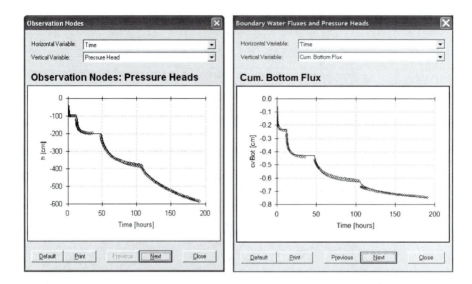

FIGURE 5.57 Observed (circles) and predicted (curves) pressure heads (left) and cumulative outflows (right) in the parameter optimization example.

vertical flow and do not account for saturation excess runoff. Numerical solutions can be used to understand complex soil water processes such as infiltration, evaporation, redistribution, transpiration, groundwater recharge, and groundwater discharge. Preferential flow is one of the most challenging aspects of predicting soil water movement, but various methods have been developed to account for the effects of macropores, fingering, funnel flow, and spatial variability.

5.12 DERIVATIONS

In this section, the Green-Ampt Equation 5.22 is derived following Jury and Horton (2004). The infiltration rate i [LT^{-1}] of a square-wave wetting front can be described using the Buckingham-Darcy equation:

$$i = -K(h_0)\frac{h_0 + L - h_f}{L}$$
$$= -K(h_0)\frac{\Delta h + L}{L},$$

(5.30)

where h_0 [L] is the pressure head at the soil surface and behind the wetting front, $K(h_0)$ [LT^{-1}] is the corresponding unsaturated hydraulic conductivity, h_f is the pressure head [L] at the wetting front, L is the distance from the soil surface to the wetting front, and $\Delta h = h_0 - h_f$. Equation 5.30 is written assuming infiltration as a downward flux will be negative (z positive in the upward direction). The infiltration rate is also

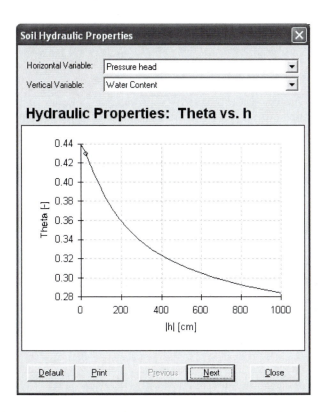

FIGURE 5.58 Predicted water retention curve and the single observed data point (circle) for $h(\theta)$ in the parameter optimization example.

equal to the rate of change of the volume of new water behind the advancing wetting front (and must also be negative):

$$i = -\frac{d}{dt}\left[\left(\theta_0 - \theta_i\right)L\right]$$

$$= -\Delta\theta \frac{dL}{dt},$$

(5.31)

where θ_i is the initial or antecedent water content, θ_0 is the water content at the surface and extends to the wetting front, $\Delta\theta = \theta_0 - \theta_i$, and dL/dt is the rate of advance of the wetting front.

These two expressions for the infiltration rate are set equal to each other:

$$\Delta\theta \frac{dL}{dt} = K\left(h_0\right)\frac{\Delta h + L}{L}.$$

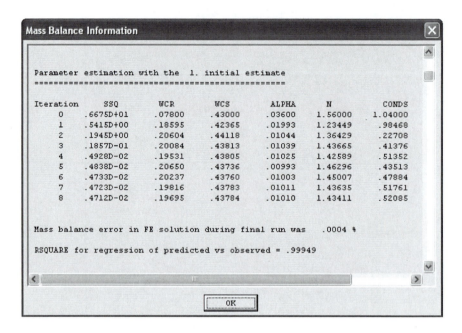

FIGURE 5.59 Iteration results from the parameter optimization example.

Separating variables and integrating the wetting front distance from 0 to L and time from 0 to t, one obtains:

$$\int_0^L \frac{L}{\Delta h + L}\, dL = \int_0^t \frac{K(h_0)}{\Delta\theta}\, dt. \tag{5.32}$$

The integral on the left side of Equation 5.32 can be found in standard tables (Lide 2000). It is:

$$\int_0^L \frac{L}{\Delta h + L}\, dL = \left[L - \Delta h \ln\left(1 + \frac{L}{\Delta h}\right)\right]\Bigg|_0^L$$

$$= L - \Delta h \ln\left(1 + \frac{L}{\Delta h}\right).$$

The integral on the right side of Equation 5.32 is:

$$\int_0^t \frac{K(h_0)}{\Delta\theta}\, dt = \frac{K(h_0)}{\Delta\theta}\, t.$$

Substituting for the integrals in Equation 5.32, the equation is

$$\frac{K(h_0)}{\Delta\theta}t = L - \Delta h \ln\left(1 + \frac{L}{\Delta h}\right)$$

$$K(h_0)t = \Delta\theta L - \Delta\theta\Delta h \ln\left(1 + \frac{\Delta\theta L}{\Delta\theta\Delta h}\right).$$

Recognizing that the cumulative infiltration $I(t) = \Delta\theta L$, Equation 5.22 is produced:

$$K(h_0)t = I(t) - \Delta\theta\Delta h \ln\left(1 + \frac{I(t)}{\Delta\theta\Delta h}\right).$$

5.13 PROBLEMS

1. Use the curve number method to calculate the runoff in centimeters for a rainfall event of 8.1 cm on July 10 for a golf course with >75% grass cover. The soil is a sandy loam. The rainfall during the 5 days previous to July 10 is 1.52 cm. The National Engineering Handbook Part 630 (Hydrology) is available at http://www.wsi.nrcs.usda.gov/products/w2q/H&H/tech_info/engHbk.html. See Chapter 7 for soil hydrologic groups; Chapter 9, Table 9-5 for ARC II CNs; and Chapter 10, Table 10-1 for adjusting CNs for ARC I or III.

2. Some computer models require only the texture of a soil to make predictions of water movement such as infiltration. Use the information in Table 3.1 for a silty clay loam soil to calculate how long it will take for 5 cm of rain to infiltrate under incipient ponding conditions at the surface, based on the Green-Ampt infiltration equation. Assume that the initial water content in the soil is $0.13 \text{ cm}^3 \text{ cm}^{-3}$. What if the initial water content is $0.30 \text{ cm}^3 \text{ cm}^3$ instead?

3. Use Excel to show the effect of antecedent moisture content on infiltration rate. Set up a spreadsheet as shown in Example 5.2 for the same soil with an initial water content of $0.10 \text{ cm}^3 \text{ cm}^{-3}$ (dry soil). In the spreadsheet add two new columns for the soil with an initial water content of $0.30 \text{ cm}^3 \text{ cm}^{-3}$ (wet soil): one column to calculate the time and the second column to calculate $i(t)$. On the same graph, plot the infiltration rate $i(t)$ vs. t for the wet and dry soils. In order to see the curve better, make the y-axis go from 0 to 30 cm h^{-1} and the x-axis go from 0 to 0.20 h.

4. In Figure 5.24, the infiltration rate from the borehole example was shown as simulated by HYDRUS (2D/3D) for a loamy sand. Use the steady infiltration rate after 2 h ($i_s = 97 \text{ cm } h^{-1}$) to see how close you can estimate the value of K_s in this soil using the borehole Equation 3.34 following Example 3.5 in Chapter 3. The value of K_s used in the HYDRUS (2D/3D) simulation

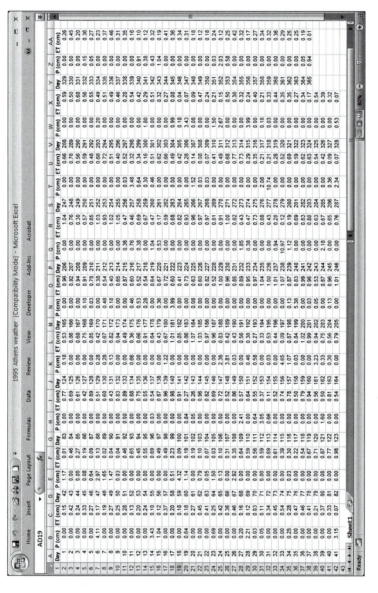

FIGURE 5.60 Daily precipitation and pan evaporation in centimeters for Athens, Georgia, in 1995.

was 4.36 cm h⁻¹. Use the value for λ_c from Table 3.1 for a loamy sand. How does your calculated value for K_s compare to the "true" value?

5. Use HYDRUS-1D to run the first infiltration example in Section 5.4.1 for a uniform loamy sand. Show a figure similar to Figure 5.7. Make a graph of the infiltration rate using Excel similar to Figure 5.9.

6. Use HYDRUS-1D to run the evaporation example in Section 5.6. Use a 50 cm-deep profile, a final time of 72 h, a time-variable boundary condition with one record of an evaporation rate of 0.042 cm h⁻¹, five print times (3, 6, 12, 24, and 48 h), an atmospheric upper boundary condition, a free drainage lower boundary condition, an evenly spaced grid with 101 nodes, and an initial pressure head distribution of −100 cm. Show a figure similar to Figure 5.29.

7. Use HYDRUS-1D to run the transpiration example in Section 5.7. Show figures similar to Figures 5.34 through 5.37. Daily precipitation and pan evaporation in centimeters for Athens, GA, in 1995 are given in Figure 5.60 and are available on the HYDRUS website as HYDRUS Simulation 5.9 - Data.

8. Use HYDRUS-1D to perform the parameter optimization process in Example 5.3. Show your results in figures similar to Figures 5.57 through 5.59.

6 Solute Transport

6.1 INTRODUCTION

Solute transport is perhaps the most complex subject in soil physics. It involves water and chemical movement, chemical reactions, and microbial transformations. It is an area of intense research due to concern over water quality. The field has developed in many disciplines besides soil science: civil engineering, environmental engineering, geophysics, and geochemistry.

6.2 CONSERVATION AND FLUX EQUATIONS

The objective is to predict the concentration of a given chemical compound (such as nitrate, various pesticides, or heavy metals) in solution where concentration is usually in units of mass of solute per volume of soil water. The first step is to develop the conservation equation. Then flux equations are developed for each of the transport processes.

6.2.1 CONSERVATION EQUATION

The solute _conservation equation_ is derived in the same manner as the heat conservation equation in Chapter 4 (Section 4.4.1) and the water conservation equation in Chapter 5 (Section 5.2.1). Only solute flow in the vertical direction (z positive up) is considered for simplification purposes. Solute flux (J_s) is the rate at which solute flows through a given cross-sectional area [$ML^{-2}T^{-1}$], much like the flux terms for water (J_w) and heat (J_H). The solute conservation equation states that the change in total solute concentration (C_T) in units of mass per volume of soil [ML^{-3}] in a representative elementary volume over time is equal to the negative of the rate of change of the solute flux with depth, minus any source/sink (S) in units of mass per volume per time [$ML^{-3}T^{-1}$]:

$$\frac{\partial C_T}{\partial t} = -\frac{\partial J_s}{\partial z} - S. \tag{6.1}$$

This is a partial differential equation (PDE) with independent variables z and t and the dependent variable C_T. The units of each term are [$ML^{-3}T^{-1}$]. The total solute concentration consists of the dissolved or liquid-phase solute concentration (c), expressed as a mass of solute per volume of liquid [ML^{-3}], and the adsorbed or solid phase solute concentration (s), expressed as a mass of solute per mass of soil [MM^{-1}]. To convert each of these concentrations to units of mass per volume of soil, c is multiplied by θ and s is multiplied by ρ_b:

249

(handwritten: dissolved $\frac{mg}{L}$)

(handwritten: C_T can be $\frac{mass}{volume}$ by multiplying ρ_b's then adding to θc)

$$C_T = \theta c + \rho_b s.$$ (6.2)

(handwritten: total concentration of solute) *(handwritten: sorbed $\frac{mg}{kg}$)*

(handwritten: solute out) *(handwritten: Δs elemental volume)* *(handwritten: solute in)*

6.2.2 TRANSPORT PROCESSES

Before proceeding with Equation 6.1, J_s must be defined. Solute flux in the liquid phase consists of three components that represent different transport processes: *advective* (also called *convective*) flux (J_{lc}), *diffusive* flux (J_{ld}), and *hydrodynamic dispersive* flux (J_{lh}):

(handwritten: concentration gradient driven)

(handwritten: total solute flux →) $$J_s = J_{lc} + J_{ld} + J_{lh}.$$ (6.3)

(handwritten: liquid phase convection) *(handwritten: diffusion)* *(handwritten: hydrodynamic dispersion)*

6.2.2.1 Advection

The first term in Equation 6.3 represents mass flow of solute at the average rate caused by the flux of water (J_w from Darcy's law). Looking at the units for the advective flux term, the correct units for solute flux are

(handwritten: concentration)

$$J_w \left[\frac{L^3 \, water}{L^2 \, T} \right] \cdot c \left[\frac{M \, solute}{L^3 \, water} \right] = J_{lc} \left[\frac{M \, solute}{L^2 \, T} \right].$$

(handwritten: water flux)

If transport mechanisms were limited to advection, all the solute molecules would travel at the same velocity as the *average* water velocity. If solutes were moving into a soil from a source of constant concentration at the surface (c_0), this would produce a square-wave distribution of c with depth (Figure 6.1). This is called *piston flow*. Note that concentrations are divided by the input concentration, c_0, so behind the solute front the relative or *reduced* concentration (c/c_0) is one.

The velocity of this front is not J_w. The Darcy flux is the macroscopic velocity or the volume of water crossing a unit cross-sectional area in a given time. The cross-sectional area is for bulk soil and includes area that is occupied by solid particles and air (in addition to water). Since water only flows through the wetted cross-sectional area, the *average pore water velocity v* at the microscopic scale is a function of the wetted cross-sectional area and greater than the Darcy velocity:

$$v = \frac{J_w}{\theta}.$$ *(handwritten: $= v_s$ seepage velocity)* (6.4)

For $v = 5$ cm h^{-1}, the piston flow front is centered at a depth of 50 cm after 10 h (Figure 6.1).

6.2.2.2 Diffusion

Another transport mechanism is due to *diffusion* or the movement of solute molecules in response to differences in concentration. Chemical diffusion in free water (not soil) is described by Fick's first law:

[handwritten: diffusion coefficient → normally 1×10^{-5} $\frac{cm^2}{s}$]

$$J_{ld} = -D_1^w \frac{\partial c}{\partial z}, \quad \text{[handwritten: concentration gradient]} \tag{6.5}$$

where D_1^w is the diffusion coefficient in free water [L^2T^{-1}]. This is given in chemical handbook tables and it is on the order of 1×10^{-5} cm^2 s^{-1} or about 1 cm^2 d^{-1} for almost all chemical compounds. Note that Fick's first law takes the same form as Darcy's law (Equation 3.5) and Fourier's law (Equation 4.6). In soil, the reduced cross-sectional area and more tortuous path lengths due to the presence of the solid and air phases (compared to diffusion in a bulk solution) must be accounted for. This will slow diffusion in soil, so the diffusion coefficient must be reduced. To account for tortuosity, D_1^w is multiplied by a *tortuosity factor* ξ_1 [–] that varies with θ (Millington and Quirk 1961):

$$\xi_1(\theta) = \frac{\theta^{7/3}}{\theta_s^2}. \quad \text{[handwritten: a measure of how diffusion is effected by the path length]} \tag{6.6}$$

In Figure 6.2, the effect of water content on $\xi_1(\theta)$ is shown for a value of $\theta_s = 0.50$. As water content increases, the tortuosity factor increases (diffusion paths are more direct and shorter). To account for the reduced cross-sectional area, D_1^w is multiplied by θ so that diffusive flux in soil is described as follows:

$$J_{ld} = -\theta \xi_1(\theta) D_1^w \frac{\partial c}{\partial z} = -\theta D_1^s \frac{\partial c}{\partial z}, \tag{6.7}$$

[handwritten: If we cut off solute and clean water was moving through it would look like:]

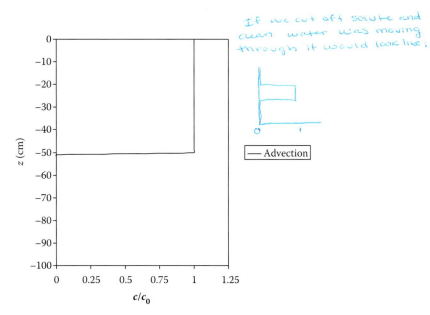

FIGURE 6.1 Relative concentration as a function of depth for a hypothetical solute front moving purely by advection at $t = 10$ h with a mean pore water velocity of 5 cm h^{-1}.

[handwritten: saturated:]

$$\xi = \frac{0.5^{7/3}}{0.5^2} = 0.79$$

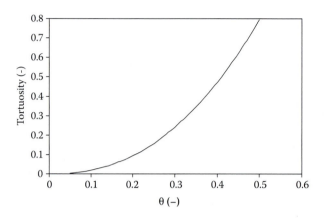

FIGURE 6.2 Tortuosity as a function of water content for a soil with $\theta_s = 0.50$.

where D_l^s is the effective diffusion coefficient in soil $[L^2T^{-1}]$ and equal to $\xi_l(\theta) \cdot D_l^w$. Note that the values for $\xi_l(\theta)$ will always be less than one. Hence, D_l^s will often be < 1 cm² d⁻¹. Diffusion causes some molecules to diffuse ahead of the average velocity of the piston flow front (Figure 6.3). This causes spreading or *dispersion* of the solute front.

6.2.2.3 Hydrodynamic Dispersion

In soil, one must also account for *hydrodynamic dispersion*. Three processes contribute to this type of dispersion. Not all pores are the same size (Figure 6.4, top). Pores

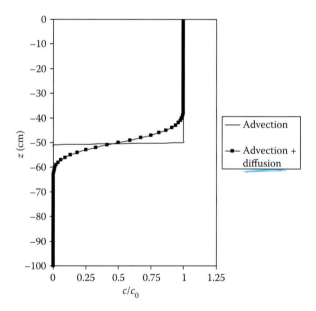

FIGURE 6.3 Effect of diffusion in dispersing a solute front at $t = 10$ h with a mean pore water velocity of 5 cm h⁻¹. *[handwritten: due to concentration gradient]*

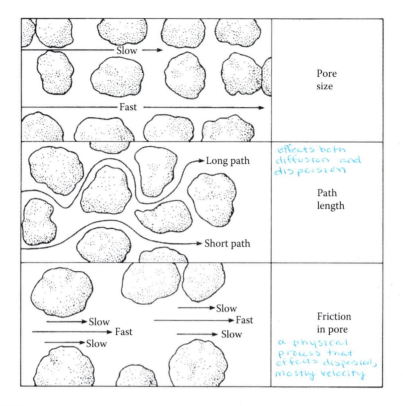

FIGURE 6.4 Sources of hydrodynamic dispersion: differences in pore size (top), differences in flow path length and mixing between pores (middle), and differences in velocity within a pore (bottom). (Šimůnek and van Genuchten, 2006.)

with larger diameters have larger *mean* velocities, as we know also from Poiseuille's law (Section 3.2.1). Not all streamlines (the paths followed by a solute particle) parallel the mean water flow direction and this causes differences in path lengths and velocities (Figure 6.4, middle). Also, pores interconnect and there is mixing between pores, which affects solute velocity. Finally, within each pore, velocities are not uniform (as shown in the derivation of Poiseuille's law, Section 3.6.1). Rather, they display a parabolic distribution with zero velocities at the pore walls and maximum velocities in the center of the pore (Figure 3.25 and Figure 6.4, bottom).

All these processes contribute to diversity in solute velocities and cause the solute front to be even more dispersed (much greater effect than diffusion), as shown in Figure 6.5. Since the effect of hydrodynamic dispersion is to cause spreading, much like diffusion, a Fick's law type equation can be used to describe hydrodynamic dispersive flux:

$$J_{\text{lh}} = -\theta D_{\text{lh}} \frac{\partial c}{\partial z}, \tag{6.8}$$

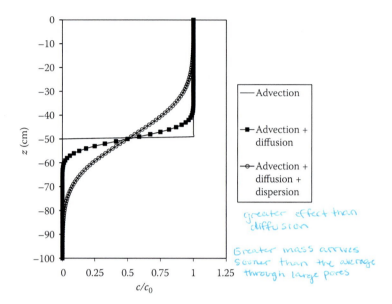

FIGURE 6.5   Effect of hydrodynamic dispersion and diffusion in dispersing a solute front at $t = 10$ h with a mean pore water velocity of 5 cm h^{-1}.

where D_{lh} is the coefficient of hydrodynamic dispersion [L^{2}T^{-1}]. Experiments have often shown that D_{lh} is proportional to v (Bresler 1973):

$$D_{lh} = \lambda v, \quad \leftarrow \text{ pore velocity} \tag{6.9}$$

where the proportionality constant λ is known as *dispersivity* [L]. The larger the value of λ, the more dispersed the solute front will be. It is affected by, and hence a measure of, pore-size distribution. A large λ indicates a wide range in pore-size distribution.

Another interesting aspect of this parameter is that values of λ vary depending on the scale of measurement. For laboratory experiments with packed columns, λ is typically 0.5–2 cm (Jury and Horton 2004). For intact columns and field soils, values are much higher, in the 5–20 cm range. For groundwater flow in aquifers, λ is higher still, often greater than 1 m. This phenomenon is known as *scale-dependent dispersivity*. There will be more on this in Section 6.3.9.

Using these equations for each component, the total solute flux (Equation 6.3) can be written:

$$J_s = J_w c - \theta D_l^s \frac{\partial c}{\partial z} - \theta D_{lh} \frac{\partial c}{\partial z}. \tag{6.10}$$

The diffusion and dispersion coefficients can be combined into a single *effective dispersion coefficient, D_e* [L^{2}T^{-1}]:

$$D_e = D_l^s + D_{lh}. \tag{6.11}$$

Then one can write: *Because diffusion is negligable*

$$J_s = J_w c - \theta D_e \frac{\partial c}{\partial z}. \tag{6.12}$$

Example 6.1

Calculate the effective dispersion coefficient (D_e) in square centimeters per hour for a soil with a steady water content of 0.38 cm³ cm⁻³, a steady Darcy flux of 1.4 cm h⁻¹, dispersivity of 4.2 cm, and a bulk density of 1.40 g cm⁻³. Assume a diffusion coefficient in free water of 10⁻⁵ cm² s⁻¹.

First, use ρ_b to calculate porosity (Equation 1.9):

$$\phi = 1 - \frac{\rho_b}{\rho_s} = 1 - \frac{1.40 \dfrac{g}{cm^3}}{2.65 \dfrac{g}{cm^3}} = 0.472.$$

$\phi = \theta_s$

Then calculate the tortuosity factor (Equation 6.6):

$$\xi_l(\theta) = \frac{\theta^{7/3}}{\theta_s^2} = \frac{0.38^{7/3}}{0.47^2} = \frac{0.105}{0.221} = 0.475.$$

Calculate D_l^s:

$$D_l^s = \xi_l(\theta) D_l^w = 0.475 \left(10^{-5} \frac{cm^2}{s} \right) \times \left(\frac{3600\,s}{h} \right) = 0.017 \frac{cm^2}{h}.$$

Calculate v (Equation 6.4):

$$v = \frac{J_w}{\theta} = \frac{1.4 \dfrac{cm}{h}}{0.38} = 3.68 \frac{cm}{h}.$$

Calculate D_{lh} (Equation 6.9):

$$D_{lh} = \lambda v = (4.2\,cm) \left(3.68 \frac{cm}{h} \right) = 15.45 \frac{cm^2}{h}.$$

Calculate D_e (Equation 6.11):

$$D_e = D_l^s + D_{lh} = 0.017 \frac{cm^2}{h} + 15.45 \frac{cm^2}{h} = 15.47 \frac{cm^2}{h}.$$

Note that 99.9% of the value of D_e comes from the effect of hydrodynamic dispersion. Diffusion has a negligible effect on D_e and dispersion of the solute front, in this case.

As shown in the example, the effect of diffusion on the effective dispersion coefficient can often be ignored. In this case:

$$D_e \approx D_{lh} = \lambda v. \tag{6.13}$$

6.3 ADVECTION DISPERSION EQUATION

Now the equation for solute flux and the solute conservation equation are used to develop the equation for solute transport. Substituting Equations 6.2 and 6.12 into Equation 6.1 results in the following equation:

$$\frac{\partial(\theta c + \rho_b s)}{\partial t} = -\frac{\partial}{\partial z}\left(J_w c - \theta D_e \frac{\partial c}{\partial z}\right) - S. \tag{6.14}$$

If no adsorption or sources/sinks are assumed, then the result is

$$\frac{\partial(\theta c)}{\partial t} = \frac{\partial}{\partial z}\left(\theta D_e \frac{\partial c}{\partial z}\right) - \frac{\partial(J_w c)}{\partial z}. \tag{6.15}$$

The units for each term are mass of solute divided by volume of soil and time. This is one form of the *Advection Dispersion Equation* (ADE). The first term on the right side accounts for dispersion and the second term accounts for mass flow (advection). It is also called the *Convection Dispersion Equation* (CDE).

For steady-state water conditions, J_w and θ are constant. This means that D_e is also constant (since D_e is proportional to v, which is constant for constant J_w and θ). Bringing θ and D_e outside the derivatives results in the equation:

$$\theta \frac{\partial c}{\partial t} = \theta D_e \frac{\partial^2 c}{\partial z^2} - J_w \frac{\partial c}{\partial z}. \tag{6.16}$$

Then dividing by θ and making the substitution $J_w/\theta = v$ (Equation 6.4) produces:

$$\frac{\partial c}{\partial t} = D_e \frac{\partial^2 c}{\partial z^2} - v \frac{\partial c}{\partial z}. \tag{6.17}$$

This is a PDE that is similar to the heat transport Equation (4.21), except for the additional term caused by advection. When convection is included in the heat transport equation (see Equation 4.56), it is very similar to the ADE. The solution of this equation will give concentrations as a function of depth and time: $c(z,t)$. The solution will depend on the particular initial conditions (profile concentrations at time zero) and boundary conditions (concentrations or solute fluxes at the surface and bottom of the profile). It will also depend on whether the soil is uniform or layered.

6.3.1 ANALYTICAL SOLUTION

For a uniform soil and steady water flow conditions, the ADE can usually be solved analytically. Solutions for various boundary and initial conditions are given by Skaggs and Leij (2002) and Jury and Roth (1990). The Laplace transform method can be used to find a solution for given initial and boundary conditions. For example, an initial concentration of zero everywhere in the profile can be specified. Three types of boundary conditions are possible: type-1, type-2, and type-3 (Table 6.1). For the surface, a type-3 boundary condition will be used that specifies the *flux concentration*. The other boundary condition will be that concentrations do not change deep in the soil (a type-2 boundary condition). The boundary conditions are (note that z is considered positive in the downward direction for this derivation):

$$J_w c(0,t) - \theta D_e \frac{\partial c(0,t)}{\partial z} = J_w c_0(t)$$

$$\frac{\partial c(\infty,t)}{\partial z} = 0 \tag{6.18}$$

The initial conditions are

$$c(z,0) = 0. \tag{6.19}$$

The Laplace transform method of finding the solution is shown in Section 6.8.1. The final solution is

$$c(z,t) = c_0 \left[\sqrt{\frac{v^2 t}{\pi D_e}} \exp\left(-\frac{(z-vt)^2}{4D_e t} \right) + \frac{1}{2} \operatorname{erfc}\left(\frac{z-vt}{\sqrt{4D_e t}} \right) \right. $$

$$\left. -\frac{1}{2}\left(1 + \frac{vz}{D_e} + \frac{v^2 t}{D_e} \right) \exp\left(\frac{vz}{D_e} \right) \operatorname{erfc}\left(\frac{z+vt}{\sqrt{4D_e t}} \right) \right] \tag{6.20}$$

Like the solution to the heat flow equation for similar boundary and initial conditions (Equation 4.25), this solution to the ADE involves the complementary error function. However, it is more complex due to the additional transfer mechanism of advective flux (mass flow) and the type-3 boundary condition at the surface.

TABLE 6.1

Solute transport boundary conditions

Type	Name	Example
1	Dirichlet	$c(0,t) = c_0(t)$
2	Neumann	$\dfrac{\partial c(\infty,t)}{\partial z} = 0$
3	Cauchy	$J_w\, c(0,t) - \theta D_e\, \dfrac{\partial c(0,t)}{\partial z} = J_w\, c_0(t)$

The transport parameters in this solution are v and D_e. The units used for depth, z, and time, t, will determine the units for v and D_e. Equation 6.20 was used to plot the curves for different values of D_e to simulate diffusion and hydrodynamic dispersion in Figure 6.5. A value of $v = 5$ cm h^{-1} was used for both curves (and the piston flow square wave). A small value of D_e (1 cm^2 h^{-1}) was used to simulate diffusion alone and a large value of D_e (10 cm^2 h^{-1}) was used to simulate diffusion plus hydrodynamic dispersion. Obviously, increasing D_e causes more spreading (dispersion) of the solute front. Since D_e is proportional to dispersivity (λ), a wide range in pore-size distribution causes a more dispersed solute front. The leading edge of the front represents solute traveling in the largest pores (macropores when λ is large).

The movement of the solute front can be examined by plotting $c(z,t)$ as a function of z for different times (Figure 6.6). The S-shape of the curve is apparent and

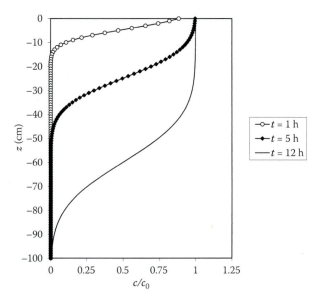

FIGURE 6.6 Concentrations as a function of depth for times of 1, 5, and 12 h based on Equation 6.20 with $v = 5$ cm h^{-1} and $D_e = 10$ cm^2 h^{-1}.

distinguishes advective-dispersive flow from purely dispersive or diffusive flow (see Figure 4.8). It is also apparent that the solute front becomes more dispersed as it travels deeper into the soil. The distribution of concentrations with depth is certainly not linear for solute flow.

Example 6.2

Use Excel to make a graph similar to Figure 6.6 for $v = 5$ cm h^{-1} and $D_e = 10$ cm^2 h^{-1} for the $t = 1$ and 5 h curves.

Enter the parameter values at the top of the spreadsheet (Figure 6.7). Enter the negative depths from -1 to -100 cm in 1 cm increments in the first column. Convert these to positive values in the second column. Since Equation 6.20 is complex, it helps to calculate the arguments of each term in separate columns (C through G) and then calculate the concentration in the final column. Excel cannot calculate the complementary error function of a negative argument, so an *if statement* must be used when the argument is negative to implement the following equality:

$$erfc(-x) = 1 + erf(|-x|).$$

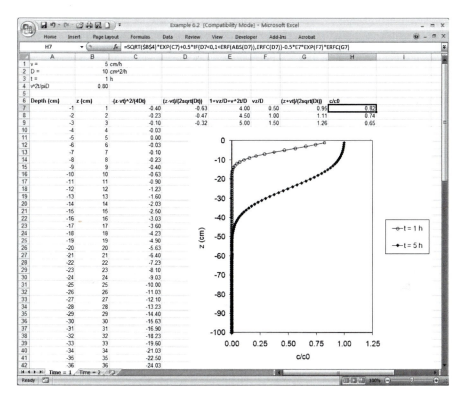

FIGURE 6.7 Excel spreadsheet for Example 6.2.

Use different worksheets (see lower left corner of spreadsheet) to calculate curves for times 1 and 5 h. Caution: In Equation 6.20, as depth increases the arguments of the exp and erfc functions in the last term increase. This causes the exp function to become very large and the erfc function to become very small. Accurate calculation of the product may require greater precision than is available in Excel (15 significant digits). For example, the $t = 10$ h curve in Figure 6.6 can not be calculated accurately with Excel.

6.3.2 Breakthrough Curves

Just as temperature was plotted as a function of time at a given depth in Figure 4.10, concentration can be plotted as a function of time at a given depth. To do this at a depth of 50 cm, for example, Equation 6.20 can be used to plot $c(L,t)$ vs. t, where $L = 50$ cm. This is done in Figure 6.8 for various values of D_e and the plots are called *breakthrough curves* (BTC). Initially, concentrations are zero (the initial conditions), then the solute front arrives and concentrations begin to rise. This is solute traveling in the largest pores. Although the concentrations in these pores may be equal to the input concentration at the surface (c_0), the average concentration is still low because the water flowing in the smaller pores at this depth is still solute free.

After about 10 h, the relative concentrations reach a value of 0.5 and half of the BTC has arrived. Eventually, concentrations reach the input concentration. At this point, solute has arrived at this depth in even the smallest pores.

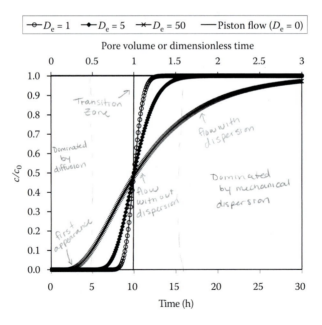

FIGURE 6.8 Breakthrough curves at a depth of 50 cm with $v = 5$ cm h^{-1} for piston flow and for the ADE solution (Equation 6.20) using $D_e = 1$, 5, and 50 cm^2 h^{-1} as a function of time (bottom axis) and pore volumes or dimensionless time (top axis).

Since dispersion of the solute front is affected by D_e, so are the BTCs. As D_e increases, the BTC becomes more dispersed. However, all the curves are "centered" at about the same time. That is, they all reach a relative concentration of 0.5 at about the same time. Note that this time corresponds to the time that the piston flow curve would arrive at this depth for this value of v. This is called the piston flow breakthrough time (shown in Figure 6.8 as the square wave), or $t_b = 10$ h. Essentially, $D_e = 0$ corresponds to piston flow when all the solute travels at the same velocity. Anything that causes velocities to vary in a soil causes the BTC to become more dispersed.

Looking carefully at Figure 6.8, it is apparent that the curves are not symmetrical about the piston flow curve when D_e is large. For the curve with $D_e = 50$ cm² h⁻¹, solute first appears about 7 h before the piston flow BTC, but concentrations don't reach $c/c_0 = 1$ until at least 20 h after the piston flow breakthrough. At high values of D_e, BTCs exhibit *early breakthrough* and *tailing*.

The time for breakthrough of the piston flow curve at a depth of $z = L$ can be calculated simply from the mean pore water velocity. Velocity is distance divided by time, so solving for time we obtain:

$$t_b = \frac{L}{v}. \tag{6.21}$$

A relative time can be developed (just as was done with concentration) by dividing t by t_b:

$$T = \frac{t}{t_b} = \frac{vt}{L}, \tag{6.22}$$

where T is *dimensionless* or *reduced* time. If the same data are plotted as a function of T, the curves in Figure 6.8 are produced (see the top axis). All the curves are now centered near a dimensionless time of one.

Depth can also be made dimensionless (Z) by dividing z by L, where L is referred to as the *characteristic length*, which is usually the maximum depth or column length in a BTC experiment:

$$Z = \frac{z}{L}. \tag{6.23}$$

BTCs can also be plotted against cumulative drainage per unit area (d_w in cubic centimeter per square centimeter) rather than time. Since steady-state water conditions are assumed, the Darcy flux is constant and the cumulative drainage per unit area is

$$d_w = J_w t. \tag{6.24}$$

Piston flow will reach the depth of interest ($z = L$) when the cumulative drainage is exactly equal to the volume of *old water* (water to be displaced by piston flow *new water*) on a unit area basis. If a soil has a cross-sectional area, A, then the total soil volume is $A \cdot L$. The total volume of old water is $A \cdot L \cdot \theta$. The volume of old water per unit area is simply $L \cdot \theta$, which is known as a *pore volume*. Hence, the cumulative drainage per unit area when piston flow breakthrough occurs at depth L is

$$PV = d_{wb} = L\theta = L\frac{J_w}{v}. = \forall_r \cdot \theta_s = \forall_r \phi \tag{6.25}$$

If a dimensionless or reduced cumulative drainage per unit area is created by dividing d_w by d_{wb}, the result is

$$\frac{d_w}{d_{wb}} = \frac{J_w t}{LJ_w / v} = \frac{vt}{L} = \frac{z}{L} = T. \tag{6.26}$$

In other words, dimensionless time and pore volumes are the same and the symbol T can be used for both. BTCs in Figure 6.8 can be thought of as being plotted as a function of dimensionless time *or* pore volumes. BTCs are often plotted in this manner. It is interesting to consider how many pore volumes are required for concentrations to reach the input concentration (a relative concentration of one). In Figure 6.8, as D_e increases, it takes more pore volumes to displace all the old water. For $D_e = 50 \text{ cm}^2$ h^{-1}, it takes about four pore volumes for the BTC to be complete. This is because for large values of D_e, and consequently large values of λ, there are many different pore sizes and solute velocities. Essentially, it is a *poorly mixed system* when λ is large and it takes more new water to flush out all the old water.

The BTCs looked at so far, and the solution to the ADE represented by Equation 6.20, are for a boundary condition at the surface where there is a constant input of solute at a given concentration, c_0. This is known as a *step input*, in that it appears as a square wave or a step function when plotted as a function of time (Figure 6.9).

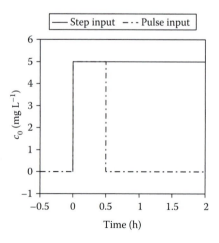

FIGURE 6.9 Input concentration ($c_0 = 5 \text{ mg L}^{-1}$) as a function of time for a step input and pulse input with a pulse duration of 0.5 h.

It would be appropriate for an animal waste lagoon at the soil surface with a steady seepage rate releasing nitrate or bacteria to the underlying soil at a constant concentration. It would also be appropriate for a bioremediation effort where nutrients were being released through an irrigation system at a constant rate and concentration to stimulate microbial breakdown of a contaminant in the soil.

6.3.3 PULSE INPUTS

Another common type of input for BTCs is a *pulse input*. In this case, solute of a given concentration, c_0, is applied at time zero, just as with the step input. However, after a certain time, t_0, solute-free water is applied (Figure 6.9). This boundary condition would be appropriate for a contaminant spill at the soil surface. Mathematically, a pulse input type-3 (see Equation 6.18) boundary condition can be expressed in two ways. One way is the more obvious approach:

$$c_0(t) = c_0 \qquad 0 < t \le t_0$$
$$c_0(t) = 0 \qquad t > t_0 \tag{6.27}$$

Another way to express a pulse input involves the use of the *Dirac delta function*, $\delta(t)$. In Chapter 4 (Section 4.7.3), it was noted that the error function is similar to the *normal* (or *Gauss*) *distribution*, the familiar bell-shaped curve from statistics. The normal distribution is defined as:

$$N(t,\sigma) = \frac{1}{\sigma\sqrt{2\pi}} \exp\left(-\frac{t^2}{2\sigma^2}\right), \tag{6.28}$$

where σ^2 is the *variance* and σ is the *standard deviation* of the distribution. The normal distribution is centered at $t = 0$, and as σ decreases, the distribution narrows (Figure 6.10).

The normal distribution is a *probability density function*, which is a plot of probability per unit t vs. t. The units of probability density are inverse of the variable (t^{-1} in this case). The area under the curve (sum of the probabilities) is always one:

$$\int_{-\infty}^{+\infty} N(t,\sigma)\, dt = 1. \tag{6.29}$$

The Dirac delta function $\delta(t)$ is defined as the limit of the normal distribution as $\sigma \to 0$ (Jury and Roth 1990):

$$\delta(t) = \lim_{\sigma \to 0} N(t,\sigma). \tag{6.30}$$

Hence, it is an infinite spike at $t = 0$. It retains the property that the area under this function is equal to one. It is a useful function for the mathematical representation

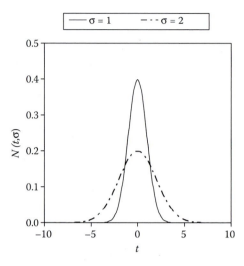

FIGURE 6.10 Normal distribution (probability density as a function of t, Equation 6.28) for standard deviations of 1 and 2.

of a sharp input, such as a hammer blow or a voltage spike. A very short duration input ($t_0 \rightarrow 0$) can be represented using the Dirac delta function as follows (Jury and Roth 1990):

$$c_0(t) = \frac{M}{J_w} \delta(t) \qquad (6.31)$$

where M is the mass of solute applied in the pulse per unit area of soil surface [ML^{-2}]. We will refer to Equation 6.31 as a *Dirac delta* input to distinguish it from a *pulse* input (Equation 6.27). It is especially useful when the input concentration of the pulse is not known, but the total mass of solute in the pulse is known. This might be the case in a spill of a solid contaminant or application of a solid fertilizer.

The analytical solution to the ADE for a pulse input boundary condition (Equation 6.27) and the same lower semi-infinite boundary condition (Equation 6.18) and initial conditions (Equation 6.19) used before, is

$$c(z,t) = c_{step}(z,t) \quad 0 < t \le t_0$$

$$c(z,t) = c_{step}(z,t) - c_{step}(z,t-t_0) \quad t > t_0, \qquad (6.32)$$

where $c_{step}(z,t)$ is the solution for the step input (Equation 6.20) and $c_{step}(z,t-t_0)$ is the same equation with $t-t_0$ substituted for t. Note that for times shorter than the pulse duration ($t \le t_0$), the solution is the same as the step input. Beyond that time, a step input term is subtracted that represents a front of solute-free water that starts at a delayed time, $t-t_0$ (Figure 6.11).

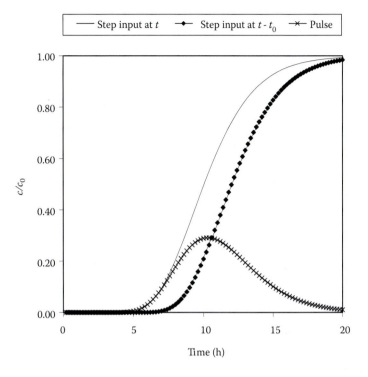

FIGURE 6.11 BTC solution for a pulse input (Equation 6.32) is the difference between a step input starting at $t = 0$ and delayed step input starting at $t = t_0$ ($= 2$ h).

Profile relative concentrations at a time of 10 h for a pulse input (where the pulse was applied for 2 h) with $v = 5$ cm h^{-1} are shown in Figure 6.12 for $D_e = 1$ and 10 cm^2 h^{-1}. Note that peak concentrations are less than the input relative concentration of one. The piston flow curve for a pulse input is also shown and it is a square pulse for which the leading edge has moved to a depth of 50 cm after 10 h. Unlike the other curves, piston flow concentrations extend to a relative concentration of one. Note that all the curves have a peak concentration (are "centered") at about the same depth of 45 cm, which corresponds to the center of the piston flow curve. Since most contaminants are not toxic below a given concentration, a common saying is *dilution is the solution to pollution*. It is apparent that as the dispersion coefficient (D_e) and dispersivity (λ) increase, the pulse becomes more dispersed and the peak concentration decreases. Hence, large values for D_e and λ favor dilution, but they also favor deeper movement of the leading edge of the pulse.

The analytical solution for a Dirac delta input (surface boundary condition specified by Equation 6.31) is given by Jury and Horton (2004):

$$c(z,t) = \frac{M}{J_w}\left\{\frac{v}{\sqrt{\pi D_e t}}\exp\left[-\frac{(z-vt)^2}{4D_e t}\right] - \frac{v^2}{2D_e}\exp\left(\frac{vz}{D_e}\right)\text{erfc}\left(\frac{z+vt}{\sqrt{4D_e t}}\right)\right\} \quad (6.33)$$

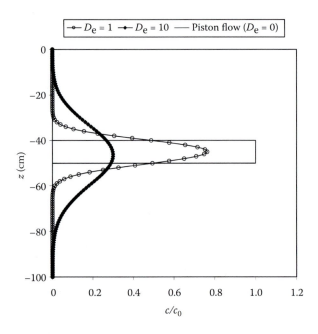

FIGURE 6.12 Relative concentrations as a function of depth predicted by Equation 6.32 after 10 h for a pulse input of 2-h duration with $v = 5$ cm h⁻¹ and for $D_e = 1$ and 10 cm² h⁻¹ compared to piston flow.

Note that c_0 does not appear in this equation (since it does not appear in the boundary condition at the surface), but M (the mass of solute applied in the pulse per unit area of soil surface) does. Profile relative concentrations at a time of 10 h for a Dirac delta input of $M = 5$ g cm⁻² with $J_w = 2.5$ cm h⁻¹ and $v = 5$ cm h⁻¹ are shown in Figure 6.13 for $D_e = 1$ and 10 cm² h⁻¹. This combination of M and J_w results in the same mass of solute being applied in Figures 6.12 and 6.13. The curves in both figures are very similar and this will be the case for small values of the pulse duration, t_0. The piston flow curve in the case of a Dirac delta input is an infinite spike at a depth of 50 cm (Figure 6.13). Again, both curves have a peak concentration at about the same depth as the piston flow spike.

One can look at the movement of a pulse input by plotting $c(z,t)$ as a function of z for different times using Equation 6.32, as was done earlier for a step input (Figure 6.6). In Figure 6.14, profile relative concentrations for a pulse input of 2 h duration with $v = 5$ cm h⁻¹ and $D_e = 10$ cm² h⁻¹ are shown after 3, 6, and 12 h. It is clear that as the pulse moves down the profile, it becomes more dispersed and the peak concentrations decrease. This is due to dilution as solute molecules with the higher velocities move farther and farther ahead of solute molecules with the lower velocities.

The BTCs for a pulse input at a depth of 50 cm for $v = 5$ cm h⁻¹ and $D_e = 1$, 10, and 50 cm² h⁻¹ are shown in Figure 6.15 using Equation 6.32. As D_e increases, the BTCs become more dispersed. All the curves are centered about the piston flow curve in that the solute mass is evenly distributed on both sides of the square pulse. The time for breakthrough of the center of the piston flow pulse (11 h) is a reasonable approximation of the time for breakthrough for the center of the other pulses. The lack of

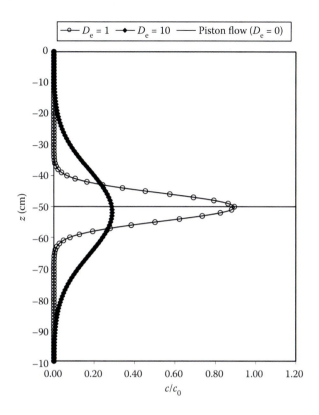

FIGURE 6.13 Relative concentrations as a function of depth predicted by Equation 6.33 after 10 h for a Dirac delta input with $v = 5$ cm h^{-1} for $D_e = 1$ and 10 cm^2 h^{-1} compared to piston flow.

symmetry in the BTCs when D_e is large is apparent and it causes early breakthrough of the solute front and tailing. If the BTCs for pulse inputs are plotted as a function of dimensionless time or pore volumes, $T = vt/L$, the curves are centered about $T = 1$ (see Figure 6.15, top axis).

6.3.4 EQUILIBRIUM ADSORPTION AND THE ADVECTION DISPERSION EQUATION

So far, it has been assumed that solutes do not undergo adsorption to minerals or organic matter in soil. The next step is to include the adsorption process because many contaminants of interest (such as heavy metals and pesticides) are adsorbed. In Equation 6.2, it was shown that the total concentration consisted of a dissolved solute concentration (c) and an adsorbed concentration (s), where the units for s are mass of solute adsorbed per mass of soil [MM^{-1}]. If s is included in the conservation Equation 6.1, the result is

$$\frac{\partial(\theta c + \rho_b s)}{\partial t} = -\frac{\partial}{\partial z}\left(J_w c - \theta D_e \frac{\partial c}{\partial z}\right). \tag{6.34}$$

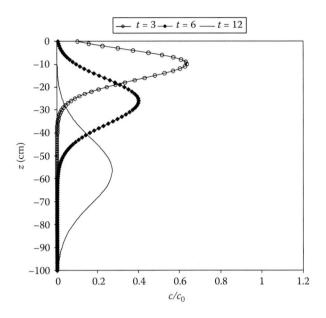

FIGURE 6.14 Relative concentrations as a function of depth for times of 3, 6, and 12 h predicted by Equation 6.32 for a pulse input with $v = 5$ cm h^{-1} for $D_e = 10$ cm^2 h^{-1}.

Under steady flow conditions, one can factor out a constant θ and (recalling that $J_w/\theta = v$) obtain:

$$\frac{\partial c}{\partial t} + \frac{\rho_b}{\theta}\frac{\partial s}{\partial t} = D_e\frac{\partial^2 c}{\partial z^2} - v\frac{\partial c}{\partial z}. \tag{6.35}$$

The ADE now includes an adsorption term, but there are two dependent variables in the ADE: c and s. To eliminate one of these variables, a relationship between c and s is sought. Adsorption can be divided into linear and nonlinear adsorption, an important distinction mathematically because the ADE can be solved analytically only for linear adsorption.

6.3.4.1 Linear Equilibrium Adsorption

Adsorption can be further subdivided into instantaneous (equilibrium) adsorption and kinetic (or nonequilibrium) adsorption. For instantaneous linear adsorption:

$$s = K_d c, \tag{6.36}$$

where K_d is the distribution coefficient [L^{3}M^{-1}] and the slope of the adsorption isotherm (Figure 6.16). K_d has units of volume of soil water over mass of soil. Adsorption of hydrophobic organic compounds has been correlated with the organic C content

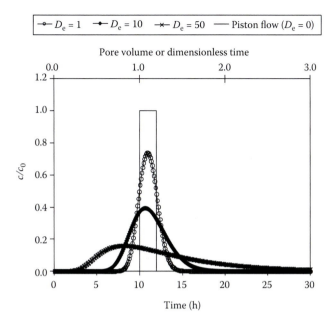

FIGURE 6.15 BTCs at a depth of 50 cm predicted by Equation 6.32 for a pulse input with $v = 5$ cm h^{-1} for $D_e = 1$, 10, and 50 cm^2 h^{-1} compared to piston flow plotted as a function of time (bottom axis) and as a function of pore volume or dimensionless time (top axis).

of the soil, f_{oc} [MM^{-1}] (e.g., Papiernik et al. 2002). For these compounds, K_d can be estimated using the following equation:

$$K_d = K_{oc}f_{oc}, \qquad (6.37)$$

where K_{oc} is the organic C partitioning coefficient [L^{3}M^{-1}]. Soil organic matter is typically about 60% organic C (Papiernik et al. 2002). Jury et al. (1984b) provided K_{oc} for 35 organic chemicals and these ranged from $2 \cdot 10^1$ to $2.3 \cdot 10^4$ cm^3 g^{-1}.

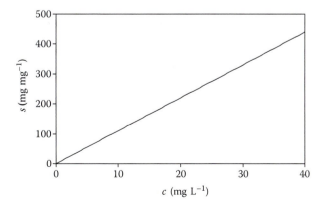

FIGURE 6.16 Linear adsorption isotherm with $K_d = 11$ L mg^{-1}.

Linear adsorption is most likely in dilute solutions with inorganic solutes and instantaneous adsorption is also most likely with these solutes. One can eliminate s from the ADE by substituting Equation 6.36 into Equation 6.35:

$$\frac{\partial c}{\partial t} + \frac{\rho_b}{\theta}\frac{\partial(K_d c)}{\partial t} = D_e\frac{\partial^2 c}{\partial z^2} - v\frac{\partial c}{\partial z}$$

$$\left[1+\frac{\rho_b K_d}{\theta}\right]\frac{\partial c}{\partial t} = D_e\frac{\partial^2 c}{\partial z^2} - v\frac{\partial c}{\partial z} \tag{6.38}$$

$$R\frac{\partial c}{\partial t} = D_e\frac{\partial^2 c}{\partial z^2} - v\frac{\partial c}{\partial z},$$

where R is a unit-less *retardation factor*:

$$R = 1 + \frac{\rho_b K_d}{\theta}. \tag{6.39}$$

If there is no adsorption ($K_d = 0$) then $R = 1$. From an adsorption isotherm, R can be calculated for a given ρ_b and θ. If Equation 6.38 is divided by R, the result is (Jury and Horton 2004):

$$\frac{\partial c}{\partial t} = \frac{D_e}{R}\frac{\partial^2 c}{\partial z^2} - \frac{v}{R}\frac{\partial c}{\partial z}$$

$$\frac{\partial c}{\partial t} = D_R\frac{\partial^2 c}{\partial z^2} - v_R\frac{\partial c}{\partial z}. \tag{6.40}$$

where *retarded dispersion coefficient* (D_R) and *retarded velocity* (v_R) are defined as:

$$D_R = \frac{D_e}{R}$$

$$v_R = \frac{v}{R}. \tag{6.41}$$

Equation 6.40 is the exact same form as the ADE for nonadsorbed solute transport (Equation 6.17), except for the new parameters, v_R and D_R. Therefore, all the solutions to the ADE are applicable, except for the substitution of v_R for v and D_R for D. The solution for a step input with a nonadsorbed solute is given in Equation 6.20. Making the substitutions for adsorbed solute transport, the result is (Skaggs and Leij, 2002):

$$c(z,t) = c_0 \left[\sqrt{\frac{\left(\frac{v}{R}\right)^2 t}{\pi \frac{D_e}{R}}} \exp\left(-\frac{\left(z-\frac{v}{R}t\right)^2}{4\frac{D_e}{R}t}\right) + \frac{1}{2}\mathrm{erfc}\left(\frac{z-\frac{v}{R}t}{\sqrt{4\frac{D_e}{R}t}}\right) \right.$$

$$\left. -\frac{1}{2}\left(1 + \frac{vz}{D_e} + \frac{\left(\frac{v}{R}\right)^2 t}{\frac{D_e}{R}}\right) \exp\left(\frac{vz}{D_e}\right) \mathrm{erfc}\left(\frac{z+\frac{v}{R}t}{\sqrt{4\frac{D_e}{R}t}}\right) \right]$$

$$= c_0 \left[\sqrt{\frac{v^2 t}{\pi D_e R}} \exp\left(-\frac{(Rz-vt)^2}{4D_e Rt}\right) + \frac{1}{2}\mathrm{erfc}\left(\frac{Rz-vt}{\sqrt{4D_e Rt}}\right) \right. \tag{6.42}$$

$$\left. -\frac{1}{2}\left(1 + \frac{vz}{D_e} + \frac{v^2 t}{D_e R}\right) \exp\left(\frac{vz}{D_e}\right) \mathrm{erfc}\left(\frac{Rz+vt}{\sqrt{4D_e Rt}}\right) \right]$$

The effect of linear adsorption and R on profile dissolved concentrations with a step input is shown in Figure 6.17 using Equation 6.42 and values of $R = 1$ (for a

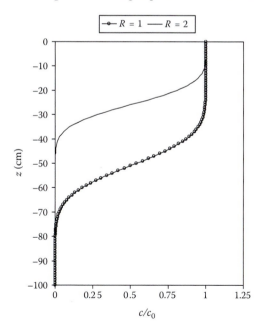

FIGURE 6.17 Relative concentrations as a function of depth after 10 h for a step input with $v = 5$ cm h^{-1}, $D_e = 5$ cm^2 h^{-1}, and $R = 1$ (nonadsorbed solute) and $R = 2$ (adsorbed solute) using Equation 6.42.

nonadsorbed solute) and $R = 2$ (for an adsorbed solute). Both curves represent the concentrations in solution (since the equation is for c). The movement of the dissolved solute front when adsorption occurs is delayed because some of the solute has been lost to adsorption sites.

The effect of adsorption on the arrival of the BTC at a depth of 50 cm is shown in Figure 6.18. Curves are shown for $R = 1$, 2, and 3. With adsorption, the arrival of the dissolved solute front is delayed as R increases (hence the term, *retardation coefficient*). The BTC also becomes more dispersed. When the same curves are plotted as a function of pore volumes ($T = vt/L$), each curve is centered around a pore volume equal to R (Figure 6.18, top axis). That is, it takes one pore volume to displace the center of the nonadsorbed front to this depth, two pore volumes to displace the center of the adsorbed solute with $R = 2$ to this depth, etc. The retardation coefficient thus tells one how much slower a particular solute travels as compared to tracers (nonreactive solutes), such as chloride or bromide, for which the retardation coefficient is equal to 1 ($K_d = 0$).

The effect of adsorption on a pulse input can be seen by making the same substitutions of v_R for v and D_R for D_e in the appropriate solutions. For a pulse input, this is just a matter of using Equation 6.42 as $c_{step}(z,t)$ and $c_{step}(z,t - t_0)$ in Equation 6.32. The effect of adsorption on BTCs for a pulse input of 2-h duration with $v = 5$ cm h^{-1}, $D_e = 5$ cm^2 h^{-1}, and $R = 1$, 2, and 3 is shown in Figure 6.19. The arrival of the BTC for a pulse input is delayed as R increases. The BTCs become more dispersed and

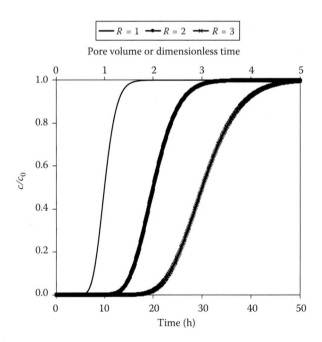

FIGURE 6.18 BTCs for a depth of 50 cm for a step input with $v = 5$ cm h^{-1}, $D_e = 5$ cm^2 h^{-1}, and $R = 1$, 2, and 3 using Equation 6.42 as a function of time (bottom axis) and pore volume or dimensionless time (top axis).

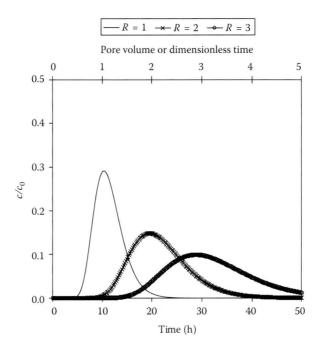

FIGURE 6.19 BTCs for a depth of 50 cm for a pulse input of 2 h duration with $v = 5$ cm h^{-1}, $D_e = 5$ cm^2 h^{-1}, and $R = 1$, 2, and 3 using Equations 6.42 and 6.32 as a function of time (bottom axis) and pore volume or dimensionless time (top axis).

the peak concentration is less when adsorption occurs. Plotting the same BTCs as a function of pore volumes makes it clear that the pulses are centered about a pore volume $T = R$ (Figure 6.19 top axis).

R may be thought of as a ratio of the mean velocity of a nonadsorbed solute v to that of the retarded solute v_R (Jury and Horton 2004):

$$R = \frac{v}{v_R}. \tag{6.43}$$

This is useful because it permits calculation of the average velocity of a solute pulse and the time it will take to reach a given depth. It was shown earlier that the time for breakthrough for nonadsorbed piston flow (t_b) was L/v (Equation 6.21). The time for breakthrough for adsorbed piston flow is

$$t_{bR} = \frac{L}{v_R} = \frac{RL}{v} = Rt_b. \tag{6.44}$$

R can also be thought of as a ratio of the total mass of solute divided by the mass in solution. For a step input, this is

$$R = \frac{J_w \int_0^t [c_0 - c(L,\tau)]d\tau}{\theta \int_0^L [c(z,t) - c_i]dz}, \tag{6.45}$$

where c_0 is the input concentration, c_i is the initial dissolved concentration (this has been zero in the problems so far), and $c(L,t)$ is the dissolved concentration at the lower boundary. The integral in the numerator subtracts outputs from inputs over time to get the total amount of solute in the column. It is multiplied by J_w to produce units of mass of solute per unit area of soil. The integral in the denominator sums the dissolved solute over the length of the column. It is multiplied by θ to produce units of mass of solute per unit area of soil, as well. If there is no adsorption, then the total mass of solute and the mass in solution are the same and $R = 1$.

6.3.4.2 Nonlinear Equilibrium Adsorption

The Freundlich (1909) equation is commonly used to describe instantaneous nonlinear adsorption isotherms:

$$s = K_f c^\beta, \tag{6.46}$$

where K_f is the Freundlich distribution coefficient [$L^{3\beta}M^{-\beta}$] and β is an exponent [–], both obtained from the isotherm for a particular soil. Adsorption isotherms for different values of β are shown in Figure 6.20. As β decreases, the isotherm becomes more nonlinear (convex). Linear adsorption (Equation 6.36) is a special case of the Freundlich equation with $\beta = 1$. For $\beta < 1$, less solute is adsorbed per unit increase in c at high concentrations compared to low concentrations. This often happens when

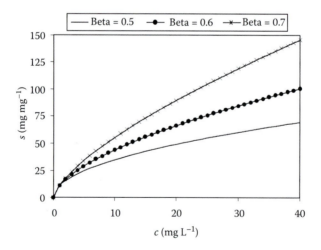

FIGURE 6.20 Nonlinear adsorption isotherm using the Freundlich Equation 6.46 with $K_f = 11$ $L^{3\beta}$ mg$^{-\beta}$ and $\beta = 0.5$, 0.6, and 0.7.

the amount of solute added is sufficient to fill most of the adsorption sites in soils. Some adsorption isotherms are concave ($\beta > 1$) and this happens with hydrophobic pesticides where adsorption is facilitated as more pesticides coat the adsorption surfaces.

Another common equation used to describe instantaneous nonlinear isotherms is the Langmuir (1918) equation:

$$s = \frac{K_d c}{1 + \eta c},\tag{6.47}$$

where K_d [L^{3}M^{-1}] and η [L^{3}M^{-1}] are Langmuir equation parameters. Adsorption isotherms for different values of η are shown in Figure 6.21. As η increases, the isotherm becomes more nonlinear and (unlike the Freundlich isotherm) approaches a maximum sorbed concentration.

If either Equation 6.46 or Equation 6.47 is substituted into the ADE (Equation 6.34), the ADE becomes a *nonlinear* PDE, since c is now raised to a power and there is no analytical solution. Hence, only numerical solutions can be used to study transport of solutes that undergo nonlinear adsorption.

6.3.5 Transformations and the Advection Dispersion Equation

Many important contaminants such as pesticides and nutrients such as nitrate undergo important microbial transformations in soils. Pesticides are broken down by microbes and ammonium is converted to nitrite and then nitrate through nitrification. Radionuclides also undergo radioactive decay. All these processes can be represented as *first-order transformation processes*. In each case, the rate of transformation is proportional to the mass of chemical present:

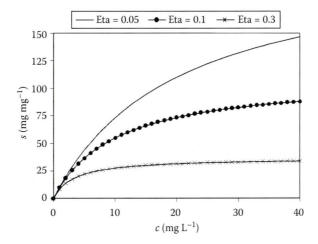

FIGURE 6.21 Nonlinear adsorption isotherm using the Langmuir Equation 6.47 with $K_d = 11$ L^3 mg^{-1} and $\eta = 0.05$, 0.1, and 0.3 L^3 mg^{-1}.

$$\frac{\partial(\theta c)}{\partial t} = -\mu\theta c, \tag{6.48}$$

where μ is the *first-order transformation rate constant* with units of inverse time [T^{-1}]. Multiplying c by θ converts concentration to mass per unit volume of soil. The units for this equation are mass loss rate per volume of soil [$ML^{-3}T^{-1}$]. The equation shows, and it is usually assumed (although not for radionuclides), that only the solution phase of the solute (c) undergoes decay. The half-life of a chemical $T_{1/2}$ (the time for the chemical concentration to decrease to one half of the original concentration) is related to the first-order transformation constant:

$$\mu = \frac{\ln(2)}{T_{1/2}}.$$

Jury et al. (1984b) provided $T_{1/2}$ for 35 organic chemicals and these ranged from 15 days to infinity.

To add transformation to the transport process under steady flow, Equation 6.48 is added as a *source/sink term* to a form of the ADE seen earlier (Equation 6.16):

$$\theta\frac{\partial c}{\partial t} = \theta D_e\frac{\partial^2 c}{\partial z^2} - J_w\frac{\partial c}{\partial z} - \mu\theta c. \tag{6.49}$$

If this equation is divided by θ and linear adsorption is assumed, one obtains the form of the ADE that includes transformation and adsorption:

$$R\frac{\partial c}{\partial t} = D_e\frac{\partial^2 c}{\partial z^2} - v\frac{\partial c}{\partial z} - \mu c. \tag{6.50}$$

Adding decay to the transport process results in the analytical solutions seen so far (Equation 6.42) being multiplied by a new exponential term that includes the transformation rate constant. The solution for a step input with decay (and adsorption) is

$$c(z,t) = c_0\exp(-\mu t)\left[\begin{array}{l}\sqrt{\frac{v^2 t}{\pi D_e R}}\exp\left(-\frac{(Rz-vt)^2}{4D_e Rt}\right) + \frac{1}{2}\mathrm{erfc}\left(\frac{Rz-vt}{\sqrt{4D_e Rt}}\right) \\ -\frac{1}{2}\left(1+\frac{vz}{D_e}+\frac{v^2 t}{D_e R}\right)\exp\left(\frac{vz}{D_e}\right)\mathrm{erfc}\left(\frac{Rz+vt}{\sqrt{4D_e Rt}}\right)\end{array}\right] \tag{6.51}$$

To get the solution for a pulse input, Equation 6.51 is substituted for c_{step} in Equation 6.32.

The effect of a positive transformation rate constant (decay) on the movement of a nonadsorbed solute pulse down a soil profile is shown in Figure 6.22. Two sets of curves are shown. One set is the concentrations after 3 h, without and with decay,

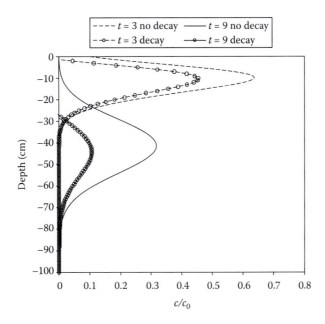

FIGURE 6.22 Relative concentration as a function of depth with and without decay for a pulse input after 3 and 9 h ($\mu = 0.1$ h^{-1}, $t_0 = 2$ h, $v = 5$ cm h^{-1}, $D_e = 10$ cm^2 h^{-1}, $R = 1$).

where $\mu = 0.1$ h^{-1}. The other set is the concentrations after 9 h, without and with decay. It's obvious that as the pulse moves deeper in the soil, decay has more of an effect (due to the additional time). Decay also diminishes the trailing edge of the pulse. Adsorption can play an important role in delaying a pulse and allowing more time for decay (even though the decay occurs in the liquid phase). Many contaminants do not make it to groundwater because of the combination of adsorption and decay.

6.3.6 VOLATILIZATION AND THE ADVECTION DISPERSION EQUATION

Volatilization is increasingly recognized as an important process affecting the fate of many organic chemicals, including pesticides, fumigants, and explosives in field soils (Jury et al. 1983, 1984a; Glotfelty and Schomburg 1989).

When the gas phase is included, the total solute concentration is

$$C_T = \theta c + \rho_b s + ag, \tag{6.52}$$

where a is the volumetric air content [L^{3}L^{-3}] and g is the solute concentration in the gas phase [ML^{-3}].

The most important process for solute transport in the gas phase is diffusion in soil air (since soil air movement is limited, gas-phase advection is usually negligible). This can be described using Fick's law:

$$J_{gd} = -a\xi_g(a)D_g^w \frac{\partial g}{\partial z} = -aD_g^s \frac{\partial g}{\partial z}, \tag{6.53}$$

where D_g^w and D_g^s are the diffusion coefficient in free gas and the effective diffusion coefficient in soil [L^2T^{-1}], respectively, and $D_g^s = \xi_g(a) \cdot D_g^w$. A typical value for D_g^w is 0.05 cm^2 s^{-1} (Jury et al. 1984a). Note that this is almost four orders larger than for liquid diffusion (see Example 6.1). Tortuosity can be defined in a manner similar to Equation 6.6:

$$\xi_l(a) = \frac{a^{7/3}}{\theta_s^2}.$$

(6.54)

When Equations 6.52 and 6.53 are included in the ADE, the following equation is produced:

$$\frac{\partial}{\partial t}(\theta c + \rho_b s + ag) = -\frac{\partial}{\partial z}\left(J_w c - \theta D_e \frac{\partial c}{\partial z} - aD_g^s \frac{\partial g}{\partial z}\right).$$

(6.55)

Henry's law describes the relationship between solute concentrations in the liquid and gas phases:

$$g = K_H c,$$

(6.56)

where K_H is the dimensionless Henry's constant. Jury et al. (1984b) provided K_H for 35 organic chemicals and these ranged from $5.5 \cdot 10^{-9}$ to $9.7 \cdot 10^1$.

Substituting Equation 6.56 into Equation 6.55 (and assuming equilibrium linear adsorption in the liquid phase and steady flow), a new form of the ADE is obtained that includes gas diffusion:

$$\frac{\partial}{\partial t}(\theta c + \rho_b K_d c + aK_H c) = -\frac{\partial}{\partial z}\left(J_w c - \theta D_e \frac{\partial c}{\partial z} - aD_g^s K_H \frac{\partial c}{\partial z}\right)$$

$$\theta \frac{\partial}{\partial t}\left[\left(1 + \frac{\rho_b K_d + aK_H}{\theta}\right)c\right] = \theta \frac{\partial}{\partial z}\left(D_e \frac{\partial c}{\partial z} + \frac{aD_g^s K_H}{\theta} \frac{\partial c}{\partial z}\right) - J_w \frac{\partial c}{\partial z},$$

$$R\frac{\partial c}{\partial t} = D_E \frac{\partial^2 c}{\partial z^2} - v\frac{\partial c}{\partial z},$$

(6.57)

where the new retardation coefficient is

$$R = 1 + \frac{\rho_b K_d + aK_H}{\theta},$$

(6.58)

and the new effective dispersion coefficient is

$$D_E = D_e + \frac{aD_g^s K_H}{\theta}. \tag{6.59}$$

Since the ADE that includes adsorption and volatilization (Equation 6.57) is the same form as the ADE for an adsorbing solute (Equation 6.38), the analytical solution seen earlier (Equation 6.42) can be used to describe adsorption and volatilization by substituting the new definition for R (Equation 6.58) and substituting D_E (Equation 6.59) for D_e.

Because volatilization increases retardation (Equation 6.58) and the dispersion coefficient (Equation 6.59), volatilization will slow liquid-phase solute movement through the soil and increase dispersion of the liquid-phase solute front or pulse.

Example 6.3

Calculate R and D_E for the soil fumigant, methyl bromide, which has a $K_{oc} = 22$ cm³ g⁻¹ and a $K_H = 1.5$ (Jury et al. 1980b). Use $D_g^w = 0.05$ cm² s⁻¹, $f_{oc} = 0.053$ g g⁻¹, $\theta = 0.38$ cm³ cm⁻³, $v = 0.1$ cm h⁻¹, $\lambda = 4.2$ cm, $\rho_b = 1.40$ g cm⁻³, $\varphi = 0.475$ cm³ cm⁻³, and $D_l^s = 0.017$ cm² h⁻¹. Compare this to R and D_e in the same soil with no volatilization.

Calculate K_d using Equation 6.37:

$$K_d = K_{oc} f_{oc} = \left(22 \frac{cm^3}{g} \right)(0.053) = 1.17 \frac{cm^3}{g}.$$

Calculate R assuming no volatilization (Equation 6.39):

$$R = 1 + \frac{\rho_b K_d}{\theta} = 1 + \frac{\left(1.40 \dfrac{g}{cm^3} \right)\left(1.17 \dfrac{cm^3}{g} \right)}{0.38} = 5.31.$$

Calculate a using Equation 1.5:

$$a = \varphi - \theta = 0.472 - 0.38 = 0.092.$$

Calculate R with volatilization (Equation 6.58):

$$R = 1 + \frac{\rho_b K_d + a K_H}{\theta} = 1 + \frac{\left(1.40 \dfrac{g}{cm^3} \right)\left(1.17 \dfrac{cm^3}{g} \right) + (0.092)(1.5)}{0.38} = 5.67.$$

Calculate D_g^s using Equation 6.54:

$$D_g^s = D_g^w \frac{a^{7/3}}{\theta_s^2} = \left(0.05 \frac{cm^2}{s} \right)\left(\frac{3,600\,s}{h} \right)\left(\frac{0.092^{7/3}}{0.38^2} \right) = 4.80 \frac{cm^2}{h}.$$

Calculate D_e using Equation 6.9 and Equation 6.11:

$$D_e = D_i^s + \lambda v = 0.017\frac{cm^2}{h} + (4.2\,cm)\left(0.1\frac{cm}{h}\right) = 0.437\frac{cm^2}{h}.$$

Calculate D_E using Equation 6.59:

$$D_E = D_e + \frac{aD_g^s K_H}{\theta} = 0.437\frac{cm^2}{h} + \frac{(0.092)\left(4.80\frac{cm^2}{h}\right)(1.5)}{0.38} = 2.18\frac{cm^2}{h}.$$

Volatilization causes a slight increase in R and a substantial increase in the effective dispersion coefficient. Gas diffusion is a dominant process for losses of methyl bromide to the atmosphere, where it depletes the stratospheric ozone layer (USEPA 2009).

6.3.7 FLUX VS. RESIDENT CONCENTRATIONS

A distinction can be made between two types of measured concentrations, depending on how a sample is taken. In taking a sample from a given volume of soil, one can extract the solution and measure the concentration. This is a volume-averaged or *resident concentration*, and what most people might think of as a concentration. However, samples are often taken under conditions where flow is occurring. For example, in a column BTC experiment, samples are usually collected with test tubes and a fraction collector. If there is a macropore in the column, more of the sample in a test tube will come from the macropore than from other pores and the concentration in the macropore will be *weighted* more than in a soil sample (which would be volume weighted). A concentration measured under conditions where flow is occurring is a *flux concentration*. The relationship between flux concentrations (c_f) and resident concentrations (c_r) is (Parker and van Genuchten 1984):

$$c_f = c_r - \frac{D_e}{v}\frac{\partial c_r}{\partial z}.$$

(6.60)

Note that if the contribution of diffusion to dispersion is ignored (see Equation 6.13), then $D_e/v = \lambda$. This shows that there is little difference between flux and resident concentrations in a soil with a small λ. Only in soils with large values for λ, such as soils with macropores, does the distinction become important.

Outflow samples from columns, wells, and tile drains are all flux concentrations. Soil samples and measurements made with time domain reflectometry (TDR) are resident concentrations. There's debate over what to consider samples from suction samplers or lysimeters (Parker and van Genuchten 1984).

The significance of this distinction emerges when comparing predicted concentrations from solutions to the ADE to measured concentrations. For example, if samples are collected in a BTC experiment and a solution to the ADE is fit to the measured concentrations to obtain transport parameters, then the ADE predicted concentrations should be flux concentrations.

How does one know if the solution to the ADE predicts flux or resident concentrations? It depends on the boundary condition used at the soil surface in obtaining the solution. If the boundary condition is given in terms of a concentration (*type-1 boundary condition*), then the solution will predict flux concentrations within the model domain. However, a type-1 boundary condition at the soil surface is not recommended because it can result in a mass balance error (Parker and van Genuchten, 1984). A type-3 boundary condition which predicts resident concentrations within the model domain is preferred. Equation 6.60 can be used to convert predicted resident concentrations to flux concentrations. In numerical solutions (Sections 6.5 and 6.6), a type-2 boundary condition at a finite depth that specifies a zero spatial gradient will result in flux concentrations being predicted at the outlet (see Equation 6.60), which is often required for a BTC experiment.

6.3.8   Nonequilibrium Solute Transport

One of the most challenging aspects of predicting solute transport is the effect of preferential flow (Flury 1996; Germann 1988; Jury and Flühler 1992; Vervoort et al., 1999). Where preferential flow occurs, BTCs can be very skewed, exhibiting the early breakthrough and tailing discussed in relation to Figures 6.8 and 6.15. In this case, it takes a very large value of λ to fit the ADE to the data. Even then, the fit of ADE is often poor. Large values of λ and poor fit of the ADE may indicate nonequilibrium conditions. In this case, the ADE needs to be modified to accurately represent additional processes.

6.3.8.1   Physical Nonequilibrium

Van Genuchten and Wierenga (1976) proposed a two-region conceptual model consisting of *mobile* (θ_{mo}) and *immobile* (θ_{im}) water. This is also called a *dual-porosity* model, as discussed in Chapter 5 in relation to Figure 5.38. Solute moves by advection with the water that flows through the mobile regions between aggregates or in macropores. The interior of aggregates or the soil matrix has such small pores that water is *relatively* immobile. Solute can enter and exit the aggregates or matrix by diffusion, but this is slow compared to advective flow in the mobile region. Early breakthrough occurs because only a portion of the solute diffuses into the immobile regions. Tailing occurs with a pulse input because solute that has entered immobile regions must diffuse back out before concentrations will drop to zero.

According to this theory, the liquid (c) and adsorbed (s) solutes reside in mobile and immobile regions. We will use the notation c_{mo} and c_{im} for mobile and immobile liquid concentrations and s_{mo} and s_{im} for the mobile and immobile adsorbed concentrations. The total concentration in soil (C_T) is then:

$$C_T = \theta_{mo}c_{mo} + f\rho_b s_{mo} + \theta_{im}c_{im} + (1-f)\rho_b s_{im}, \tag{6.61}$$

where f is the dimensionless fraction of adsorption sites in the mobile region, θ_{mo} is the mobile water content [L^3L^{-3}], θ_{im} is the immobile water content [L^3L^{-3}], and

$\theta = \theta_{mo} + \theta_{im}$. Substituting the mobile phases of this expression for the total concentration into the left side of the ADE (Equation 6.15) and assuming linear equilibrium adsorption, the mobile-phase ADE is

$$\frac{\partial}{\partial t}\left(\theta_{mo} c_{mo}\right) + \frac{\partial}{\partial t}\left(f\rho_b K_d c_{mo}\right) = \frac{\partial}{\partial z}\left(\theta_{mo} D_{mo} \frac{\partial c_{mo}}{\partial z}\right) - \frac{\partial(J_w c_{mo})}{\partial z} - \Gamma_s, \quad (6.62)$$

where D_{mo} is the effective dispersion coefficient for the mobile phase [$L^2 T^{-1}$]. Exchange with the immobile phases occurs via the last term on the right side of Equation 6.62:

$$\Gamma_s = \alpha_s(c_{mo} - c_{im}) = \frac{\partial}{\partial t}(\theta_{im} c_{im}) + \frac{\partial}{\partial t}\left[(1 - f)\rho_b K_d c_{im}\right], \quad (6.63)$$

where α_s is the *exchange rate coefficient* [T^{-1}]. It is assumed that exchange will be proportional to the difference in concentrations between the mobile and immobile regions (i.e., liquid diffusion causes exchange). The larger the α_s, the faster the exchange between the mobile and immobile regions and the nearer the system is to equilibrium conditions. The rate coefficient α_s depends on the size and shape of the basic structural unit (aggregate or ped), but typical values are 0.1–5 h^{-1}.

Van Genuchten and Wierenga (1976) used two dimensionless parameters, β and ω. The first parameter is defined as:

$$\beta = \frac{\theta_{mo} + f\rho_b K_d}{\theta + \rho_b K_d}. \quad (6.64)$$

If there is no adsorption ($K_d = 0$) or if it is assumed that adsorption sites are equally distributed in the mobile and immobile regions (commonly done) then $f = \theta_{mo}/\theta$ and Equation 6.64 reduces to:

$$\beta = \frac{\theta_{mo}}{\theta}. \quad (6.65)$$

Hence, β is simply the fraction of the total water content that is mobile and ranges between zero and one. The second dimensionless parameter includes the exchange rate coefficient and is defined as:

$$\omega = \frac{\alpha_s L}{J_w}. \quad (6.66)$$

The relationship between D_{mo} in Equation 6.62 and D_e is

$$D_e = \frac{\theta_{mo}}{\theta} D_{mo}. \quad (6.67)$$

A mobile water velocity can be defined as $v_{mo} = J_w/\theta_{mo}$ [LT^{-1}]. Similarly, the relationship between v_{mo} and v is

$$v = \frac{\theta_{mo}}{\theta}\, v_{mo}. \qquad (6.68)$$

Hence, with the two-region physical nonequilibrium model there are five transport parameters: v, D_e, R, β, and ω. Van Genuchten and Wierenga (1976) used the Laplace transform method to solve the set of equations (Equations 6.62 and 6.63). The solutions are not shown here, but they are used to plot BTCs for a pulse input to show the effect of β and ω using the CXTFIT model (Toride et al. 1999) in the STANMOD software package (further discussed in Section 6.4). In Figure 6.23, BTCs at a depth of 30 cm are shown for a pulse input with a duration (t_0) of 5 h, $v = 1.5$ cm h^{-1}, $D_e = 1.25$ cm^2 h^{-1}, and $R = 1$. An intermediate value for $\omega = 1$ was used for all curves. A value of $\beta = 1$ corresponds to physical equilibrium. CXTFIT does not allow users to simultaneously run equilibrium and nonequilibrium problems, so $\beta = 0.99$ was used to represent near equilibrium. This value produces a pulse centered at about one pore volume ($t_b = L/v = 20$ h). This is the same as the

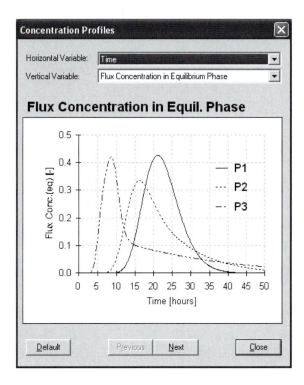

FIGURE 6.23 Effect of varying β on the BTCs at a depth of 30 cm for a pulse input using the nonequilibrium ADE. All parameter sets used $v = 1.5$ cm h^{-1}, $D_e = 1.25$ cm^2 h^{-1}, $R = 1$, and $\omega = 1$. For parameter set P1, $\beta = 0.99$; for P2, $\beta = 0.7$; and for P3, $\beta = 0.3$.

conventional ADE. But as the mobile fraction (β) decreases to 0.7 and 0.3, physical nonequilibrium prevails and the BTCs become more skewed. The early breakthrough and tailing that is typical in well-structured soils can be seen. A small value of β would simulate a soil with macropores where there is a lack of physical equilibrium between the macropores and the matrix. That is, the concentrations of solute in the macropores (c_{mo}) are very different from the concentrations in the matrix (c_{im}). It's clear from Figure 6.23 that with macropores, most of the solute may come through before 1 pore volume (see the $\beta = 0.3$ curve). This example is discussed in detail in Section 6.4.1.

The effect of varying ω is shown in Figure 6.24. The same values for v, D_e, R, and t_0 were used as in Figure 6.23. An intermediate value of $\beta = 0.7$ was used for all curves. When ω is large (5), the BTC is nearly symmetrical at about one pore volume, despite a significant fraction of immobile water. This would correspond to physical equilibrium and be well described using the conventional ADE. Equilibrium conditions prevail ($c_{mo} \approx c_{im}$) because there is rapid exchange between the mobile and immobile regions due to a large value for α_s and, consequently, a large value for ω. This might occur where a soil is well aggregated, but the aggregates are small so that chemical diffusion quickly establishes equilibrium between concentrations inside the aggregates (immobile region) and concentrations in the spaces between

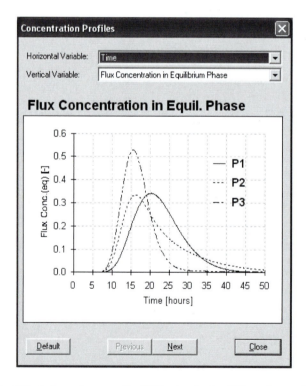

FIGURE 6.24 Effect of varying ω on the BTCs at a depth of 30 cm for a pulse input using the nonequilibrium ADE. All parameter sets used $v = 1.5$ cm h^{-1}, $D_e = 1.25$ cm^2 h^{-1}, $R = 1$, and $\beta = 0.7$. For parameter set P1, $\omega = 5$; for P2, $\omega = 1$; and for P3, $\omega = 0.1$.

the aggregates (mobile region). At an intermediate value for ω (1), the BTC becomes skewed with early breakthrough and tailing. This is because exchange between the mobile and immobile regions is slow and concentrations are very different in the two regions. It's interesting to note what happens at small values of ω. In this case, the BTC again becomes symmetrical, but it comes through earlier than the equilibrium curve ($\omega = 5$). This is because when ω is small, exchange with the immobile region is so slow that the immobile region does not participate in the transport. The transport volume is simply the mobile region, and within the mobile region, equilibrium prevails.

6.3.8.2 Chemical Nonequilibrium

Another type of nonequilibrium is chemical nonequilibrium. So far, it has been assumed that adsorption was instantaneous (linear or nonlinear). For large organic molecules, adsorption may take time. The simplest chemical nonequilibrium model assumes that adsorption is kinetic and can be described by the equation:

$$\frac{\partial s}{\partial t} = \alpha_k (K_d c - s), \tag{6.69}$$

where $K_d c$ is the final adsorbed concentration at equilibrium and α_k is a *kinetic rate constant* [T^{-1}]. This is known as a *one-site* chemical nonequilibrium model. A more complex model assumes that there are two types of adsorption sites and this is known as a *two-site model* (as opposed to the physical nonequilibrium *two-region model*). For *type-1* sites, it is assumed that adsorption is instantaneous and linear (Equation 6.36 applies). For *type-2* sites, adsorption is kinetic and can be described by Equation 6.69. In the two-site model, β and ω are defined differently from the physical nonequilibrium model. The definition for β is

$$\beta = \frac{\theta + f \rho_b K_d}{\theta + \rho_b K_d}, \tag{6.70}$$

which is the same as Equation 6.64 except that θ_{mo} does not appear. In this case, f is the fraction of the total number of adsorption sites that are type-1 sites. Chemical equilibrium would be indicated by a value of β (and f) close to one. Smaller values of β (between zero and one) would indicate more type-2 sites and chemical nonequilibrium. The definition for ω with the two-site chemical nonequilibrium model is

$$\omega = \frac{\alpha_k (1 - \beta) R L}{v}. \tag{6.71}$$

Large values for ω are caused by large values for α_k. This indicates rapid adsorption kinetics (Equation 6.69) and corresponds to chemical equilibrium. Small values of ω indicate slow kinetics and chemical nonequilibrium.

Nkedi-Kizza et al. (1984) showed that the traditionally used physical and chemical nonequilibrium transport models (i.e., the dual-porosity model with mobile and immobile flow regions and the two-site sorption model, respectively) are mathematically identical when applied to solute BTCs. Hence, by analyzing the BTC of a single chemical, it is often not possible to discriminate which process (physical or chemical) is responsible for the observed nonequilibrium process. In that case, one needs additional information, such as BTCs measured simultaneously for a tracer and a reactive chemical, to better analyze the underlying transport processes (see Section 6.4.2). Since the overall equation is the same, the effect of varying β and ω for chemical nonequilibrium is exactly the same as that shown in Figures 6.23 and 6.24.

6.3.9 STOCHASTIC ADVECTION DISPERSION EQUATION

Transport models can be classified as *deterministic* or *stochastic* (Addiscott and Wagenett 1985). Various definitions have been given for these terms. A deterministic model always produces the same results, given a set of inputs. A stochastic model *may* produce different results under these circumstances. One way to distinguish between the models is by looking at the input parameters. Deterministic models use a single (or mean) value for each parameter, such as saturated hydraulic conductivity at a given point. Stochastic models use a distribution of values for *some* parameters (usually specified as a mean and a variance, assuming a population distribution—normal or lognormal, for example). A Monte Carlo approach would be a stochastic model, but there are simpler, less time consuming ways to do this (van Genuchten and Jury 1987). Stochastic models are commonly used in soil physics to simulate field-scale solute transport and account for spatial variability.

The ADE assumes that spreading caused by hydrodynamic dispersion can be described as a Fick's law process (see Equation 6.8). This assumes a high degree of mixing occurs between flow paths such that when solute molecules traveling in fast pores move ahead of solutes in slow pores, the pores interconnect frequently and the molecules in the fast pores diffuse into the slow pores. This slows solute movement in the fast pores and limits the spread (dispersion) of the solute front or pulse.

When prediction of *field-scale* average values of $c(z,t)$ are attempted, areas of the field may differ greatly in v and mixing may not be complete until the solute moves much deeper than the depths being simulated. For example, see the discussion in Chapter 5 on preferential flow caused by field-scale variability (Section 5.8.1) and Figures 5.39 and 5.40. This is indicated by a scale-dependent λ (see the discussion in Section 6.2.2.3).

Under these conditions, a *stream-tube model* may be appropriate. It is one type of stochastic approach. The field is conceptualized as a system of parallel tubes (Figure 6.25). The *local-scale* solute transport in each tube is described by a transport model such as piston flow or the ADE. Some transport variables vary among the tubes based on their mean and variance (they are stochastic). The other transport variables are the same in all stream tubes (they are deterministic). There is no connection (mixing) between the tubes. The field-scale concentration $\hat{c}(z,t)$ or *ensemble average* $<c>$ [ML^{-3}] is the average of all the concentrations in the tubes over the area

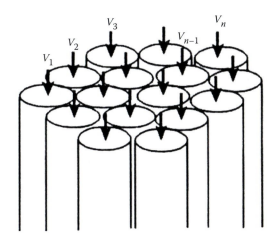

FIGURE 6.25 Stream-tube model with stochastic mean pore water velocity v. (Adapted from Toride, N., et al., *Research Report No. 137*, U.S. Salinity Laboratory, USDA, ARS, Riverside, CA, 1999.)

A of the field. It is also the average of the stream-tube concentrations as the number of stream tubes becomes very large.

Toride and Leij (1996) developed a stream-tube model where the model describing transport in each tube was based on the ADE. If only v is considered a stochastic variable, then the field-scale mean concentration is

$$\hat{c}(z,t) = \int_0^\infty c(z,t;v)f(v)\,dv. \tag{6.72}$$

This shows that the local-scale concentration is summed and weighted by the frequency distribution of v or its *probability density function* (pdf) $f(v)$. The most familiar pdf is the *normal* or bell-shaped distribution for an arbitrary variable Z (Jury and Horton 2004):

$$f(Z) = \frac{1}{\sigma\sqrt{2\pi}}\exp\left[\frac{(Z-\mu)^2}{2\sigma^2}\right], \tag{6.73}$$

where μ is the mean and σ is the standard deviation of the distribution. The *expected value* of Z or ensemble average of Z is defined as:

$$E(Z) = \int_{-\infty}^{+\infty} Z f(Z)\,dZ. \tag{6.74}$$

Essentially, this is the sum of possible values of the variable multiplied by their probabilities. It can be shown that the expected value of Z is μ.

A less familiar pdf is the *lognormal* distribution:

$$f(Z) = \frac{1}{\sigma Z\sqrt{2\pi}} \exp\left[-\frac{\left(\ln[Z]-\mu\right)^2}{2\sigma^2}\right],\qquad(6.75)$$

where μ and σ are the mean and standard deviation of the log transformed Z. In the case of the lognormal distribution, the expected value is a function of μ and σ (see Section 6.8.2):

$$E(Z) = \exp\left(\mu + \frac{\sigma^2}{2}\right).\qquad(6.76)$$

In Equation 6.72, it is usually assumed that v is lognormally distributed (as most research shows):

$$f(v) = \frac{1}{\sigma_v v\sqrt{2\pi}} \exp\left(-\frac{\left[\ln(v)-\mu_v\right]^2}{2\sigma_v^2}\right),\qquad(6.77)$$

where μ_v and σ_v are the mean and standard deviation of the log transformed v.

Three types of field-scale concentrations can be defined: (1) the field-scale flux concentration ($<c_f>$), (2) the field-scale resident concentration ($<c_r>$), and (3) the field-scale flux-averaged concentration ($<vc_r>/<v>$). Note that because v is a stochastic variable, $<vc_r>/<v> \neq <c_f>$ (Leij and van Genuchten 2002).

In Figure 6.26, the field-scale resident concentration of a nonadsorbed solute predicted using the CXTFIT model (Toride et al. 1999) in the STANMOD software package by the stochastic stream-tube model is compared to the deterministic ADE. Equations 6.72 and 6.77 were used for both models. Since the CXTFIT model cannot run both deterministic and stochastic models simultaneously, to simulate a deterministic ADE model (parameter set P1), σ_v was given a very low value of 0.001 cm d^{-1} ($\sigma_v = 0$ cannot be used with Equation 6.77). For the stochastic model (parameter set P2), σ_v was assigned a value of 0.5 cm d^{-1}. A value of $D_e = 20$ cm^2 d^{-1} was used in both models. The graphs are shown for 3 days after applying a pulse input of 2-day duration. Both models have peak concentrations at about the same depth, but the stream-tube model predicts much deeper movement of the solute front and a more skewed distribution. Both models recognize that velocities vary. The ADE assumes that all the variation occurs at the local scale (within a stream tube). The variation in pore water velocities is quantified in λ, which in this case is $D_e/v = 12.5$ cm (assuming Equation 6.13 applies), a relatively large value. The ADE ignores variation in velocities between stream tubes. The stream-tube model recognizes variation between stream tubes and

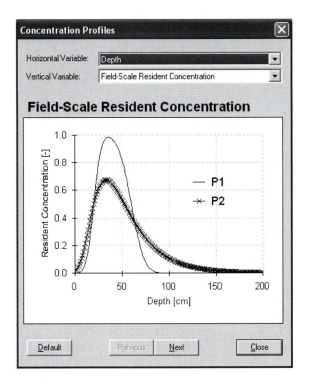

FIGURE 6.26 The field-scale resident concentration predicted by the deterministic ADE (parameter set P1) and the stream-tube model with stochastic v (parameter set P2). The parameter values for P1 are: $v = 20$ cm d^{-1}, $\sigma_v = 0.001$ cm d^{-1}, and $D_e = 20$ cm^2 d^{-1}. The parameter values for P2 are the same except $\sigma_v = 0.5$ cm d^{-1}. Graphs are for 3 days after applying a pulse input of 2-day duration.

within stream tubes. It assumes that areas with different local velocities are so far apart that there is no mixing between stream tubes. Because of the additional source of variation and the lack of mixing, the solute front moves much deeper with the stream-tube model.

If transport variables v and K_d are considered stochastic, then the field-scale mean concentration is (Toride and Leij 1996):

$$\hat{c}(z,t) = \int_0^\infty \int_0^\infty c(z,t;v,K_d)f(v,K_d)\,dv\,dK_d, \tag{6.78}$$

where the bivariate lognormal joint probability density function $f(v,K_d)$ for v and K_d is given by:

$$f(v,K_d) = \frac{1}{2\pi\sigma_v\sigma_{K_d}vK_d\sqrt{1-\rho_{vK_d}^2}}\exp\left(-\frac{Y_v^2 - 2\rho_{vK_d}Y_vY_{K_d} + Y_{K_d}^2}{2\left(1-\rho_{vK_d}^2\right)}\right), \tag{6.79}$$

with

$$\rho_{vK_d} = \int_0^\infty \int_0^\infty Y_v Y_{K_d} f(v, K_d)\, dv\, dK_d,$$

$$Y_v = \frac{\ln(v) - \mu_v}{\sigma_v}, \quad Y_{K_d} = \frac{\ln(K_d) - \mu_{K_d}}{\sigma_{K_d}},$$

where μ_{K_d} and σ_{K_d} are the mean and standard deviations of the log transformed K_d, and ρ_{vK_d} is the correlation coefficient between Y_v and Y_{K_d}. For positive ρ_{vK_d}, K_d tends to increase with v, while K_d tends to decrease with v for negative ρ_{vK_d}. A negative correlation might apply to a field with variable texture. The more sandy areas might have a higher v (due to a higher K_s) and (due to the lower clay content) a lower K_d (Leij and van Genuchten 2002).

The mass of solute in each tube with a pulse input can be considered constant among tubes or proportional to the local-scale velocity. Since some mixing is likely locally, it may be more realistic to consider a constant mass among tubes.

6.4   STANMOD AND CXTFIT

A large number of analytical models for one-, two-, and three-dimensional solute transport problems have been incorporated into the public domain software package STANMOD (STudio of ANalytical MODels) (Šimůnek et al. 1999b). It includes not only programs for equilibrium advective-dispersive transport (i.e., the CFITM of van Genuchten (1980b) for one-dimensional transport and 3DADE (Leij and Bradford 1994) for three-dimensional problems), but also programs for more complex problems. For example, STANMOD also incorporates the CFITIM and N3DADE (Leij and Toride 1997) programs for nonequilibrium transport (i.e., the two-region mobile-immobile model for physical nonequilibrium and the two-site sorption model for chemical nonequilibrium) in one and multiple dimensions, respectively. One of the models in STANMOD is the widely used CXTFIT code (Toride et al. 1999). In this section, the focus is on CXTFIT. The model can be run in direct mode or inverse mode.

6.4.1   DIRECT MODE

Here we show an example of using the nonequilibrium ADE model in direct mode. This is STANMOD Simulation 6.1 – Effect of Beta. When STANMOD is first opened, a New Project window will appear where a name and description can be assigned to the project. The next window to appear is the Program Selection window (Figure 6.27) where the various analytical models available in STANMOD are shown. In this example, the CXTFIT code is selected.

To set up a problem in CXTFIT, one must go through each of the windows listed in the Pre-processing panel on the left side of Figure 6.28. The first window to appear after the Program Selection window (Figure 6.27) is the Type of Problem window

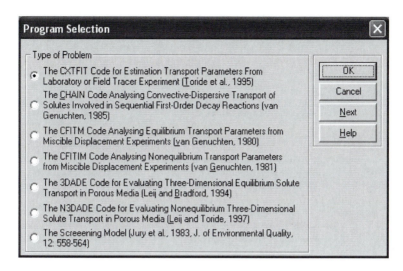

FIGURE 6.27   Program Selection window in STANMOD.

(Figure 6.29). Here the direct problem is chosen and three parameter sets are specified in order that the effect of varying the nonequilibrium ADE parameter β can be shown. Next is the Type of Model window (Figure 6.30). The deterministic nonequilibrium CDE is selected.

Next is the Input and Output Data Code window (Figure 6.31). Time and position are chosen to be dimensional. In the next window, length units are set to centimeters, time units are set to hours, and concentration units are set dimensionless (Figure 6.32). In this example, solute concentrations will be simulated at a depth of 30 cm as might be observed in a 30 cm long column BTC experiment. So flux concentrations are chosen and a characteristic length (L, see Equation 6.23) of 30 cm is set (Figure 6.33). In the Transport and Reaction Parameters window (Figure 6.34), there are three rows of parameter values since three parameter sets were specified in Figure 6.30. For all

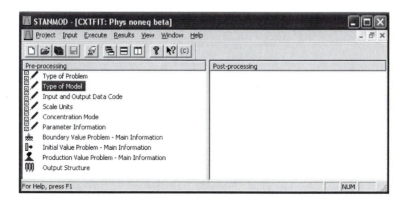

FIGURE 6.28   Pre-processing panel (left side) of CXTFIT.

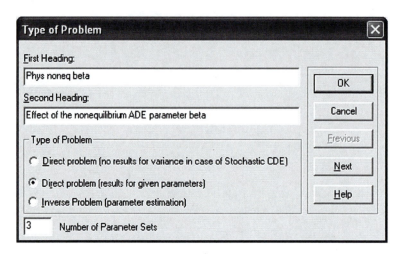

FIGURE 6.29 Type of Problem window in CXTFIT.

three parameter sets, $v = 1.5$ cm d^{-1} (the units are determined by the selections in Figure 6.32), $D_e = 1.25$ cm^2 d^{-1}, and $R = 1$. Three separate values for β are chosen: 0.99, 0.7, and 0.3. For all three parameter sets, an intermediate value of $\omega = 1$ is used. A value of zero is specified for the first-order transformation parameter in the mobile (μ_1) and immobile (μ_2) phases for physical nonequilibrium (or type-1 and type-2 sites for chemical nonequilibrium).

In the Boundary Value Problem window, a pulse input of unit concentration with an application time of 5 h is specified (Figure 6.35). In the Initial Value Problem window (not shown), zero initial concentration is chosen. In the Production Value Problem window (not shown), zero production is specified. The settings in the Output Structure window (Figure 6.36) control graphing of the results. At the bottom, concentration vs. time is chosen. For Space Discretization, one output position at a depth of 30 cm is indicated. For Time Discretization, a time increment of 0.1 h is chosen (to

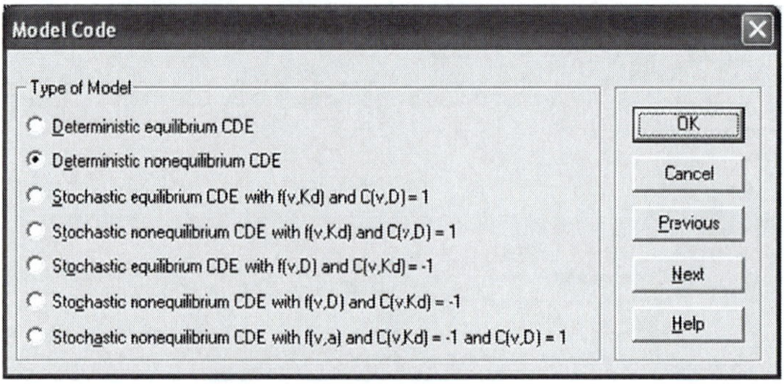

FIGURE 6.30 Type of Model window in CXTFIT.

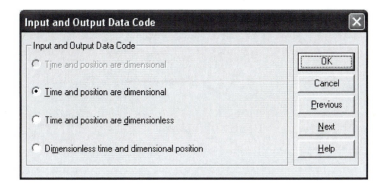

FIGURE 6.31   Input and Output Data Code window in CXTFIT.

get a very smooth curve) and the curve is extended for 500 increments (a final time of 50 h) with a start time of zero.

At this point, the output structure is defined, all the input parameters have been set, and the input file to the program code is complete. Once the program has run (the DOS window has appeared), two new entries will appear in the Post-processing panel (right side of Figure 6.37).

Clicking on the Graphical Output (Line Graphs) will produce the display shown earlier in Figure 6.23.

6.4.2   INVERSE MODE AND PARAMETER OPTIMIZATION

Similar to the discussion in Chapter 5 on inverse solutions and parameter optimization (Section 5.10), inverse methods can be used to estimate solute transport parameters from data sets such as column BTC experiments.

An example program from a boron BTC experiment of van Genuchten (1974) will be used to show CXTFIT in the inverse mode. This is STANMOD Simulation

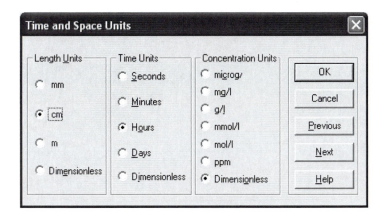

FIGURE 6.32   Time and Space Units window in CXTFIT.

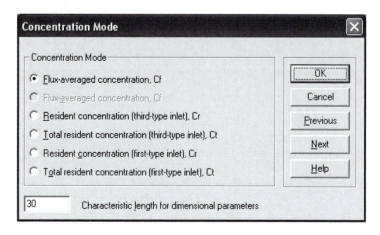

FIGURE 6.33 Concentration Mode window in CXTFIT.

6.2 – Parameter Optimization. It is also in the STANMOD project directory under the title FIG79B. This can be found by clicking on the Project Manager icon on the toolbar in STANMOD and selecting the Inverse Workspace directory. With the inverse mode, there are additional preprocessing steps that don't appear in the direct mode (Figure 6.38).

In the Type of Problem window, inverse problem is selected (Figure 6.39). In the Model Code window, the deterministic nonequilibrium model is chosen. Time and position are dimensionless. Flux concentrations and a characteristic length of 30 are chosen, since BTCs were measured using 30 cm long columns. In the inverse mode, one can specify constraints on parameter estimations, include the total mass applied as part of the optimization process, and specify the maximum number of iterations to be used in finding optimal parameter (and mass) values (Figure 6.40). In this case, no constraints are used, the total mass is not included in the estimation process, and the maximum number of iterations is set at 20. The two-region physical nonequilibrium model is chosen with independent degradation rates for the solution and adsorbed phase (although degradation was assumed to be zero in this experiment with boron, thus this choice is irrelevant) (Figure 6.41).

Transport and Reaction Parameters

	v	D	R	Beta	Omega	Mu1	Mu2
Parameter Name	v	D	R	Beta	Omega	Mu_1	Mu_2
Parameter Set 1	1.5	1.25	1	0.99	1	0	0
Parameter Set 2	1.5	1.25	1	0.7	1	0	0
Parameter Set 3	1.5	1.25	1	0.3	1	0	0

OK   Cancel   Previous ...   Next ...   Help

FIGURE 6.34 Transport and Reaction Parameters window in CXTFIT.

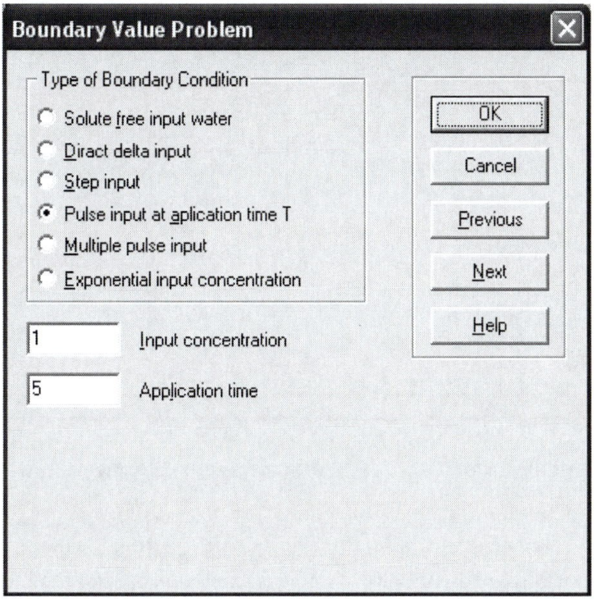

FIGURE 6.35 Boundary Value Problem window in CXTFIT.

The Transport and Reaction Parameters window is shown in Figure 6.42. In the BTC experiment, the column volumetric water content was measured and found to be 0.45 cm^3 cm^{-3}. Using the steady flow rate through the column and θ, v was calculated as 38.5 cm d^{-1}. From an earlier BTC experiment on the same column using tritium, D_e was found through optimization to be 15.5 cm^2 d^{-1}. It was assumed that

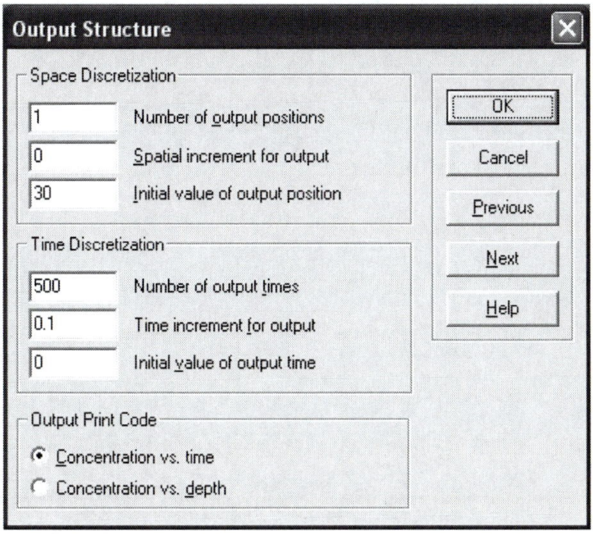

FIGURE 6.36 Output Structure window in CXTFIT.

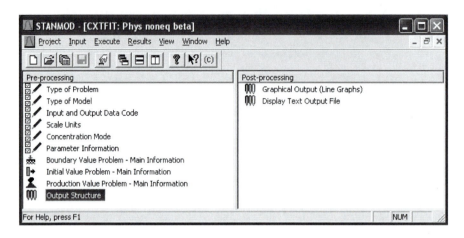

FIGURE 6.37   Main window of CXTFIT with the post-processing panel (right side).

D_e would be the same for tritium and boron (v was nearly the same for both experiments). It's important to note that these input parameters are not the mean pore water velocity and dispersion coefficient for the mobile phase (v_{mo} and D_{mo}), but rather v and D_e. The relationships between these parameters are given in Equations 6.67 and 6.68. Using the water content and bulk density ($\rho_b = 1.22$ g cm^{-3}) of the column and K_d for boron, a retardation coefficient of $R = 3.9$ was calculated. The two physical nonequilibrium parameters, β and ω, are identified as fitted parameters by checking the box below the values for initial estimates of the parameters (0.5 and 0.2). Transformation rate constants are specified as zero.

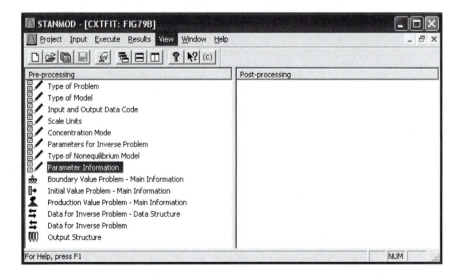

FIGURE 6.38   Main window with the pre-processing panel (left side) in the inverse mode of CXTFIT.

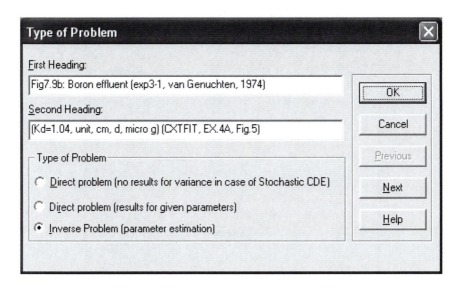

FIGURE 6.39 Type of Problem window for the project FIG79B (selected from the STANMOD Inverse workspace).

In the Boundary Value Problem window (not shown), a pulse input is selected. The pulse was applied for 6.494 pore volumes, so $T = 6.494$ is entered as the "Application time" (time is dimensionless in this project). An input concentration of $c/c_0 = 1$ is used, since concentrations are also dimensionless. In the Initial Value Problem window (not shown), an initial concentration of zero is specified. In the Production Value Problem window (not shown), zero production is specified.

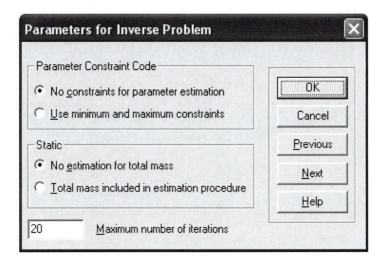

FIGURE 6.40 Parameters for Inverse Problem window in project FIG79B.

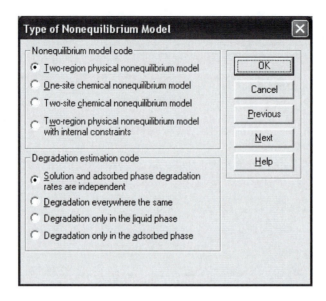

FIGURE 6.41 Type of Nonequilibrium Model window in project FIG79B.

In the Inverse Data Structure window (Figure 6.43), time and concentration for fixed depth (BTC) is chosen. A value of 30 is entered for the number of data points as there were 30 measurements of concentration for the boron BTC experiment. A dimensionless position for the depth at which the BTC was measured of $Z = z/L = 1$ is used. With the characteristic length, $L = 30$ cm, this corresponded to the length of the column.

In the Data for Inverse Problem window, the 30 measured data pairs from the BTC are entered. Time is entered as dimensionless values (pore volumes = T) as are concentrations (c/c_0). In Figure 6.44, the first 11 values are shown.

The settings in the Output Structure window are shown in Figure 6.45. The output position is specified in dimensionless units ($Z = 1$). A dimensionless time increment

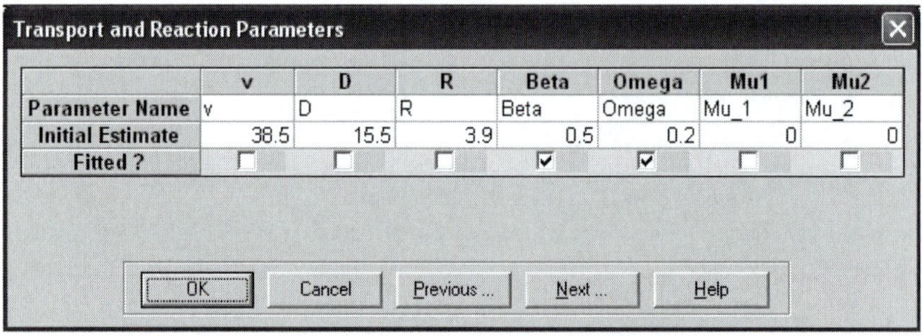

Transport and Reaction Parameters

	v	D	R	Beta	Omega	Mu1	Mu2
Parameter Name	v	D	R	Beta	Omega	Mu_1	Mu_2
Initial Estimate	38.5	15.5	3.9	0.5	0.2	0	0
Fitted ?	☐	☐	☐	☑	☑	☐	☐

FIGURE 6.42 Transport and Reaction Parameters window in project FIG79B.

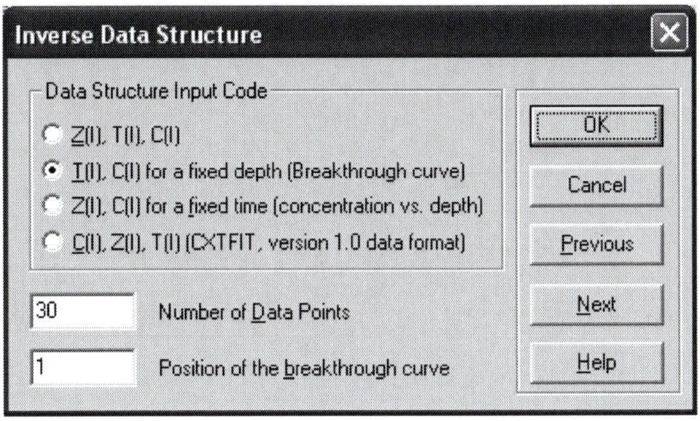

FIGURE 6.43 Inverse Data Structure window in project FIG79B.

of 0.1 pore volumes is specified with 201 increments and a start time of zero. This will extend the display of the BTC from 0 to 20 pore volumes.

When the program is executed, the DOS window display shown in Figure 6.46 appears. It shows that optimal values were found for the parameters β and ω after five iterations. The error sum of squares between the observed and predicted dimensionless concentrations was reduced from 1.5 to 0.08 in this process.

The graphical output from the Post-processing panel is shown in Figure 6.47. The optimized physical nonequilibrium ADE (solid line) is a very good fit to the measured data (circles). Because boron was applied for an extended time (more than six pore volumes), peak concentrations are close to $c/c_0 = 1$. The BTC is skewed, implying that nonequilibrium conditions occurred.

The Display Text Output tab entry in the Post-processing panel can be selected to view additional information on the optimization results. Scrolling down in this window, the display shown in Figure 6.48 can be seen. The r^2 for the fit of the model

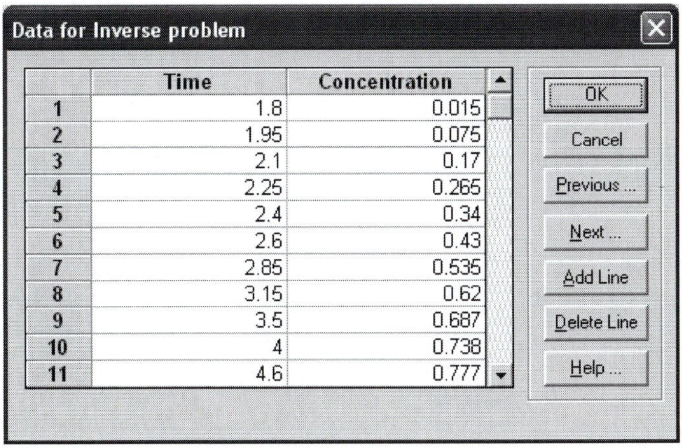

FIGURE 6.44 Data for Inverse Problem window in project FIG79B.

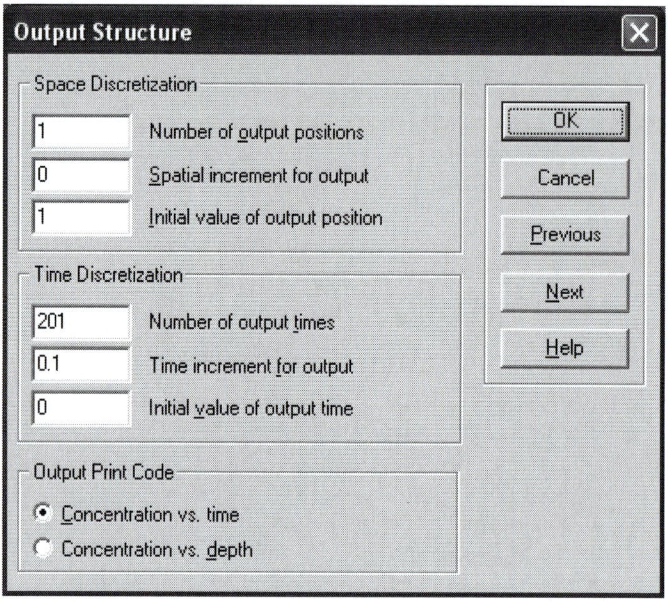

FIGURE 6.45   Output Structure window in project FIG79B.

to the data is very good (0.97). The nonequilibrium parameters are inversely correlated ($r = -0.756$), indicating that if one of the parameters (β) is reduced and the other (ω) increased by a corresponding amount, the results will be similar. The optimized values are $\beta = 0.578$ and $\omega = 0.70$. Since $\beta < 1$, nonequilibrium conditions were prevalent. Using the value for β (0.822) obtained with the tritium BTC, θ_{mo} can be calculated using Equation 6.65. Then (assuming that θ_{mo} was the same for tritium and boron) using Equation 6.64, the fraction of adsorption sites in the mobile region

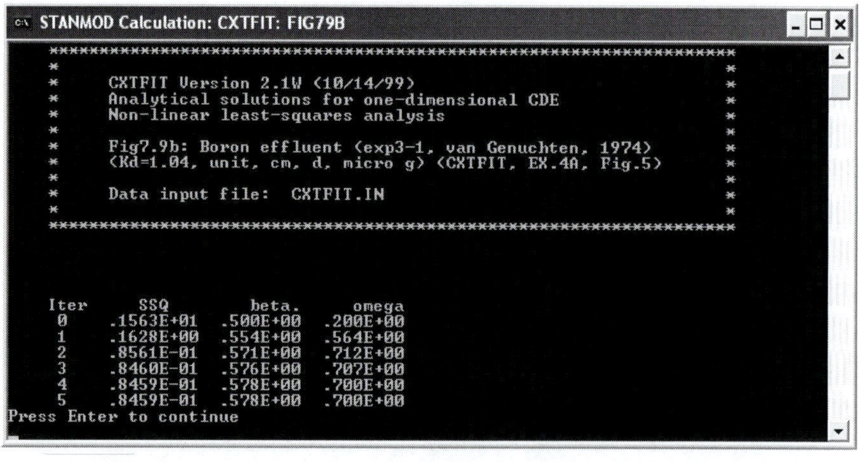

FIGURE 6.46   DOS window display after executing project FIG79B.

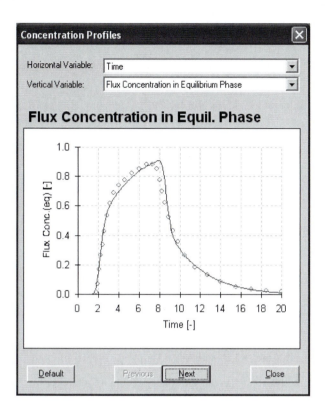

FIGURE 6.47 Predicted (line) and observed (circles) flux concentrations in project FIG79B.

f can be calculated and it is 0.49. Using Equation 6.66, α_s can also be calculated and it is 0.40 d^{-1} (Toride et al. 1999).

The solution is unique in that using different initial estimates of β or ω does not change the fitted values. This is not the case if D_e is fitted (instead of using the value obtained from a fit of the tritium BTC on the same column) as well as β and ω. Nonequilibrium transport of an adsorbed solute is best studied with experiments that measure BTCs of both the adsorbed and a nonadsorbed solute on the same column where some of the transport parameters can be determined using the nonadsorbed solute (Toride et al. 1999).

6.5 NUMERICAL APPROACHES FOR SOLUTE TRANSPORT

Many solute transport problems cannot be solved using analytical approaches. These include transport of a solute that experiences nonlinear adsorption, transport under variable water contents and fluxes (transient flow conditions), and transport in irregularly shaped domains. However, numerical methods can be used to solve the ADE under these circumstances. Common numerical approaches are the finite difference approach and the finite element approach.

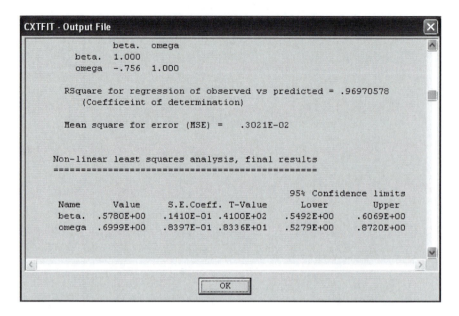

```
CXTFIT - Output File                                                    ☒

              beta.  omega
       beta.  1.000
       omega  -.756  1.000

     RSquare for regression of observed vs predicted = .96970578
        (Coefficeint of determination)

     Mean square for error (MSE) =     .3021E-02


     Non-linear least squares analysis, final results
     =================================================

                                                95% Confidence limits
       Name      Value     S.E.Coeff.  T-Value    Lower       Upper
       beta.   .5780E+00   .1410E-01  .4100E+02  .5492E+00   .6069E+00
       omega   .6999E+00   .8397E-01  .8336E+01  .5279E+00   .8720E+00
```

```
                         ┌──────OK──────┐
```

FIGURE 6.48   Output File window in project FIG79B.

6.5.1 FINITE DIFFERENCE APPROACH

In this section, a finite difference numerical approach to solving the ADE for transport in one dimension is briefly described. As with heat flow and water flow, the finite difference approach is used to develop a set of algebraic equations from the ADE in its various forms. For the sake of simplicity, the form for equilibrium transport of a nonadsorbed solute under steady flow (e.g., Equation 6.17) is used. For the same reason, the boundary conditions are both type-1 (known concentrations) and the initial conditions are concentrations of zero (Equation 6.19).

The soil profile is discretized into N depth intervals, Δz, such that $z = i\Delta z$ and $i = 0...N$. Time is discretized into intervals Δt such that $t = j\Delta t$. The depth and time intervals can vary, but they are kept constant in this case. Then the derivatives in Equation 6.17 are written as discrete differences divided by the appropriate interval:

$$\frac{c_i^{j+1}-c_i^j}{\Delta t} = D_e\left(\frac{\left.\frac{\partial c}{\partial z}\right|_{i+1/2} - \left.\frac{\partial c}{\partial z}\right|_{i-1/2}}{\Delta z}\right) - v\left.\frac{\partial c}{\partial z}\right|_i$$

$$= D_e\left(\frac{\frac{c_{i+1}-c_i}{\Delta z} - \frac{c_i-c_{i-1}}{\Delta z}}{\Delta z}\right) - v\left(\frac{c_{i+1}-c_{i-1}}{2\Delta z}\right) \qquad (6.80)$$

$$= D_e\left(\frac{c_{i+1}-2c_i+c_{i-1}}{\Delta z^2}\right) - v\left(\frac{c_{i+1}-c_{i-1}}{2\Delta z}\right).$$

Subscripts denote depth nodes and superscripts denote time steps. If all terms on the right side of Equation 6.80 are evaluated at the known (j) time step, then there is only one unknown in the equation and the equation can be solved explicitly for c_i^{j+1} at each node ($i = 1...N-1$, with c_0^{j+1} and c_N^{j+1} known from the boundary conditions):

$$c_i^{j+1} = c_{i-1}^{j}\left(\frac{\Delta t\,v}{2\Delta z} + \frac{\Delta t\,D_e}{\Delta z^2}\right) + c_i^{j}\left(1 - \frac{2\Delta t\,D_e}{\Delta z^2}\right) + c_{i+1}^{j}\left(-\frac{\Delta t\,v}{2\Delta z} + \frac{\Delta t\,D_e}{\Delta z^2}\right) \quad (6.81)$$

This is the *explicit* finite difference method and has the disadvantage of being unstable for large time steps.

If all terms on the right side of Equation 6.80 are evaluated at the unknown ($j + 1$) time step, then there are three unknowns: c_{i-1}^{j+1}, c_i^{j+1}, and c_{i+1}^{j+1}. Collecting coefficients of these terms on the left side gives:

$$bc_{i-1}^{j+1} + dc_i^{j+1} + ec_{i+1}^{j+1} = c_i^{j}, \quad (6.82)$$

and:

$$b = -\frac{\Delta t\,v}{2\,\Delta z} - \frac{\Delta t\,D_e}{\Delta z^2}$$

$$d = 1 + \frac{2\Delta t\,D_e}{\Delta z^2} \quad (6.83)$$

$$e = \frac{\Delta t\,v}{2\,\Delta z} - \frac{\Delta t\,D_e}{\Delta z^2}$$

Equation 6.82 can be written for each node so there are $N + 1$ equations with $N-1$ unknowns (c_i^{j+1}) to be solved simultaneously (the concentrations at the boundaries are known). This system of equations can be written as a combination of a matrix and two vectors:

$$
\begin{bmatrix}
1 & 0 & 0 & & & & & 0 \\
b & d & e & & & & & 0 \\
0 & b & d & e & & & & 0 \\
 & & . & . & . & & & \\
0 & & & b & d & e & & 0 \\
0 & & & & b & d & e \\
0 & & & & & 0 & 0 & 1
\end{bmatrix}
\cdot
\begin{bmatrix}
c_0^{j+1} \\
c_1^{j+1} \\
c_2^{j+1} \\
. \\
c_{N-2}^{j+1} \\
c_{N-1}^{j+1} \\
c_N^{j+1}
\end{bmatrix}
=
\begin{bmatrix}
c_0^{j} \\
c_1^{j} \\
c_2^{j} \\
. \\
c_{N-2}^{j} \\
c_{N-1}^{j} \\
c_N^{j}
\end{bmatrix}
\quad (6.84)
$$

This can be written in matrix notation as:

$$[A] \cdot \{c\}^{j+1} = \{c\}^j, \tag{6.85}$$

where $[A]$ is the square symmetrical tridiagonal matrix in Equation 6.84 with the elements b, d, and e along the subdiagonal, diagonal, and superdiagonal, $\{c\}^{j+1}$ is the vector of the $N-1$ unknown values of c and the two known values at the boundaries, and $\{c\}^j$ is a similar vector containing the values of c at the known time step. Computer algorithms are used to invert $[A]$ and solve Equation 6.85 for $\{c\}^{j+1}$ at a given time step.

The system shown so far (specifying all terms on the right side of Equation 6.80 at the $j + 1$ time step) is a *fully implicit* finite difference method. In the *Crank-Nicolson* approach, the terms on the right side of Equation 6.80 are evaluated at the $j + 1/2$ time step (by taking an average of the right side at the j and $j + 1$ time steps).

It can be shown (Section 6.8.3) that the advection term (which is not present in the heat flow equation if we only consider conduction) introduces additional error into the finite difference approximation. This is referred to as "numerical dispersion" and methods have been developed to correct for this error (see Section 6.8.3).

6.5.2 FINITE ELEMENT APPROACH

Finite element methods have the advantage that it is easier to discretize complex two- and three-dimensional problems (Figure 6.49). The finite element approach assumes that concentration can be approximated by a finite series $c'(x,t)$ of the form:

$$c'(x,t) = \sum_{m=1}^{N} \phi_m(x) c_m(t), \tag{6.86}$$

where ϕ_m are linear basis functions, c_m are the unknown time-dependent coefficients that represent solutions to the ADE at the finite element nodal points, and N is the total number of nodal points.

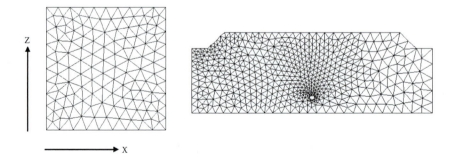

FIGURE 6.49 Examples of the triangular finite element grids for regular (left) and irregular (right) two-dimensional transport domains. (Šimůnek and van Genuchten, 2006.)

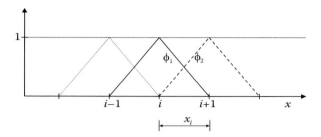

FIGURE 6.50 One-dimensional linear basis functions. ϕ_i is the basis function for the node i. (Šimůnek and van Genuchten, 2006.)

For example, for the one-dimensional finite element $x_i < x < x_{i+1}$, linear basis functions have the following triangular form (Figure 6.50):

$$\phi_1 = 1 - \frac{x - x_i}{\Delta x}$$

$$\phi_2 = \frac{x - x_i}{\Delta x},$$

(6.87)

where Δx $(= x_{i+1} - x_i)$ is the size of the finite element [L], that is, the distance between two neighboring nodal points. The product of the triangular basis elements and the coefficients approximate $c'(x,t)$ over the interval (Figure 6.51).

Application of the Galerkin finite element method leads to a system of N time-dependent equations. The system of equations can be written in matrix form and then solved, much like the finite difference approach.

6.5.3 Stability and Oscillations

Numerical solutions of the transport equation often exhibit oscillatory behavior and/or excessive numerical dispersion near relatively sharp concentration fronts (Huyakorn

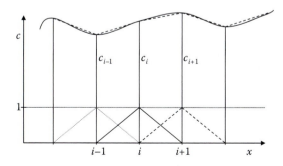

FIGURE 6.51 The sum of the product of local basis functions and coefficient concentrations provide an estimate of $c(x,t)$ for a particular time. (Šimůnek and van Genuchten, 2006.)

and Pinder 1983; Šimůnek et al. 1998). These problems can be especially serious for advection-dominated transport characterized by small dispersivities. One way to partially circumvent numerical oscillations is to use *upstream weighing*. In this case, the second (flux) term of the ADE is not weighted by regular linear basis functions, but instead using nonlinear functions (Yeh and Tripathi 1990). The nonlinear weighting functions ensure that more weight is placed on flow velocities of nodes located at the upstream side of an element. Undesired oscillations can often be prevented also by selecting an appropriate combination of the space and time discretizations. Two dimensionless numbers may be used to characterize the discretizations in space and time. One of these is the grid *Peclet number*, *Pe*, which defines the predominant type of solute transport (notably the ratio of the advective and dispersive transport terms) in relation to coarseness of the finite element grid:

$$Pe = \frac{v \, \Delta x}{D_e}, \tag{6.88}$$

where Δx is the length of a finite element [L]. The Peclet number increases when the advective part of the transport equation dominates the dispersive part (i.e., when a relatively steep concentration front is present). To achieve acceptable numerical results, the spatial discretization must be kept relatively fine to maintain a low Peclet number. Numerical oscillations can be virtually eliminated when the local Peclet numbers do not exceed about 5. However, acceptably small oscillations may be obtained with local Peclet numbers as high as 10 (Huyakorn and Pinder 1983). Undesired oscillations for higher Peclet numbers can be effectively eliminated by using upstream weighing.

A second dimensionless number, the Courant number, *Cr*, may be used to characterize the relative extent of numerical oscillations in the numerical solution. The Courant number is associated with the time discretization, Δt [T], as follows:

$$Cr = \frac{v \, \Delta t}{\Delta x}. \tag{6.89}$$

Given a certain spatial discretization, the time step must be selected such that the Courant number remains less than or equal to 1, except for a fully implicit discretization scheme.

Perrochet and Berod (1993) developed a criterion for minimizing or eliminating numerical oscillations that is based on the product of the Peclet and Courant numbers:

$$Pe \cdot Cr \leq \omega_s \, (= 2), \tag{6.90}$$

where ω_s is the stability criterion [–]. This indicates that advection-dominated transport problems having large *Pe* numbers can be safely simulated provided ω_s is reduced to ≤ 2.

6.6 HYDRUS EXAMPLES OF SOLUTE TRANSPORT

HYDRUS-1D (Šimůnek et al. 2008), HYDRUS-2D (Šimůnek et al. 1999a), and HYDRUS (2D/3D) (Šimůnek et al. 2006) simulate the one-, two-, and three-dimensional movement of multiple solutes in variably saturated media. All programs use finite elements in space and finite differences in time to numerically solve the ADE for solute transport. The solute transport equations assume advective-dispersive transport in the liquid phase (see Section 6.3), and diffusion in the gaseous phase (see Section 6.3.6). The transport equations also include provisions for nonlinear (see Section 6.3.4.2) and/or nonequilibrium reactions between the solid and liquid phases (see Section 6.3.8.2), linear equilibrium reactions between the liquid and gaseous phases (see Section 6.3.6), zero-order production, and two first-order degradation reactions (see Section 6.3.5): one that is independent of other solutes, and one that provides the coupling between solutes involved in the sequential first-order decay reactions. In addition, physical nonequilibrium solute transport can be accounted for by assuming a two-region, dual-porosity type formulation, which partitions the liquid phase into mobile and immobile regions (see Section 6.3.8.1). Alternatively, the transport equations include provisions for kinetic attachment/detachment of solute to the solid phase and it can be thus used to simulate transport of viruses, colloids, or bacteria (Schijven and Šimůnek 2002; Bradford et al. 2003).

HYDRUS-1D includes modules for simulating movement of carbon dioxide (Šimůnek and Suarez 1994) and major ions and their mutual interactions (Šimůnek et al. 1996). In addition, a special module, HP1 (Jacques and Šimůnek 2005; Jacques et al. 2008), which was recently developed by coupling HYDRUS-1D with the PHREEQC geochemical code (Parkhurst and Appelo 1999), can simulate a broad range of low-temperature biogeochemical reactions in water, the vadose zone, and in groundwater systems, including interactions with minerals, gases, exchangers, and sorption surfaces based on thermodynamic equilibrium, kinetics, or mixed equilibrium-kinetic reactions. However, these topics are beyond the scope of this textbook.

The HYDRUS programs implement a Marquardt-Levenberg parameter estimation technique (Marquardt 1963; Šimůnek and Hopmans 2002) for inverse estimation of solute transport and reaction parameters from measured transient or steady-state flow and/or transport data (Šimůnek and Hopmans 2002).

In the following sections, several HYDRUS examples of solute transport are demonstrated.

6.6.1 Nonlinear Adsorption and Transport

In this example with HYDRUS-1D (HYDRUS Simulation 6.1 – Nonlinear Adsorption and Transport), one-dimensional transport of a solute that undergoes nonlinear sorption (for which there is no analytical solution) is shown. This is also TEST4 in the HYDRUS-1D Direct Projects directory. It is based on an experiment described in Selim et al. (1987). Steady flow was established in a 10.75-cm long loam soil column using a 10 mmol$_c$ L^{-1} CaCl$_2$ solution. Then the input solution was changed to 10 mmol$_c$ L^{-1} MgCl$_2$. After 14.92 days, the input solution was changed back to the original CaCl$_2$ solution. The output concentrations were analyzed for magnesium (Mg).

In this example, only solute transport is selected in the Main Processes window (Figure 6.52). When this is done, HYDRUS-1D maintains the initial pressure head distribution constant throughout the problem. The General Solute Transport option is selected. Other solute transport options allow simulating transport of major ions or considering complex biogeochemical reactions.

In the Geometry Information window (not shown), length units of centimeters are selected and the depth of the profile is set to 10.75 cm. Days are selected for time units in the Time Information window and the time discretization settings are as shown in Figure 6.53. The maximum time step is set to 0.1 days to eliminate oscillations in time that occur in the predicted concentrations near the soil surface with a larger time step. Five print times are selected in the Print Information window (not shown) and the default print times are selected (every 5 days). In the Iteration Criteria window (not shown), the default values for water simulation are used. The default Soil Hydraulic Model window settings are taken (not shown). A saturated hydraulic conductivity of 6.495 cm d^{-1} and saturated water content of 0.633 cm^3 cm^{-3} are used in the Water Flow Parameters window (not shown). The other parameters in this window are not important because only saturated flow is simulated. In the Water Flow Boundary Conditions window (not shown), a constant pressure head condition is selected for the upper and lower boundary conditions and the initial conditions are in terms of pressure heads (the default setting).

The Solute Transport window settings are shown in Figure 6.54. The default settings for the time weighting scheme (Crank-Nicolson) and space weighting scheme (Galerkin Finite Elements) are used. For mass units, "mmol" is entered in the space provided. This is only used as a label for graphs produced by HYDRUS-1D. The equilibrium ADE model is selected. The iteration criteria for nonlinear problems

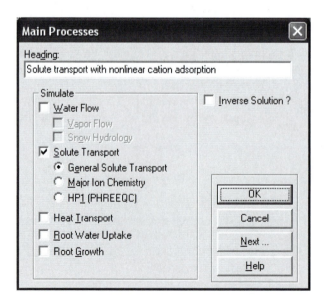

FIGURE 6.52 Main Processes window in the TEST4 example.

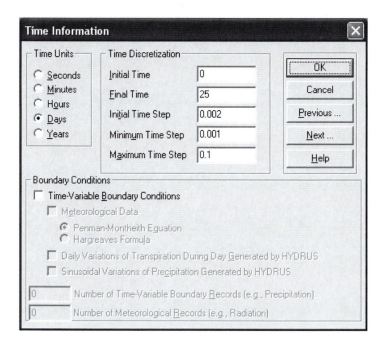

FIGURE 6.53 Time Information window in the TEST4 example.

(required since adsorption is nonlinear in this example) are changed from the default values to those shown in Figure 6.54. A pulse duration of 14.919 days is entered. Since this is less than the total simulation time, this is a pulse input problem, but a rather wide pulse. Note that in this window, users can select whether or not their transport and reaction parameters are temperature- and water content-dependent (which many reactions are) and various nonequilibrium solute transport models that are available in HYDRUS-1D.

The Solute Transport Parameters window settings are shown in Figure 6.55. Bulk density is set to a value of 0.884. The units of bulk density must be the inverse of the units used for the Freundlich distribution coefficient (in the next window). No units for mass are specified in HYDRUS-1D (the "mass units" in Figure 6.54 are only used as a graph label). The product of bulk density and the sorption distribution coefficient is used to determine the dimensionless retardation coefficient (R, see Equation 6.39), which appears in the numerical solution of the transport equation. The physical/chemical nonequilibrium transport model parameter f (see Equations 6.64 and 6.70, respectively), called "Frac" in this window, is set to 1, confirming that this simulation is for equilibrium conditions. The water and gas diffusion coefficients are set to zero, indicating that it is assumed that diffusion will have a minimal effect on dispersion. Dispersivity ("Disp.") is set to 2.727 cm.

In the Reaction Parameters window (Figure 6.56), values for the Freundlich distribution coefficient ("Kd") and exponent ("Beta") parameters (see Equation 6.46)

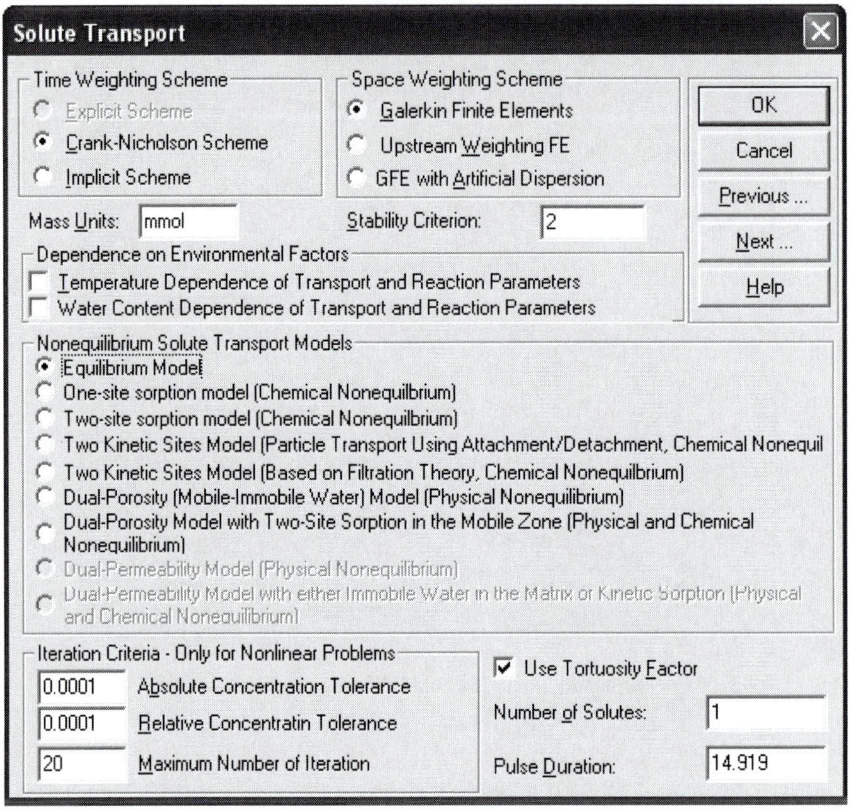

FIGURE 6.54 Solute Transport window in the TEST4 example.

are entered for Mg. As mentioned above, the units for the distribution coefficient must be the inverse of the bulk density specified in the Solute Transport Parameters window (Figure 6.55). "Nu" corresponds to the Langmuir adsorption parameter η (see Equation 6.47). When it is set to zero, the Freundlich equation is used. A nonzero value results in use of the Langmuir equation. All the other reaction parameters (scroll to the right to see these) are set to zero.

The Solute Transport Boundary Conditions window is shown in Figure 6.57. "Concentration Flux BC" is selected for the upper boundary. This boundary condition is recommended over the "Concentration BC" in almost all cases. It corresponds to a type-3 solute boundary condition (see Table 6.1) that results in the most accurate solute mass balance and is a physically more appropriate boundary condition for most practical problems (see the Help button for this window). For the lower boundary condition, "Zero Concentration Gradient" is selected. This boundary condition is appropriate for both semi-infinite and finite systems in that it assumes that the concentration is continuous across the lower boundary (Toride et al. 1999). This also results in the prediction of flux concentrations

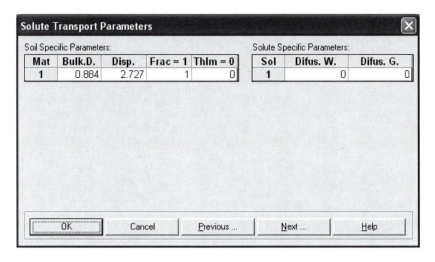

FIGURE 6.55 Solute Transport Parameters window in the TEST4 example.

at the lower boundary (see Equation 6.60). The Mg concentration at the upper boundary is set at 10 mmol cm^{-3}.

The Profile Information window is shown in Figure 6.58. The model domain is divided into 44 evenly spaced depth increments of 0.25 cm. Four observation nodes are inserted at 0, 2.5, 5, 7.5, and 10.75 cm. The initial conditions for pressure head and concentrations are set to zero at all depths.

After the model has been run, the user can select the "Observation Points" display from the Post-processing panel (Figure 6.59). The curves represent predicted

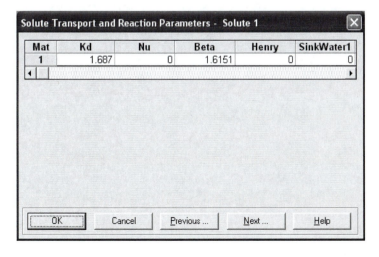

FIGURE 6.56 Solute Transport and Reaction Parameters window in the TEST4 example.

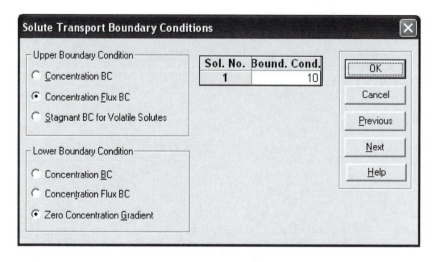

FIGURE 6.57   Solute Transport Boundary Conditions window in the TEST4 example.

Mg concentrations at depths of 0, 2.5, 5, 7.5, and 10.75 cm. The BTCs become more dispersed and less skewed with depth and it is clear that this is a (wide) pulse input.

If the "Solute Transport – Actual and Cumulative Boundary Fluxes" display is selected from the Post-processing panel and "All Solutes" is chosen from the pull-down menu for the vertical variable, the solute fluxes into and out of the model

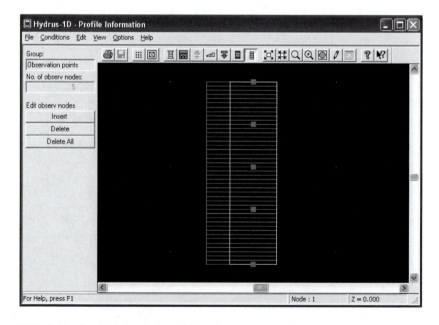

FIGURE 6.58   Profile Information window in the TEST4 example.

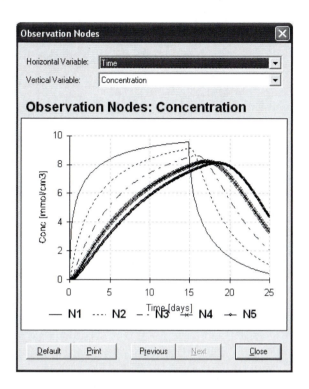

FIGURE 6.59 Observation Nodes window in the TEST4 example. The curves represent predicted Mg concentrations at depths of 0, 2.5, 5, 7.5, and 10.75 cm (N1–N5). The 0 cm curve has the highest peak concentration and the 10.75 cm curve has the lowest peak concentration with the other depths intermediate and in order of depth.

domain as a function of time can be seen (Figure 6.60). The input at the top is a square wave (wide pulse). Flux out the bottom (a negative flux) starts very soon after the solute is applied and is not complete by the end of the simulation. Units for solute flux are per unit-area of surface and bottom boundaries (mmol cm^{-2} d^{-1}). If the Mass Balance display is selected from the Post-processing panel, the user can scroll down to the last print time (25 days) and see that concentration mass balance error (inputs-outputs vs. mass present in the model domain) is less than 1% (not shown).

6.6.2 TRANSPORT OF NITROGEN SPECIES

The next example is a nitrogen chain model consisting of three species (HYDRUS Simulation 6.2 – Transport of Nitrogen Species). This is also TEST3 in the HYDRUS-1D Direct Project directory. First-order transformation constants control the conversion of ammonium to nitrite and then nitrate:

$$NH_4^+ \rightarrow NO_2^- \rightarrow NO_3^-.$$

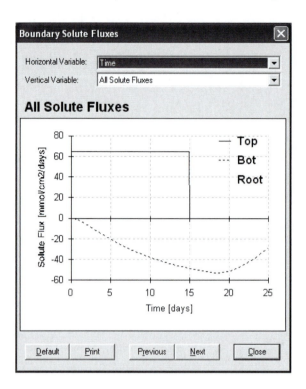

FIGURE 6.60 Boundary Solute Fluxes window in the TEST4 example.

The simulation is based on an example described by van Genuchten (1985), where NH_4 was added to a soil profile. Van Genuchten derived an analytical solution for this transport problem. The results of this analytical solution were used to verify the correctness of the numerical scheme in HYDRUS-1D (see Section 6.6.3).

In the Main Processes window, only solute transport is selected using the General Solute Transport option (same as in the first example in Section 6.6.1). Centimeters are chosen for the length units and the depth of the soil profile is 200 cm. Hours are chosen for the time units and the final time is 200 h. The initial time step is 1 h, the minimum time step 0.0001 h, and the maximum time step 100 h. Three print times (50, 100, and 200 h) are selected. The default iteration criteria for water flow and default soil hydraulic model are used. The water flow parameters are shown in Figure 6.61. Since the water flow option was not chosen in the Main Processes window, the only parameters from this window that are important in the simulation are the saturated hydraulic conductivity (which is set to 1 cm h⁻¹) and the saturated water content (which is set to 1 to be in accord with the analytical solution). The saturated hydraulic conductivity is the flow rate through the profile (unit gradient flow, see Equation 3.29) and the saturated water content is used to calculate the retardation coefficient R.

A constant pressure head is selected for the upper and lower boundary conditions and the initial conditions are in terms of pressure heads (the default setting),

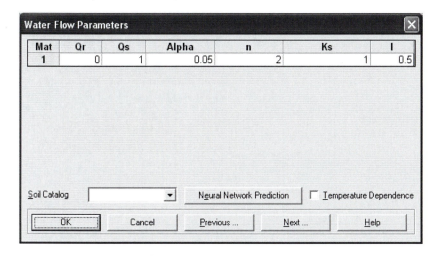

FIGURE 6.61 Water Flow Parameters window in example TEST3.

as in the first example. The settings in the Solute Transport window are the default settings for the most part (Figure 6.62). No label is chosen for mass units as relative concentrations are used. The number of solutes is set to 3 and the pulse duration is set to time equal to or larger than the maximum time of the simulation, indicating a step input. The Solute Transport Parameters window is shown in Figure 6.63. Bulk density is set to 1000. This would be appropriate for units of kilogram per cubic meter. Dispersivity is set to zero and rather large values are given for the diffusion coefficients and water for all three solutes (0.18 cm^2 h^{-1}) to be in accord with the analytical solution. Note that in steady-state flow problems, the dispersion coefficient D_e is constant (Equation 6.11), thus it does not matter whether it is input using dispersivity or the diffusion coefficient.

In the Solute Transport and Reaction Parameters window for Solute 1 (Figure 6.64), K_d is set to a value of 0.001. Since $\beta = 1$ and $\eta = 0$, this represents linear adsorption. This results in a value of $R = 2$ for NH_4, given the settings for ρ_b and θ_s (see Equation 6.39). Scrolling to the right in the Solute Transport and Reaction Parameters window for Solute 1, one can see the rates (0.005 h^{-1}) for first-order transformation of NH_4 to NO_2 in the water and solid phases (Figure 6.65). For Solute 2 (NO_2), K_d is set to zero (no adsorption) and the first-order transformation rate for NO_2 to NO_3 in the water phase is set to 0.1 h^{-1} (not shown). Since there is no adsorption of NO_2, the transformation rate in the solid phase is zero. For Solute 3 (NO_3), K_d is zero as are the transformation rates (not shown).

The solute boundary conditions are shown in Figure 6.66. A type-3 boundary condition ("Concentratin Flux BC") is chosen for the upper boundary and a type-2 boundary condition ("Zero Concentration Gradient") is chosen for the lower boundary. An input concentration at the upper boundary of $c/c_0 = 1$ is set for NH_4, with no inputs for NO_2 and NO_3.

Two hundred and one evenly spaced nodes ($\Delta z = 1$ cm) are used in the soil profile model space and three observation nodes are placed at depths of 50, 100, and 150 cm

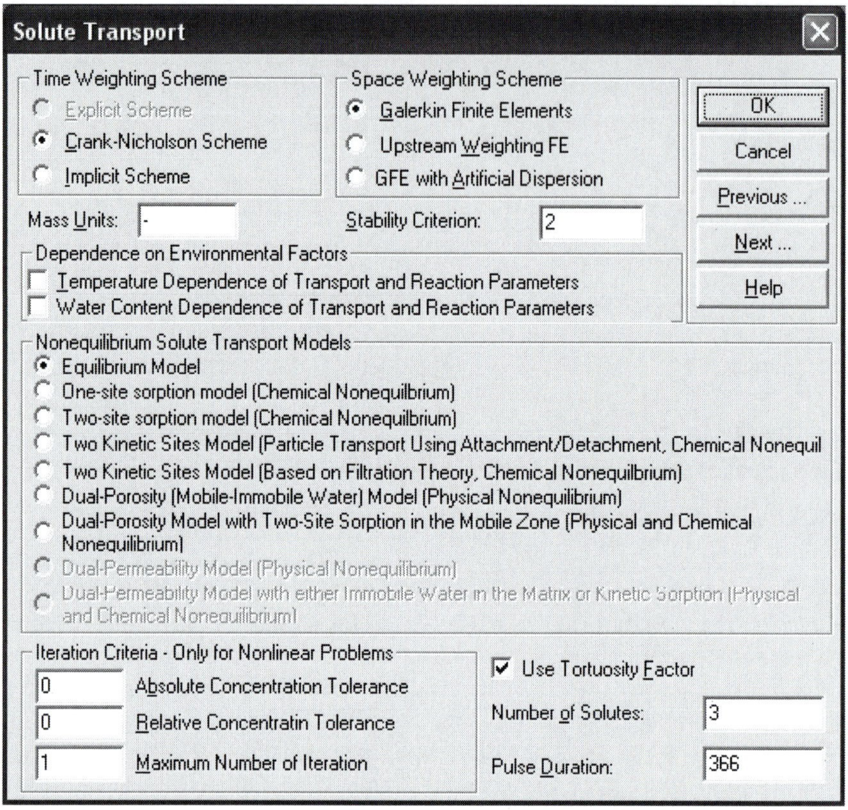

FIGURE 6.62 The Solute Transport window in example TEST3.

(not shown). The initial conditions for pressure head are zero, as are the concentrations for the three nitrogen species.

After the model has run, the Profile Information display is selected from the Post-processing panel (Figure 6.67). This shows that after 200 h, NH_4 ("Concentration - 1") has reached a depth of about 110 cm. Relative concentrations decrease steadily from 1 at the surface to about 0.45 at a depth of 100 cm behind the solute front. This decrease of soil solution NH_4 concentrations is due to the nitrification process (transformation of NH_4 to NO_2, and subsequently to NO_3) and NH_4 adsorption. By contrast, soil profile concentrations of NO_3 show that a pulse has moved beyond the deepest depth after 200 h (Figure 6.68). Peak concentrations of about 0.6 occur at a depth near 100 cm. At earlier times, the nitrate pulse occurs at shallower depths and the total mass of NO_3 (area under the curve) is much less. It's clear from these graphs that NH_4 movement into the soil profile is restricted by both adsorption and transformation losses to NO_2 (and subsequently to NO_3). Nitrate movement is not restricted by adsorption and the nitrate mass grew with the transformation from NO_2. This simulation just begins to show the complexity of modeling nitrogen species in the soil environment for

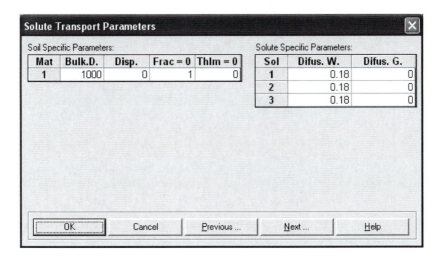

FIGURE 6.63 The Solute Transport Parameters window in example TEST3.

transformations are affected by temperature, pH, oxygen, carbon, and microbial dynamics (Tiedje 1988).

6.6.3 Parameter Optimization of Nitrogen Species Model

In this example (HYDRUS Simulation 6.3 – Parameter Optimization of Nitrogen Species Model), use of the inverse mode for the nitrogen chain model simulation described above is shown. This is FITTEST3 in the HYDRUS-1D Inverse Projects directory.

An analytical solution was used to generate synthetic "observed" data for ammonium concentrations at the depth of the shallowest node (50 cm). Most of the model

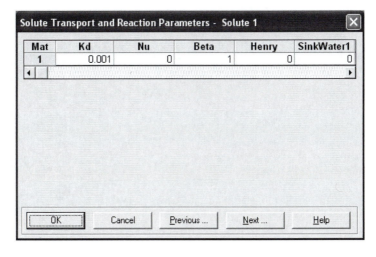

FIGURE 6.64 Solute Transport and Reaction Parameters window for Solute 1 (NH$_4$) in the TEST3 example showing the values for sorption.

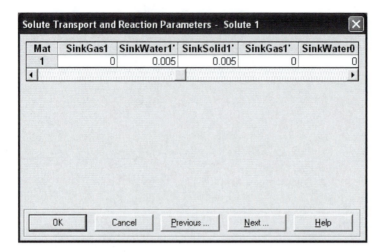

FIGURE 6.65   Scrolling to the right in the Solute Transport and Reaction Parameters window for Solute 1 (NH_4) in the TEST3 example, the values for first-order transformation to Solute 2 (NO_2) in water and solid phases can be seen.

setup is the same as in Section 6.6.2 and only the changes that are required to perform the inverse mode are described here. In the Main Processes window (not shown), the box for inverse solution is checked. The Inverse Solution window is shown in Figure 6.69. Depending on how the data to be compared with the simulation are measured, resident concentrations or flux concentrations should be chosen (see Section 6.3.7). In this example, they are resident concentrations. The number of observed data points is entered (93 pairs of time and concentration "observations"). Weights can be assigned to the data pairs individually in a later window or weights can be assigned internally. The default weighting scheme is weighting by standard deviation. This method avoids giving heavier weights to data with higher values

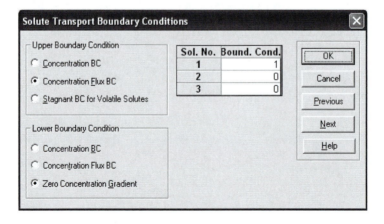

FIGURE 6.66   Solute Transport Boundary Conditions window in the TEST3 example.

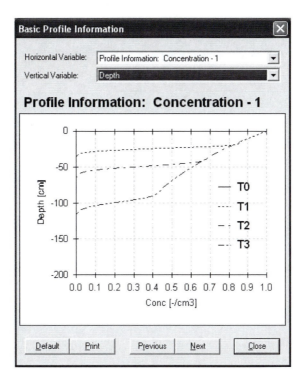

FIGURE 6.67 Soil profile concentrations for NH_4 at various times (T0 = 0, T1 = 50, T2 = 100, and T3 = 200 h) in the TEST3 example.

when different types of observed data (concentrations and fluxes for example) are used. The maximum number of iterations that will be attempted to find a best fit is set at 50.

In the Print Information window (not shown), the check mark next to Screen Output is removed. This makes it easier to see how far the optimization process has progressed while the program is running. In the Solute Transport window (not shown), the number of solutes is changed from three (in the direct nitrogen chain example above) to one, so that only ammonium transport is simulated (but losses to nitrite are included via the transformation parameters). The Reaction Parameters window is shown in Figure 6.70. Beneath each transport parameter, minimum and maximum values to constrain the optimization process can be entered (if the values are zero there are no constraints). A checkbox is provided to indicate which parameters are to be fitted. In this simulation, K_d is fitted as are the first-order transformation constants for the solid (μ_s) and liquid (μ_l) phase: "SinkWater1'" and "SinkSolid1'" (scroll to the right in this window to see these). Note the hyphen at the end of SinkWater1 and SinkSolid1 used to distinguish these chain reaction transfers from conventional sinks (see the Help file). No constraints are set for any of the fitted parameters. Entered values for optimized parameters are used by the model as initial estimates in the iteration process to find the optimal values.

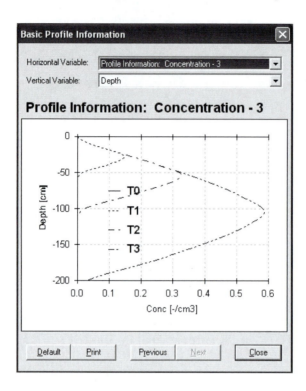

FIGURE 6.68 Soil profile concentrations for NO_3 at various times (T0 = 0, T1 = 50, T2 = 100, and T3 = 200 h) in the TEST3 example.

The Data for Inverse Solution window is shown in Figure 6.71. The Help button can be used to understand the column labels X, Y, Type, and Position. When Type = 4, the data are concentration measurements at one of the observation points, X is the time (in hours since that has been chosen in the Time Information window), and Y is the concentration (in the same units used for the Boundary Conditions window, dimensionless in this case). Position is the observation point number (assigned in the Soil Profile Graphical Editor), in this case the observation point located at a depth of 50 cm.

After the program has run, the predicted and observed NH_4 BTCs at a depth of 50 cm can be seen in the Observation Nodes display in the Post-processing panel (Figure 6.72). The inverse method found a set of parameter values that resulted in an excellent fit to the observed data. The Inverse Solution Information window is shown in Figure 6.73. Scrolling down this window, it can be seen that a satisfactory set of parameter values was found after eight iterations. The r^2 for the fit to the observed data is very good (1). Standard errors and 95% confidence limits on the fitted parameter values are also given. Although K_d is well defined in the inverse process (by the movement of the solute pulse), the two degradation coefficients (μ_i and μ_s) are fully (and inversely) correlated ($\rho = -1$) and the confidence limits are very large (ranging from negative to positive values) (Figure 6.73). This is because there is insufficient information in the observed data to determine whether degradation

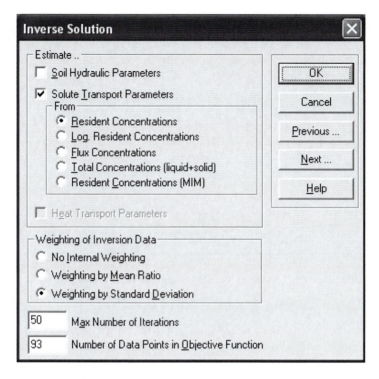

FIGURE 6.69   Inverse Solution window for the FITTEST3 example.

has taken place in the solid or liquid phase. In this case, the transformation term in the ADE (see Equation 6.49) can be written as:

$$\mu_l \theta c + \mu_s \rho_b K_d c = (\mu_l \theta + \mu_s \rho_b K_d)c. \tag{6.91}$$

Since all the terms inside the parenthesis on the right side are constants, one degradation coefficient can be decreased (even to zero) and the other increased to get the same overall value and fit. In this case, the inverse problem is *ill-posed* (Hopmans et al. 2002) for differentiating between these parameters, but not for K_d or when $\mu_l = \mu_s$. When it can be assumed that the degradation rate is the same in both phases, its value can still be easily estimated using HYDRUS-1D. Since HYDRUS-1D does not directly implement constraints such as equality among parameters, μ_s can be set equal to zero, and HYDRUS-1D rerun to estimate a new μ_l^*. Then μ can be calculated for both phases as follows:

$$\mu_l^* \theta c = \mu(\theta + \rho_b K_d)c = \mu\left(\frac{\theta + \rho_b K_d}{\theta}\right)\theta c = \mu\theta R c$$

$$\mu = \frac{\mu_l^*}{R} = \mu_l^* \frac{\theta}{\theta + \rho_b K_d} \tag{6.92}$$

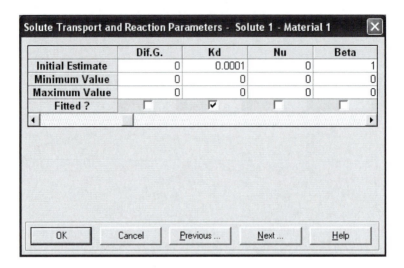

FIGURE 6.70   Reaction Parameters window for the FITTEST3 example.

6.6.4   SOLUTE TRANSPORT IN THE CAPILLARY FRINGE

Abit et al. (2008) described a field experiment that showed lateral solute transport occurred in the capillary fringe of a shallow water table. To mimic conditions that might occur beneath an onsite wastewater system with a nitrogen plume, a bromide solution was applied at the bottom of a borehole positioned about 30 cm above the water table (at a depth of about 60 cm) in a soil of the North Carolina coastal plain. Suction lysimeters at different depths and distances were used to follow the

	X	Y	Type	Position	Weight
1	10.234	1E-030	4	1	1
2	12.664	1.29E-030	4	1	1
3	15.364	1.4E-028	4	1	1
4	18.334	8.93E-026	4	1	1
5	21.304	3.47E-023	4	1	1
6	25.084	2.86E-020	4	1	1
7	30.754	1.29E-016	4	1	1
8	35.074	2.63E-014	4	1	1
9	40.204	5.62E-012	4	1	1
10	45.064	4.17E-010	4	1	1
11	50	1.76E-008	4	1	1
12	55.153	4.97E-007	4	1	1
13	60.283	8.46E-006	4	1	1
14	65.413	9.45E-005	4	1	1
15	70.003	0.000597	4	1	1

FIGURE 6.71   Data for Inverse Solution window for the FITTEST3 example.

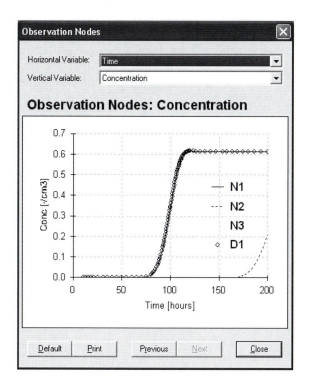

FIGURE 6.72 Observation Nodes window for the FITTEST3 example. N1 is the predicted NH$_4$ breakthrough curve at a depth of 50 cm and D1 are the data to which the curve has been fit. N2 is the predicted NH$_4$ breakthrough curve at a depth of 100 cm.

movement of bromide for up to 60 days. Several rains occurred during the period. The bromide moved downward beneath the borehole until it reached the capillary fringe and water table and then moved laterally in the direction of groundwater flow. Approximately half of the bromide detected 3 m from the borehole 59 days after application was in the capillary fringe.

In this example (HYDRUS Simulation 6.4 – Capillary Fringe), HYDRUS (2D/3D) is used to simulate similar conditions. The model space is shown in Figure 6.74. A regular grid spacing is used with $\Delta z = 4$ cm and $\Delta x = 6$ cm to create 5151 nodes. The upper boundary of the model space represents the soil surface. The lower boundary is 200 cm below the soil surface. The model space extends 800 cm in the lateral direction and the lateral boundaries are drawn at an angle of 10° to the horizontal axis (slope of 17.6%). The upper and lower boundaries are no-flux conditions for both solute and water flow, except for three nodes at the soil surface near the upslope end (the source is indicated by an arrow in Figure 6.74). At these nodes, the boundary conditions are a constant pressure head of −20 cm (sufficient to provide a steady flow of water into the soil under unsaturated conditions) and a solute concentration of 50 μg cm^{-3} (equivalent to 50 mg L^{-1}). The boundary conditions for water flow on the vertical boundaries are a pressure head distribution such that the water table is maintained at a depth of

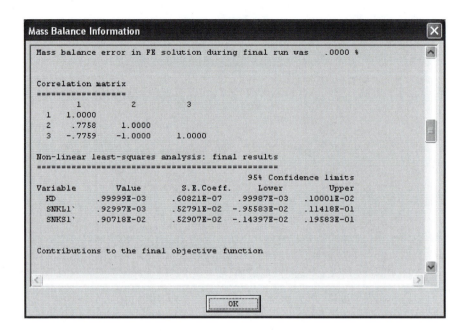

Mass Balance Information

```
Mass balance error in FE solution during final run was    .0000 %


Correlation matrix
==================
           1           2          3
   1    1.0000
   2     .7758      1.0000
   3    -.7759     -1.0000     1.0000


Non-linear least-squares analysis: final results
================================================
                                      95% Confidence limits
Variable        Value      S.E.Coeff.     Lower        Upper
  KD           .99999E-03   .60821E-07   .99987E-03   .10001E-02
  SNKL1`       .92997E-03   .52791E-02  -.95583E-02   .11418E-01
  SNKS1`       .90718E-02   .52907E-02  -.14397E-02   .19583E-01


Contributions to the final objective function
```

OK

FIGURE 6.73 Inverse Solution Information window in the FITTEST3 example.

100 cm below the soil surface at the boundaries. Due to the slope of the model domain, this causes lateral flow of groundwater from right to left. The capillary fringe extends above the water table. The vertical boundaries are designated as solute flux concentrations of zero.

The soil in this example is a sandy loam from the Rosetta database ($\theta_r = 0.038$, $\theta_s = 0.387$, $\alpha = 0.0267$ cm^{-1}, $n = 1.448$, $K_s = 38.25$ cm d^{-1}, and $l = 0.5$). Longitudinal dispersivity was set to 5 cm and transverse dispersivity to 0.5 cm. To set the solute boundary condition at the source, a concentration of 50 is entered in the Reaction Parameters window for "cBnd1" (Figure 6.75). The location i in the vector of boundary conditions cBnd$i$ that is used as the boundary concentration value is indicated by a pointer for the solute boundary. The pointer number assigned to boundaries can be seen when the solute transport boundary conditions are displayed (Figure 6.74). If the user checks "display codes" on the right panel, the pointer number will appear on the boundaries that have water fluxes. The default is to assign all these boundaries a pointer value of 1. To set a zero-concentration boundary condition to the vertical boundaries, a value of zero (the default) is assigned to "cBnd2" in Figure 6.75. Then the pointer for these boundaries is changed to 2 by selecting the nodes and, when a window appears for the "pointer to boundary vector," entering a value of 2 (Figure 6.74).

The simulated solute concentrations after 49 days are shown in Figure 6.76. Bromide moved downward below the source until it encountered the capillary fringe and water table and then moved laterally in a narrow stream. The cross-section chart tool (right panel) is used to make a graph of the solute concentrations along a vertical transect approximately 2 m downslope of the source. The graph shows that the plume

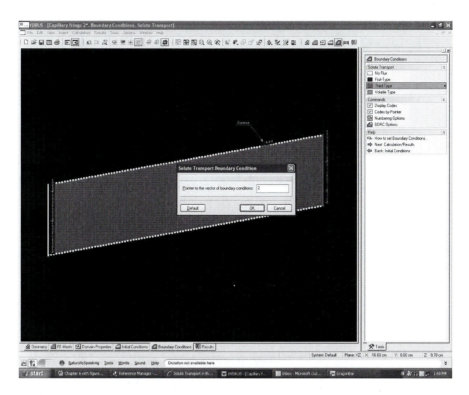

FIGURE 6.74 Two-dimensional model space in the capillary fringe example. The Solute Transport Boundary Conditions display is shown with the pointer numbers for the boundaries. The window insert appears when the boundary nodes are selected and a type-3 boundary condition is selected from the right panel.

is centered at a depth of about 100 cm below the soil surface (the depth of the water table is about 105 cm due to mounding beneath the source). Hence, a little less than half of the lateral plume occurs in the capillary fringe, as found by Abit et al. (2008). Decreasing the slope in the simulation reduced the amount of transport in the capillary fringe. Abit et al. (2008) point out that using shallow wells to quantify lateral movement of a plume could underestimate the lateral flux in that the capillary fringe component would be missed.

6.7 SUMMARY

Solute transport is complex in that it involves water and chemical movement, adsorption, and transformations. In this chapter, the various forms of the ADE have been derived. These forms include physical nonequilibrium that addresses preferential flow and chemical nonequilibrium that addresses kinetic adsorption. Stochastic forms of the ADE may be appropriate for field-scale transport problems where spatial variability in soil hydraulic properties is extensive. In many cases, analytical solutions to the ADE exist and the STANMOD software package can be used to

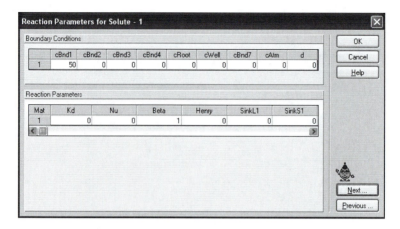

FIGURE 6.75   Reaction Parameters window for the capillary fringe example.

solve these problems, including inverse problems. Some transport problems, such as nonlinear adsorption or transient flow conditions, can only be solved using numerical methods. The two most common numerical methods are finite differences and finite elements. The HYDRUS software package uses a finite element approach in space and a finite difference approach in time to solve the ADE in its various forms,

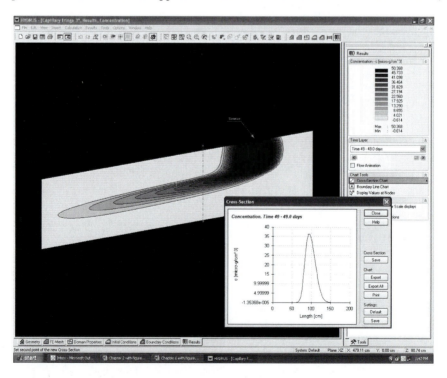

FIGURE 6.76   Predicted bromide distribution after 49 days in the capillary fringe example.

including nonequilibrium transport and inverse solutions. In this chapter, several examples of STANMOD and HYDRUS simulations of solute transport have been provided.

6.8 DERIVATIONS

In this section, the Laplace transform solution to the ADE, the expected value for a lognormal distribution, and the numerical dispersion error for an explicit finite difference solution to the ADE are derived.

6.8.1 LAPLACE TRANSFORM SOLUTION TO ADE

An analytical solution of the advection-dispersion equation (ADE) describing one-dimensional transport of a nonadsorbed solute is sought (Jury and Roth, 1990). This is the partial differential equation (PDE) (Equation 6.17):

$$\frac{\partial c}{\partial t} = D_e \frac{\partial^2 c}{\partial z^2} - v \frac{\partial c}{\partial z} \tag{6.93}$$

The Laplace transform of the PDE must be taken. The term on the left side of the ADE and the first term on the right side are essentially the same as the heat flow equation, so again these transforms are known from Chapter 4. From Equation 4.65, the left side of the ADE transforms as follows:

$$\mathscr{L}\left[\frac{\partial c(z,t)}{\partial t}\right] = s\hat{c}(z) \tag{6.94}$$

From Equation 4.66, the first term on the right side of the ADE transforms as follows:

$$\mathscr{L}\left[D_e \frac{\partial^2 c(z,t)}{\partial z^2}\right] = D_e \frac{d^2\hat{c}(z)}{dz^2} \tag{6.95}$$

The Laplace transform of the second term on the right side of the ADE must be determined (in the heat flow equation a first derivative in space did not appear). So the definition of the Laplace transform is used to find this (although it is apparent as an intermediate step in the derivation of the second derivative in space in Chapter 4). Using the definition, the transform is:

$$\mathscr{L}\left[-v\frac{\partial c(z,t)}{\partial z}\right] = -v\int\limits_{0}^{\infty} \frac{\partial c(z,t)}{\partial z} e^{-st} dt$$

where v is brought outside the integral because it is a constant parameter. As seen in Chapter 4, the order of integration/differentiation doesn't matter so one can write this as:

$$\mathcal{L}\left[-v\frac{\partial c(z,t)}{\partial z}\right] = -v\frac{d}{dz}\int_0^\infty c(z,t)e^{-st}dt$$

The differentiation notation ceases to be a partial derivative because t is lost as a variable once the transform is taken (and the only independent variable is z). The integral term is the definition of the Laplace transform of $c(z,t)$, so the transform can be written:

$$\mathcal{L}\left[-v\frac{\partial c(z,t)}{\partial z}\right] = -v\frac{d\hat{c}(z)}{dz} \tag{6.96}$$

The Laplace transform of the ADE, using Equation 6.94 through 6.96 is then :

$$s\hat{c}(z) = D_e\frac{d^2\hat{c}(z)}{dz^2} - v\frac{d\hat{c}(z)}{dz} \tag{6.97}$$

The Laplace transform of the boundary conditions (Equation 6.18) must also be taken. Taking the transform of both sides of the surface boundary condition:

$$\mathcal{L}\left\{\left[J_w c(z,t) - \theta D_e\frac{\partial c(z,t)}{\partial z}\right]\Big|_{z=0}\right\} = \mathcal{L}[J_w c_0]$$

$$\left[J_w \hat{c}(z) - \theta D_e\frac{d\hat{c}(z)}{dz}\right]\Big|_{z=0} = \frac{J_w c_0}{s} \tag{6.98}$$

The transform of the bottom boundary condition is:

$$\mathcal{L}\left[\frac{\partial c(\infty,t)}{\partial z}\right] = \frac{d\hat{c}(\infty)}{dz} = 0 \tag{6.99}$$

Transforming the PDE has eliminated time as a variable and Equation 6.97 is now an ODE, on which standard methods for solving ordinary differential equations can be used. Comparing this ODE to the one obtained after transforming the heat equation

(Equation 4.67), it can be seen that a second term has been gained on the right side due to the convection term in the ADE. However, the situation is similar to the solution in Chapter 4 in that a function is sought, which when differentiated once or twice, only differs from the original function by constants. The exponential function is such a function, so it will be used as the same trial function as before (Equation 4.70). Assume that the function sought is of the form:

$$\hat{c}(z) = e^{mz} \tag{6.100}$$

where m is undefined at this point. The first derivative with respect to z of this function is me^{mz} and the second derivative is $m^2 e^{mz}$, so substituting into Equation 6.97 with the trial function, one obtains:

$$se^{mz} = D_e m^2 e^{mz} - vme^{mz}$$

Then the equation is rearranged and e^{mz} is factored out:

$$\left(D_e m^2 - vm - s\right)e^{mz} = 0$$

Either the first term in parentheses must be zero or m must be negative ∞. The later possibility would lead to zero values for the function everywhere, so it must be true that:

$$D_e m^2 - vm - s = 0 \tag{6.101}$$

The objective is to find m. For a quadratic equation, the solution for m is:

$$m = \frac{v \pm \sqrt{v^2 + 4D_e s}}{2D_e}$$

One can rearrange this equation:

$$m = \frac{v}{2D_e}\left(1 \pm \varepsilon\right) \tag{6.102}$$

where:

$$\varepsilon = \sqrt{1 + \frac{4D_e s}{v^2}} \tag{6.103}$$

Because of the $\pm$ sign, there are two possible values for m and two functions that will satisfy Equation 6.103. Using Equation 6.100 and 6.102, the two solutions are:

$$\hat{c}(z) = \exp\left[\frac{vz}{2D_e}(1-\varepsilon)\right]$$

$$\hat{c}(z) = \exp\left[\frac{vz}{2D_e}(1+\varepsilon)\right]$$

(6.104)

Since there are two *particular* solutions to the ODE, the *general* solution is a linear combination of the particular solutions:

$$\hat{c}(z) = A\exp\left[\frac{vz}{2D_e}(1-\varepsilon)\right] + B\exp\left[\frac{vz}{2D_e}(1+\varepsilon)\right]$$

(6.105)

where A and B are undefined, but the boundary conditions can be used to find their values.

Using the bottom transformed boundary condition at $z = $ infinity (Equation 6.99), the equation is:

$$\frac{d\hat{c}(\infty)}{dz} = 0 = \frac{v}{2D_e}(1-\varepsilon)A\exp\left[\frac{v\cdot(\infty)}{2D_e}(1-\varepsilon)\right] + \frac{v}{2D_e}(1+\varepsilon)B\exp\left[\frac{v\cdot(\infty)}{2D_e}(1+\varepsilon)\right]$$

Since v and D_e are positive (s can't be negative), $\varepsilon > 1$ (see Equation 6.103). Therefore, the equation above can be written as:

$$0 = \frac{v}{2D_e}(1-\varepsilon)A\exp[-\infty] + \frac{v}{2D_e}(1+\varepsilon)B\exp[\infty]$$

$$= \frac{v}{2D_e}(1+\varepsilon)B\exp[\infty]$$

The first term is zero because $e^{-\infty} = 0$. For the second term to be zero, B must be zero. So the solution is reduced to the first term in Equation 6.105:

$$\hat{c}(z) = A\exp\left[\frac{vz}{2D_e}(1-\varepsilon)\right]$$

To find A, the solution is substituted into the surface transformed boundary condition (Equation 6.98):

$$\left\{ J_w A \exp\left[\frac{v \cdot z}{2D_e}(1-\varepsilon)\right] - \theta D_e A \exp\left[\frac{v \cdot z}{2D_e}(1-\varepsilon)\right]\frac{v}{2D_e}(1-\varepsilon)\right\}\Bigg|_{z=0} = \frac{J_w c_0}{s}$$

$$J_w A \exp\left[\frac{v \cdot 0}{2D_e}(1-\varepsilon)\right] - \theta D_e A \exp\left[\frac{v \cdot 0}{2D_e}(1-\varepsilon)\right]\frac{v}{2D_e}(1-\varepsilon) = \frac{J_w c_0}{s}$$

$$J_w A - \theta D_e A \frac{v}{2D_e}(1-\varepsilon) = \frac{J_w c_0}{s}$$

$$J_w A - \frac{J_w D_e A}{2D_e}(1-\varepsilon) = \frac{J_w c_0}{s}$$

$$A\left[1 - \frac{(1-\varepsilon)}{2}\right] = \frac{c_0}{s}$$

$$A = \frac{2c_0}{(1+\varepsilon)s}$$

So the solution to the ODE (Equation 6.97) is:

$$\hat{c}(z) = \frac{2c_0}{(1+\varepsilon)s} \exp\left[\frac{vz}{2D_e}(1-\varepsilon)\right]$$

Substituting into this equation the definition of ε (Equation 6.103), one obtains:

$$\hat{c}(z) = \frac{2c_0}{\left(1 + \sqrt{1+\frac{4D_e s}{v^2}}\right)s} \exp\left[\frac{vz}{2D_e}\left(1 - \sqrt{1+\frac{4D_e s}{v^2}}\right)\right]$$

$$= \frac{c_0}{\left(1 + \sqrt{1+\frac{4D_e s}{v^2}}\right)s} \exp\left[\frac{vz}{2D_e}\right] \exp\left[-\frac{vz}{2D_e}\sqrt{1+\frac{4D_e s}{v^2}}\right]$$

(6.106)

To get the solution to the PDE (Equation 6.93), the back transform (inverse Laplace transform) must be taken of the above equation. In other words, the solution to the PDE is:

$$c(z,t) = \mathscr{L}^{-1} \left\{ \frac{2c_0}{\left(1+\sqrt{1+\dfrac{4D_e s}{v^2}}\right)s} \exp\left[\frac{vz}{2D_e}\right] \exp\left[-\frac{vz}{2D_e}\sqrt{1+\frac{4D_e s}{v^2}}\right] \right\}$$

$$= 2c_0 \exp\left[\frac{vz}{2D_e}\right] \mathscr{L}^{-1} \left\{ \frac{\exp\left[-\dfrac{vz}{2D_e}\sqrt{1+\dfrac{4D_e s}{v^2}}\right]}{\left(1+\sqrt{1+\dfrac{4D_e s}{v^2}}\right)s} \right\}$$

(6.107)

Selected terms on the right side of Equation 6.107 coould come outside the inverse Laplace transform operator because they are coefficients that do not contain the parameter s. Hence, the crux of the problem is finding the inverse Laplace transform of the last term. Table 4.1 will be used for the inverse. There are several equations in this table for the exponential function raised to a power of the square root of s, for which the inverse Laplace transformation is given. However, all of these have just s under the radical (see transforms # 7 through 16). So in the argument of the exponential function in the numerator of Equation 6.107, one must factor out a term from the radical:

$$-\frac{vz}{2D_e}\sqrt{1+\frac{4D_e s}{v^2}} = -\frac{vz}{2D_e}\sqrt{\frac{4D_e}{v^2}}\sqrt{\frac{v^2}{4D_e}+s}$$

$$= -\frac{z}{\sqrt{D_e}}\sqrt{\frac{v^2}{4D_e}+s}$$

(6.108)

The same term must be factored out of the radical in the denominator of Equation 6.107:

$$\sqrt{1+\frac{4D_e s}{v^2}} = \sqrt{\frac{4D_e}{v^2}}\sqrt{\frac{v^2}{4D_e}+s}$$

(6.109)

Substituting Equation 6.108 and 6.109 into Equation 6.107, the new equation is:

$$c(z,t) = 2c_0 \exp\left[\frac{vz}{2D_e}\right] \mathscr{L}^{-1} \left\{ \frac{\exp\left[-\dfrac{z}{\sqrt{D_e}}\sqrt{\dfrac{v^2}{4D_e}+s}\right]}{\left(1+\sqrt{\dfrac{4D_e}{v^2}}\sqrt{\dfrac{v^2}{4D_e}+s}\right)s} \right\}$$

A property of the inverse Laplace transform called the *shifting operation* (Jury and Roth, 1990) can be used to isolate s under the radical:

$$\mathcal{L}^{-1}\left[\hat{f}(b+as)\right]=\exp\left(-\frac{bt}{a}\right)\mathcal{L}^{-1}\left[\hat{f}(s)\right]$$

If one lets $b = v^2/(4D_e)$ and $a = 1$, then using the shifting operation the equation is:

$$\mathcal{L}^{-1}\left[\frac{\exp\left(-\dfrac{z}{\sqrt{D_e}}\sqrt{b+as}\right)}{(b+as-b)\left(1+\sqrt{\dfrac{4D_e}{v^2}}\sqrt{b+as}\right)}\right]=\exp\left(-\frac{bt}{a}\right)\mathcal{L}^{-1}\left[\frac{\exp\left(-\dfrac{z}{\sqrt{D_e}}\sqrt{s}\right)}{(s-b)\left(1+\sqrt{\dfrac{4D_e}{v^2}}\sqrt{s}\right)}\right]$$

where b has been added and subtracted in the denominator of the first term on the left side in order to get a $(b + as)$ term. Substituting $v^2/(4D_e)$ back into this expression for b and 1 for a, the equation is:

$$\mathcal{L}^{-1}\left[\frac{\exp\left(-\dfrac{z}{\sqrt{D_e}}\sqrt{b+as}\right)}{(b+as-b)\left(1+\sqrt{\dfrac{4D_e}{v^2}}\sqrt{b+as}\right)}\right]$$

$$=\exp\left(-\frac{v^2t}{4D_e}\right)\mathcal{L}^{-1}\left[\frac{\exp\left(-\dfrac{z}{\sqrt{D_e}}\sqrt{s}\right)}{\left(s-\dfrac{v^2}{4D_e}\right)\left(1+\sqrt{\dfrac{4D_e}{v^2}}\sqrt{s}\right)}\right]$$

(6.110)

Now s is isolated under the radical and it is clear that inverse transform # 16 from Table 4.1 is the same form as the last term. If one substitutes x for $z/D_e^{\frac{1}{2}}$ and y^2 for $v^2/(4D_e)$ in the last term in the above equation, the equation is:

$$\mathcal{L}^{-1}\left[\frac{\exp\left(-\dfrac{z}{\sqrt{D_e}}\sqrt{s}\right)}{\left(s-\dfrac{v^2}{4D_e}\right)\left(1+\sqrt{\dfrac{4D_e}{v^2}}\sqrt{s}\right)}\right]=y\mathcal{L}^{-1}\left[\frac{\exp\left(-x\sqrt{s}\right)}{(s-y^2)(\sqrt{s}+y)}\right]$$

$$=y\left[t\mathscr{A}+\frac{\mathscr{C}}{4y}-\frac{\mathscr{D}}{4y}\left(1+2yx+4y^2t\right)\right]$$

where $\mathscr{A}$, $\mathscr{C}$, and $\mathscr{D}$ are defined in Chapter 4 just above Table 4.1. Substituting these definitions, the equation is:

$$\mathscr{L}^{-1}\left[\frac{\exp\left(-\dfrac{z}{\sqrt{D_e}}\sqrt{s}\right)}{\left(s-\dfrac{v^2}{4D_e}\right)\left(1+\sqrt{\dfrac{4D_e}{v^2}}\sqrt{s}\right)}\right]$$

$$=y\left[\begin{array}{l}\dfrac{t}{\sqrt{\pi t}}\exp\left(-\dfrac{x^2}{4t}\right)+\\[2mm]\dfrac{1}{4y}\exp\left(y^2t-yx\right)\mathrm{erfc}\left(\dfrac{x}{2\sqrt{t}}-y\sqrt{t}\right)-\\[2mm]\dfrac{1}{4y}\left(1+2yx+4y^2t\right)\exp\left(y^2t+yx\right)\mathrm{erfc}\left(\dfrac{x}{2\sqrt{t}}+y\sqrt{t}\right)\end{array}\right]$$

The overall solution then, substituting into Equation 6.107 is:

$$c(z,t)=2yc_0\exp\left(yx\right)\exp\left(-y^2t\right)\left[\begin{array}{l}\dfrac{t}{\sqrt{\pi t}}\exp\left(-\dfrac{x^2}{4t}\right)+\\[2mm]\dfrac{1}{4y}\exp\left(y^2t-yx\right)\mathrm{erfc}\left(\dfrac{x}{2\sqrt{t}}-y\sqrt{t}\right)-\\[2mm]\dfrac{1}{4y}\left(1+2yx+4y^2t\right)\exp\left(y^2t+yx\right)\\[2mm]\mathrm{erfc}\left(\dfrac{x}{2\sqrt{t}}+y\sqrt{t}\right)\end{array}\right]$$

Multiplying through by the first terms, the equation is:

$$c(z,t)=c_0\left[\begin{array}{l}\dfrac{2yt}{\sqrt{\pi t}}\exp\left(yx-y^2t-\dfrac{x^2}{4t}\right)+\dfrac{1}{2}\mathrm{erfc}\left(\dfrac{x}{2\sqrt{t}}-y\sqrt{t}\right)-\\[2mm]\dfrac{1}{2}\left(1+2yx+4y^2t\right)\exp\left(2yx\right)\mathrm{erfc}\left(\dfrac{x}{2\sqrt{t}}+y\sqrt{t}\right)\end{array}\right]$$

Substituting back into the above equation $x = z/D_e^{1/2}$ and $y = v/(2\sqrt{D_e})$, one obtains:

$$c(z,t) = c_0 \left[\begin{array}{l} \dfrac{v\sqrt{t}}{\sqrt{\pi D_e}} \exp\left(\dfrac{vz}{2D_e} - \dfrac{v^2 t}{4D_e} - \dfrac{z^2}{4D_e t} \right) + \dfrac{1}{2} \mathrm{erfc}\left(\dfrac{z}{2\sqrt{D_e t}} - \dfrac{v\sqrt{t}}{2\sqrt{D_e}} \right) - \\[3mm] \dfrac{1}{2}\left(1 + \dfrac{vz}{D_e} + \dfrac{v^2 t}{D_e} \right) \exp\left(\dfrac{vz}{D_e} \right) \mathrm{erfc}\left(\dfrac{z}{2\sqrt{D_e t}} + \dfrac{v\sqrt{t}}{2\sqrt{D_e}} \right) \end{array} \right]$$

Finding common denominators for the arguments, the equation is:

$$c(z,t) = c_0 \left[\begin{array}{l} \sqrt{\dfrac{v^2 t}{\pi D_e}} \exp\left(\dfrac{2vzt - v^2 t^2 - z^2}{4D_e t} \right) + \dfrac{1}{2}\mathrm{erfc}\left(\dfrac{z - vt}{\sqrt{4D_e t}} \right) - \\[3mm] \dfrac{1}{2}\left(1 + \dfrac{vz}{D_e} + \dfrac{v^2 t}{D_e} \right) \exp\left(\dfrac{vz}{D_e} \right) \mathrm{erfc}\left(\dfrac{z + vt}{\sqrt{4D_e t}} \right) \end{array} \right]$$

If the quadratic term in the argument of the first exponential function is expressed in terms of its roots, the final solution of Equation 6.93 is:

$$c(z,t) = c_0 \left[\sqrt{\dfrac{v^2 t}{\pi D_e}} \exp\left(-\dfrac{(z - vt)^2}{4D_e t} \right) + \dfrac{1}{2}\mathrm{erfc}\left(\dfrac{z - vt}{\sqrt{4D_e t}} \right) \right.$$
$$\left. - \dfrac{1}{2}\left(1 + \dfrac{vz}{D_e} + \dfrac{v^2 t}{D_e} \right) \exp\left(\dfrac{vz}{D_e} \right) \mathrm{erfc}\left(\dfrac{z + vt}{\sqrt{4D_e t}} \right) \right] \qquad (6.111)$$

6.8.2 Expected Value for the Lognormal Distribution

For the lognormal distribution (Equation 6.75), the expected value is

$$E(Z) = \int_{-\infty}^{+\infty} Z f(Z) dZ$$

$$= \int_{0}^{+\infty} \dfrac{Z}{\sigma Z \sqrt{2\pi}} \exp\left[-\dfrac{(\ln[Z] - \mu)^2}{2\sigma^2} \right] dZ$$

$$= \int_{0}^{+\infty} \dfrac{1}{\sigma \sqrt{2\pi}} \exp\left[-\dfrac{(\ln[Z] - \mu)^2}{2\sigma^2} \right] dZ.$$

The lower limit of the integral is zero because lognormal distributions can only be used for $Z \geq 0$. To simplify the exponential argument, make a substitution. Let:

$$y = \frac{\ln(Z) - \mu}{\sigma\sqrt{2}}.$$

Then solving for Z and finding dZ:

$$\ln(Z) = \sigma y\sqrt{2} + \mu$$

$$Z = \exp(\sigma y\sqrt{2} + \mu)$$

$$dZ = \exp(\sigma y\sqrt{2} + \mu)\sigma\sqrt{2}dy.$$

Considering the integral limits, when $Z = 0$, $y = -\infty$, so the lower limit changes (but not the upper limit). Making these substitutions in the integral above:

$$E(Z) = \int_{-\infty}^{+\infty} \frac{1}{\sigma\sqrt{2\pi}} \exp\left[-y^2\right] \exp(\sigma y\sqrt{2} + \mu)\sigma\sqrt{2}dy$$

$$= \frac{1}{\sqrt{\pi}} \exp(\mu) \int_{-\infty}^{+\infty} \exp\left[-y^2 + \sigma y\sqrt{2}\right]dy.$$

A term is sought that will complete the square of the exponential argument inside the integral:

$$E(Z) = \frac{1}{\sqrt{\pi}} \exp(\mu) \int_{-\infty}^{+\infty} \exp\left[-\left(y^2 - \sigma y\sqrt{2} + \frac{\sigma^2}{2}\right) + \frac{\sigma^2}{2}\right]dy$$

$$= \frac{1}{\sqrt{\pi}} \exp\left(\mu + \frac{\sigma^2}{2}\right) \int_{-\infty}^{+\infty} \exp\left[-\left(y - \frac{\sigma}{\sqrt{2}}\right)^2\right]dy.$$

One must show that the value of the integral is $\pi^{1/2}$ to get Equation 6.76. Make another substitution to simplify the exponential argument inside the integral. Let:

$$x = y - \frac{\sigma}{\sqrt{2}}.$$

Then $dy = dx$ and the limits of the integral stay the same. With these substitutions, the equation is

$$E(Z) = \frac{1}{\sqrt{\pi}} \exp\left(\mu + \frac{\sigma^2}{2}\right) \int_{-\infty}^{+\infty} \exp\left[-x^2\right] dx$$

$$= \exp\left(\mu + \frac{\sigma^2}{2}\right).$$

The value of the integral is $\pi^{1/2}$ (see Equation 4.76), so this is Equation 6.76.

6.8.3 NUMERICAL DISPERSION

In this section, the approach described by Moldrup et al. (1994) is followed. Starting with Equation 6.80, set the time superscript for terms on the right side to the known time level (j) and multiply through by Δt:

$$c_i^{j+1} - c_i^j = \frac{\Delta t\, D_e}{\Delta z^2}\left(c_{i+1}^j - 2c_i^j + c_{i-1}^j\right) - \frac{\Delta t\, v}{2\Delta z}\left(c_{i+1}^j - c_{i-1}^j\right). \qquad (6.112)$$

This will lead to the explicit finite difference scheme seen earlier (Equation 6.81). The error that is introduced by using a finite difference equation to express the differential terms can be seen if a Taylor expansion is used around the central node c_i^j for each term in Equation 6.112, neglecting terms with third derivatives and higher as these terms are likely to be very small:

$$c_i^{j+1} = c_i^j + \Delta t\, \frac{\partial c}{\partial t} + \frac{\Delta t^2}{2} \frac{\partial^2 c}{\partial t^2} + \cdots$$

$$c_{i+1}^j = c_i^j + \Delta z\, \frac{\partial c}{\partial z} + \frac{\Delta z^2}{2} \frac{\partial^2 c}{\partial z^2} + \cdots$$

$$c_{i-1}^j = c_i^j - \Delta z\, \frac{\partial c}{\partial z} + \frac{\Delta z^2}{2} \frac{\partial^2 c}{\partial z^2} + \cdots$$

Substituting these expansions into Equation 6.112:

$$\Delta t\, \frac{\partial c}{\partial t} + \frac{\Delta t^2}{2} \frac{\partial^2 c}{\partial t^2} = \frac{\Delta t\, D_e}{\Delta z^2}\left(\Delta z^2 \frac{\partial^2 c}{\partial z^2}\right) - \frac{\Delta t\, v}{2\Delta z}\left(2\Delta z\, \frac{\partial c}{\partial z}\right)$$

$$\frac{\partial c}{\partial t} + \frac{\Delta t}{2} \frac{\partial^2 c}{\partial t^2} = D_e\, \frac{\partial^2 c}{\partial z^2} - v\, \frac{\partial c}{\partial z}.$$

(6.113)

This is not the ADE (due to the addition of the second term on the left side), so there is clearly some error in the explicit finite difference expression for the ADE. To determine this error, another expression, $\partial^2 c/\partial t^2$, is sought. This can be found by using the ADE to substitute space derivatives of c for $\partial c/\partial t$:

$$\frac{\partial^2 c}{\partial t^2} = \frac{\partial}{\partial t}\left(\frac{\partial c}{\partial t}\right)$$

$$= \frac{\partial}{\partial t}\left(D_e \frac{\partial^2 c}{\partial z^2} - v\frac{\partial c}{\partial z}\right)$$

$$= D_e \frac{\partial}{\partial t}\left(\frac{\partial^2 c}{\partial z^2}\right) - v\frac{\partial}{\partial t}\left(\frac{\partial c}{\partial z}\right).$$

The order of differentiation can be changed:

$$\frac{\partial^2 c}{\partial t^2} = D_e \frac{\partial^2}{\partial z^2}\left(\frac{\partial c}{\partial t}\right) - v\frac{\partial}{\partial z}\left(\frac{\partial c}{\partial t}\right).$$

Then the ADE is again used to substitute the space derivatives of c for the time derivative and terms with derivatives of third and higher order are dropped:

$$\frac{\partial^2 c}{\partial t^2} = D_e \frac{\partial^2}{\partial z^2}\left(D_e \frac{\partial^2 c}{\partial z^2} - v\frac{\partial c}{\partial z}\right) - v\frac{\partial}{\partial z}\left(D_e \frac{\partial^2 c}{\partial z^2} - v\frac{\partial c}{\partial z}\right)$$

$$= D_e^2 \frac{\partial^4 c}{\partial z^4} - vD_e \frac{\partial^3 c}{\partial z^3} - vD_e \frac{\partial^3 c}{\partial z^3} + v^2 \frac{\partial^2 c}{\partial z^2}$$

$$= v^2 \frac{\partial^2 c}{\partial z^2}.$$

This is the new expression for $\partial^2 c/\partial t^2$ that was sought, so this is substituted into Equation 6.113:

$$\frac{\partial c}{\partial t} + \frac{\Delta t}{2}v^2 \frac{\partial^2 c}{\partial z^2} = D_e \frac{\partial^2 c}{\partial z^2} - v\frac{\partial c}{\partial z}$$

$$\frac{\partial c}{\partial t} = \left(D_e - \frac{\Delta t}{2}v^2\right)\frac{\partial^2 c}{\partial z^2} - v\frac{\partial c}{\partial z}.$$

This shows that the explicit finite difference scheme will be equivalent to the ADE if instead of using D_e as the dispersion coefficient, $D_e - \Delta t v^2/2$ is used. It also shows that the error is due to the inclusion of the convection term in the ADE (there would be no error if $v = 0$).

The numerical dispersion error (or correction term) varies depending on the type of numerical scheme (explicit, implicit, Crank-Nicolson, etc.) and whether sorption and transformations are included (Moldrup et al. 1994; Ataie-Ashtiani et al. 1996).

6.8   PROBLEMS

1. Using Excel, make a plot of tortuosity in the liquid phase and gas phase (two curves) as a function of θ for $\theta = 0.05$ to 0.50 ($\theta_s = 0.50$) using Equation 6.6.

2. Calculate the effective dispersion coefficient (D_e), effective liquid diffusion coefficient in soil (D_l^s), and hydrodynamic dispersion coefficient (D_{lh}) in square centimeters per hour for a soil with a steady water content of 0.38 cm^3 cm^{-3}, a steady Darcy flux of 1.4 cm h^{-1}, dispersivity (λ) of 4.2 cm, and a bulk density of 1.40 g cm^{-3}. Assume a liquid diffusion coefficient in water (D_l^w) of 10^{-5} cm^2 s^{-1}. What percentage of D_e is accounted for by diffusion vs. hydrodynamic dispersion?

3. Use Excel to do Example 6.2.

4. Show that Equation 6.106 is a solution to the ordinary differential Equation 6.99. That is, show that if you multiply Equation 6.106 by s, or if you take the second derivative of Equation 6.106 with respect to z, multiply by D_e, and subtract v times the first derivative of Equation 6.106, you get the same equation.

5. Use Excel to make a graph similar to Figure 6.18 but use a depth of 20 cm instead of 50 cm. Use different worksheets (see lower left corner of Figure 6.77) to calculate curves for different values of R. Use pore volumes for the horizontal axis. If you increase the depth, Excel precision becomes a problem when depth exceed about 28 cm (try this).

6. Use Excel to find the value of D_e and v that results in the best fit of the ADE for a step input of an adsorbed solute with $R = 5.3$, given the data in Table 6.2 from a BTC experiment on a soil column that is 30 cm long (the input concentration is 50 mg L^{-1}). What is the value of λ, assuming diffusion has a negligible effect? Do this by making a plot of the observed and predicted c/c_0 and manually adjusting D_e and v to get the best fit.

7. Use STANMOD to show the effect of v and D_e on a pulse input BTC at a depth of 50 cm with $c_0 = 100$ mg L^{-1} and $t_0 = 2$ h. To do this, make four plots on the same graph that represent combinations v and D_e. Specify a direct problem and four parameter sets. Use the deterministic equilibrium ADE. Use dimensional time and position with length in centimeters and time in hours. Choose flux-averaged concentrations. In the Transport and Reaction Parameters window, use the parameters shown in Figure 6.78. Choose a

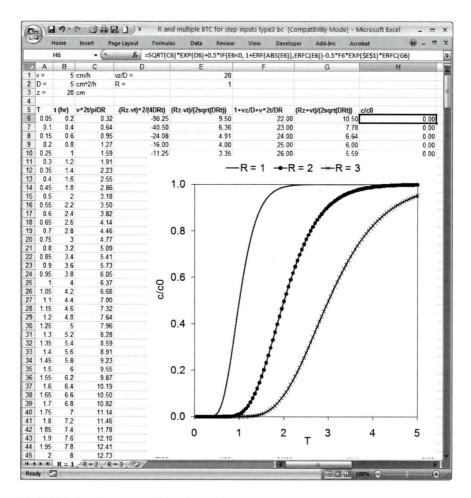

FIGURE 6.77 Excel spreadsheet for Problem 5.

pulse input with an input concentration of 100 and an application time of 2 h. The initial conditions are zero concentration and there are no transformations (zero production). For the Output Structure window, use the settings shown in Figure 6.79. Explain your results.

8. Use STANMOD to show the effect of retardation, R, and decay, μ, on a pulse input BTC at 50 cm (choose your own set of values for all parameters). Explain your results.

TABLE 6.2
Breakthrough curve data for Problem 6

t (h)	10	15	20	25	30	35	40	45	50
c (mg L^{-1})	0.2	0.5	8.1	21.0	35.7	44.7	48.6	49.0	49.5

9. Use STANMOD to show the effect of β and ω in the physical nonequilibrium ADE on a flux concentration BTC with a *step input*. Explain your results.

10. Use STANMOD to show transport (including the effect of volatilization, adsorption, and decay) of atrazine, which has transport parameters (Jury et al. 1984b):

$$K_{oc} = 160\,\text{cm}^3\,\text{g}^{-1} \qquad T_{1/2} = 71\,\text{days} \qquad K_H = 2.5\cdot 10^{-7}.$$

Use:

$$D_g^w = 0.05\,\text{cm}^2\,\text{s}^{-1} \qquad D_l^w = 1.0\cdot 10^{-7}\,\text{cm}^2\,\text{s}^{-1} \qquad J_w = 2.1\,\text{cm}\,\text{h}^{-1}.$$

$$\theta = 0.32\,\text{cm}^3\,\text{cm}^{-3} \qquad \rho_b = 1.52\,\text{g}\,\text{cm}^{-3} \qquad \lambda = 3.9\,\text{cm} \qquad f_{oc} = 0.39\,\text{g}\,\text{g}^{-1}.$$

Plot a BTC using flux concentrations at a depth of 30 cm for a pulse input of $c_0 = 100$ mg L^{-1} that lasted for 2 h. Show your calculation of v, D_E, R, and μ.

11. Show that the expected value of the normal distribution of an arbitrary variable Z (with mean μ and standard deviation σ) is μ. You will need to use the substitution $y^2 = (Z - \mu)^2/(2\sigma^2)$ and integration by parts where $u = y$ and $dv = \exp(y^2)$. Remember from Chapter 4 (based on Equation 4.76):

$$\sqrt{\pi} = \int_{-\infty}^{+\infty} \exp(-y^2)dy.$$

12. Use STANMOD to show the effect of correlation between v and K_d in the stochastic stream-tube model. In the Model Code window, select "Stochastic equilibrium CDE with $f(v, K_d)$ and $C(v, D) = 1$" (the third button). For concentration mode, use field-scale resident concentration (third-type inlet). Use three parameter sets. For all parameter sets, use:

$$v = 50\,\text{cm}\,\text{d}^{-1} \qquad D_e = 20\,\text{cm}^2\,\text{d}^{-1} \qquad K_d = 1\,\text{cm}^3\,\text{g}^{-1} \qquad \mu = 0 \qquad \sigma_v = 0.5.$$

$$\sigma_{K_d} = 0.2 \qquad \sigma_D = 0 \qquad \rho_b/\theta = 4\,\text{g}\,\text{cm}^{-3}.$$

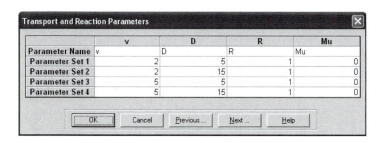

Transport and Reaction Parameters				☒
	v	D	R	Mu
Parameter Name	v	D	R	Mu
Parameter Set 1	2	5	1	0
Parameter Set 2	2	15	1	0
Parameter Set 3	5	5	1	0
Parameter Set 4	5	15	1	0

OK   Cancel   Previous ...   Next ...   Help

FIGURE 6.78 STANMOD Transport and Reaction Parameters window for Problem 7.

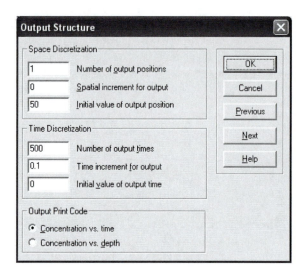

FIGURE 6.79   STANMOD Output Structure window for Problem 7.

Use three values for $\rho(v, K_d)$:
  −1 (perfect inverse correlation)
  0 (no correlation)
  1 (perfect correlation)
Use a pulse input with an application time of 2 days and dimensionless concentrations. Specify that amount of solute in each tube is the same (not proportional to velocity). Make a plot of concentration vs. depth from 0 to 200 cm. Show the results at time equal 5 days. Explain your results.

TABLE 6.3
BTC data for tritium from the experiment by van Genuchten (1974)

PV	c/c_0	PV	c/c_0	PV	c/c_0
0.512	0.001	1.558	0.915	4.125	0.353
0.599	0.016	1.646	0.923	4.255	0.236
0.686	0.082	1.754	0.947	4.386	0.166
0.730	0.138	2.016	0.967	4.516	0.118
0.817	0.296	2.604	0.981	4.777	0.066
0.904	0.465	3.125	1.000	5.037	0.038
0.992	0.593	3.342	0.986	5.385	0.018
1.079	0.685	3.516	1.015	5.818	0.008
1.166	0.764	3.712	0.971	6.251	0.004
1.253	0.806	3.842	0.838	6.791	0.002
1.340	0.850	3.951	0.638	7.331	0.006
1.428	0.901	4.038	0.480	7.439	0.000

13. Use STANMOD to fit the physical nonequilibrium ADE to the column BTC data in Table 6.3 for tritium (a nonadsorbed tracer) from the experiment by van Genuchten (1974). Use dimensionless time and position. The length of the column was 30 cm. Fit D_e, β, and ω. The steady flow rate was 17.03 cm d^{-1} and θ was 0.454 cm^3 cm^{-3}. The authors used a pulse input that lasted for 3.102 pore volumes. Run the problem at least three times with different initial estimates of D_e, β, and ω to see if you are getting a unique solution. Show the fitted parameter values and BTC for each run.

14. Use STANMOD to fit the physical nonequilibrium ADE to the BTC data for boron (Table 6.4) on the same column as in Problem 13 from the experiment by van Genuchten (1974). The steady flow rate through the column was 17.13 cm d^{-1} and the volumetric water content was 0.445 cm^3 cm^{-3}. Calculate R from their adsorption isotherm $K_d = 1.04$ cm^3 g^{-1} and column $\rho_b = 1.22$ g cm^3. The authors used a pulse input that lasted for 6.494 pore volumes. Fit D_e, β, and ω as before. Run the problem at least three times with different initial estimates of D_e, β, and ω to see if you are getting a unique solution. Then use the value for D_e that you found in Problem 1 to calculate λ and a new D_e for this problem and fit just two parameters: β and ω. Does this produce a more unique solution for β and ω? Show the fitted parameter values and BTC for each run. Assuming that θ_{mo}/θ is the same for boron and tritium, calculate f and α_s.

15. Use an explicit finite difference scheme (Equation 6.81) to solve the ADE for a step input of a nonadsorbed solute with $v = 5$ cm h^{-1} and $D_e = 10$ cm^2 h^{-1}. Use $\Delta t = 0.01$ h and $\Delta z = 1$ cm. Use dimensionless concentrations with zero initial concentrations in the profile and the type-1 boundary conditions concentration of 1 at the surface and zero at a depth of 30 cm. The depths of the nodes are shown in cells B18–B48 of Figure 6.80. The initial time and concentrations are shown in cells C16–C48. Create a macro to copy the initial time and concentrations over into the known time and concentrations column, cells E16–E48, to initialize the program. The new value for time is calculated in

---

TABLE 6.4
BTC data for boron from the experiment by van Genuchten (1974)

PV	c/c_0	PV	c/c_0	PV	c/c_0
1.80	0.015	4.60	0.777	9.30	0.433
1.95	0.075	5.30	0.819	9.80	0.357
2.10	0.170	6.00	0.852	10.50	0.269
2.25	0.265	6.70	0.880	11.50	0.186
2.40	0.340	7.30	0.882	12.70	0.133
2.60	0.430	7.75	0.852	14.00	0.090
2.85	0.535	8.00	0.776	15.50	0.054
3.15	0.620	8.25	0.699	17.00	0.040
3.50	0.687	8.55	0.621	18.50	0.029
4.00	0.738	8.90	0.527	20.00	0.025

cell F16 by using a formula to add Δt to the value in cell E16. The concentrations at the next time step are calculated in cells F19–F48 using the equation shown on the formula bar. The concentration at the boundary in the next time step is simply the boundary condition (cell F18 is equal to the value and cell E18). Use a macro to copy the calculated values for the $j + 1$ time level from cells F16–F48 back into the known time step, cells E16–E48. Do not copy the formulas from the cells, just the values. To do this, use "copy," then "paste special" and select the radio button for "values." Make a graph that uses the values of depth (cells B18–B48) and concentrations (cells F18–F48). Run the prediction out to a time of 2 h (see Figure 6.80). Try increasing Δt. At what value of Δt does the numerical solution start to show oscillations?

16. Use HYDRUS-1D to show the effect of linear vs. nonlinear adsorption on BTCs of NH_4. A typical K_d for NH_4 adsorption is about 2 cm^3 g^{-1}. Some studies have shown linear sorption for NH_4; others have shown nonlinear sorption with Freundlich β approximately 0.5. Use these values to develop BTCs for linear and nonlinear adsorption of NH_4 (you will have to make two model runs). Set up a column that is 30 cm long. Use the loam soil from the Rosetta database and set pressure heads to zero at both boundaries (gravity flow under saturated conditions). Simulate only solute transport (do not check water flow; this keeps the pressure heads constant, see the Help menu). Use a step input

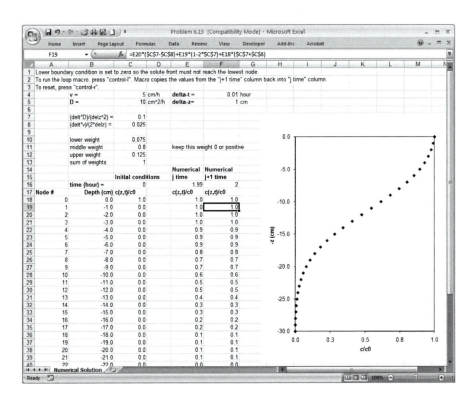

FIGURE 6.80   Excel spreadsheet for Problem 15.

concentration of 1 (relative concentrations), zero initial concentrations, and a bulk density of 1.5 g cm^3. For nonlinear adsorption, change the iteration criteria for nonlinear problems to 0.0001 for absolute and relative concentration tolerance and set the maximum number of iterations to 20. Run the problem for 20 days. Use the "Concentration Flux BC" option. Put an observation point at the bottom of the column and print out the concentration BTCs. Make sure the axes appear the same. What effect does nonlinear adsorption have?

Appendix: Unit Conversions

Concentration Conversion

From → To	mg L⁻¹	g L⁻¹	kg m⁻³	lb in⁻³	lb ft⁻³	lb gal⁻¹	grain gal⁻¹
mg L⁻¹	1	1×10^{-3}	1×10^{-3}	3.61×10^{-8}	6.24×10^{-5}	8.35×10^{-6}	5.84×10^{-2}
g L⁻¹	1×10^{3}	1	1	3.61×10^{-5}	6.24×10^{-2}	8.35×10^{-3}	58.42
kg m⁻³	1×10^{3}	1	1	3.61×10^{-5}	6.24×10^{-2}	8.35×10^{-3}	58.42
lb in⁻³	2.77×10^{7}	2.77×10^{4}	2.77×10^{4}	1	1.73×10^{3}	2.31×10^{2}	1.62×10^{6}
lb ft⁻³	1.60×10^{4}	16.02	16.02	5.79×10^{-4}	1	0.134	9.35×10^{2}
lb gal⁻¹	1.20×10^{5}	1.20×10^{2}	1.20×10^{2}	4.33×10^{-3}	7.48	1	7.0×10^{3}
grain gal⁻¹	17.12	1.71×10^{-2}	1.71×10^{-2}	6.19×10^{-7}	1.07×10^{-3}	1.43×10^{-4}	1

Length Conversion

From → To	micron	mm	cm	m	km	in	ft	mile
micron	1	1×10^{-3}	1×10^{-4}	1×10^{-6}	1×10^{-9}	3.94×10^{-5}	3.28×10^{-6}	6.21×10^{-10}
mm	1×10^{3}	1	1×10^{-1}	1×10^{-3}	1×10^{-6}	3.94×10^{-2}	3.28×10^{-3}	6.21×10^{-7}
cm	1×10^{4}	10	1	1×10^{-2}	1×10^{-5}	0.394	3.28×10^{-2}	6.21×10^{-6}
m	1×10^{6}	1×10^{3}	1×10^{2}	1	1×10^{-3}	39.37	3.28	6.21×10^{-4}
km	1×10^{9}	1×10^{6}	1×10^{5}	1×10^{3}	1	3.94×10^{4}	3.28×10^{3}	.0621
in	2.54×10^{4}	25.40	2.54	2.54×10^{-2}	2.54×10^{-5}	1	8.33×10^{-2}	1.58×10^{-5}
ft	3.05×10^{5}	3.05×10^{2}	30.48	0.305	3.05×10^{-4}	12	1	1.89×10^{-4}
mile	1.61×10^{9}	1.61×10^{6}	1.61×10^{5}	1.61×10^{3}	1.61	6.34×10^{4}	5280	1

Area Conversion

From → To	mm²	cm²	m²	km²	ha	in²	ft²	acre	mile²
mm²	1	1×10^{-2}	1×10^{-6}	1×10^{-12}	1×10^{-10}	1.55×10^{-3}	1.08×10^{-5}	2.47×10^{-10}	3.86×10^{-13}
cm²	1×10^{2}	1	1×10^{-4}	1×10^{-10}	1×10^{-8}	0.155	1.08×10^{-3}	2.47×10^{-8}	3.86×10^{-11}
m²	1×10^{6}	1×10^{4}	1	1×10^{-6}	1×10^{-4}	1.55×10^{3}	10.76	2.47×10^{-4}	3.86×10^{-7}
km²	1×10^{12}	1×10^{10}	1×10^{6}	1	1×10^{2}	1.55×10^{9}	1.08×10^{7}	2.47×10^{2}	0.386
ha	1×10^{10}	1×10^{8}	1×10^{4}	1×10^{-2}	1	1.55×10^{7}	1.08×10^{5}	2.47	3.86×10^{-3}
in²	6.45×10^{2}	6.45	6.45×10^{-4}	6.45×10^{-10}	6.45×10^{-8}	1	6.94×10^{-3}	1.59×10^{-7}	2.49×10^{-10}
ft²	9.29×10^{4}	9.29×10^{2}	9.29×10^{-2}	9.29×10^{-8}	9.29×10^{-6}	144	1	2.30×10^{-5}	3.59×10^{-8}
acre	4.05×10^{9}	4.05×10^{7}	4.05×10^{3}	4.05×10^{-3}	0.405	6.27×10^{6}	4.36×10^{4}	1	1.56×10^{-3}
mile²	2.59×10^{12}	2.59×10^{10}	2.59×10^{6}	2.59	2.59×10^{2}	4.01×10^{9}	2.79×10^{7}	640	1

Volume Conversion

From → To	cm³	L	m³	in³	ft³	pint	qt	gal	ac-ft
cm³	1	1×10^{-3}	1×10^{-6}	6.10×10^{-2}	3.53×10^{-5}	2.11×10^{-3}	1.06×10^{-3}	2.64×10^{-4}	8.11×10^{-10}
L	1×10^{3}	1	1×10^{-3}	61.02	3.53×10^{-2}	2.11	1.06	0.264	8.11×10^{-7}
m³	1×10^{6}	1×10^{3}	1	6.10×10^{4}	35.31	2.11×10^{3}	1.06×10^{3}	2.64×10^{2}	8.11×10^{-4}
in³	16.39	1.64×10^{-2}	1.64×10^{-5}	1	5.79×10^{-4}	3.46×10^{-2}	1.73×10^{-2}	4.33×10^{-3}	1.33×10^{-8}
ft³	2.83×10^{4}	28.32	2.83×10^{-2}	1.73×10^{3}	1	59.84	29.92	7.48	2.30×10^{-5}
pint	4.73×10^{2}	0.473	4.73×10^{-4}	28.87	1.67×10^{-2}	1	0.500	0.125	3.84×10^{-7}
qt	9.46×10^{2}	0.946	9.46×10^{-4}	57.74	3.34×10^{-2}	2	1	0.250	7.67×10^{-7}
gal	3.79×10^{3}	3.785	3.79×10^{-3}	2.31×10^{2}	0.134	8	4	1	3.07×10^{-6}
ac-ft	1.23×10^{6}	1.23×10^{3}	1.23×10^{3}	7.52×10^{7}	4.35×10^{4}	2.61×10^{6}	1.30×10^{6}	3.26×10^{5}	1

(Continued)

Mass Conversion

From → To	mg	g	kg	metric ton	oz	lb	short ton	grain
mg	1	1×10^{-3}	1×10^{-6}	1×10^{-9}	3.53×10^{-5}	2.21×10^{-6}	1.10×10^{-9}	1.54×10^{-2}
g	1×10^{3}	1	1×10^{-3}	1×10^{-6}	3.53×10^{-2}	2.21×10^{-3}	1.10×10^{-6}	15.43
kg	1×10^{6}	1×10^{3}	1	1×10^{-3}	35.27	2.21	1.10×10^{-3}	1.54×10^{4}
metric ton	1×10^{9}	1×10^{6}	1×10^{3}	1	3.53×10^{4}	2.21×10^{3}	1.10	1.54×10^{7}
oz	2.84×10^{4}	28.35	2.84×10^{-2}	2.84×10^{-5}	1	6.25×10^{-2}	3.13×10^{-5}	437.5
lb	4.54×10^{5}	453.6	0.454	4.54×10^{-4}	16	1	5.00×10^{-4}	7.00×10^{3}
short ton	9.07×10^{5}	9.07×10^{5}	9.07×10^{2}	0.907	3.20×10^{4}	2000	1	1.40×10^{7}
grain	64.80	6.48×10^{-2}	6.48×10^{-5}	6.48×10^{-8}	2.29×10^{-3}	1.43×10^{-4}	7.14×10^{-8}	1

Pressure Conversion

From → To	dyne cm⁻²	kg m⁻²	Pa	cm Hg	ft H₂O	bar	atm	psi	lb ft⁻²
dyne cm^{-2}	1	1.01×10^{-2}	0.100	7.50×10^{-5}	3.35×10^{-5}	1×10^{-6}	9.87×10^{-7}	1.45×10^{-5}	2.09×10^{-3}
kg m^{-2}	98.07	1	9.81	7.36×10^{-3}	3.28×10^{-3}	9.81×10^{-5}	9.68×10^{-5}	1.42×10^{-3}	0.205
Pa	10.00	0.102	1	7.50×10^{-4}	3.35×10^{-4}	1×10^{-5}	9.87×10^{-6}	1.45×10^{-5}	2.09×10^{-2}
cm Hg	1.33×10^{4}	1.36×10^{2}	1.33×10^{3}	1	0.466	1.33×10^{-2}	1.32×10^{-2}	0.193	27.85
ft H$_2$O	2.99×10^{4}	3.05×10^{2}	2.99×10^{3}	2.241	1	2.99×10^{-2}	2.95×10^{-2}	0.434	62.42
bar	1×10^{6}	1.02×10^{4}	1×10^{5}	75.01	33.46	1	0.987	14.50	2.09×10^{3}
atm	1.01×10^{6}	1.03×10^{4}	1.01×10^{5}	76.00	33.90	1.013	1	14.70	2.12×10^{3}
psi	6.90×10^{4}	7.03×10^{2}	6.90×10^{3}	5.171	2.307	6.90×10^{-2}	6.80×10^{-2}	1	144
lb ft^{-2}	4.79×10^{2}	4.882	47.88	3.59×10^{-2}	1.60×10^{-2}	4.79×10^{-4}	4.73×10^{-4}	6.94×10^{-3}	1

(Adapted from Tindall and Kunkel, 1999)

References

Abeele, W. V. 1984. Hydraulic testing of crushed Bandelier Tuff. Report No. LA-10037-MS, Los Alamos National Laboratory, Los Alamos, NM, 21.

Abit, S. M., A. Amoozegar, M. J. Vepraskas, and C. P. Niewoehner. 2008. Solute transport and the capillary fringe and shallow groundwater: field evaluation. *Vadose Zone Journal* 7:890–898.

Abramowitz, M. and I. A. Stegun. 1970. *Handbook of Mathematical Functions*. Dover, NY.

Addiscott, T. M. and R. J. Wagenet. 1985. Concepts of solute leaching in soils: a review of modeling approaches. *J. Soil Sci.* 36:411–424.

Ahuja, L. R. and C. Hebson. 1992. Root Zone Water Quality Model. GPSR Tech. Rep. No. 2, USDA, ARS, Fort Collins, CO.

Ahuja, L. R., D. L. Nofziger, D. Swartzendruber, and J. D. Ross. 1989. Relationship between Green and Ampt parameters based on scaling concepts and field-measured hydraulic data. *Water Resour. Res.* 25:1766–1770.

Andraski, B. J., and B. R. Scanlon. 2002. Thermocouple psychrometry. In *Methods of Soil Analysis: Part 4, Physical Methods*, ed. J. H. Dane and G. C. Topp. SSSA, Madison, WI. 609–642.

Arya, L. M. 2002. Wind and hot-air methods. In *Methods of Soil Analysis: Part 4, Physical Methods*, ed. J. H. Dane and G. C. Topp. SSSA, Madison, WI. 916–926.

Ashcroft, G., D. D. Marsh, D. D. Evans, and L. Boersma. 1962. Numerical methods for solving the diffusion equation: I. Horizontal flow in semi-infinite media. *Soil Sci. Soc. Proc.* 26:522–525.

Ataie-Ashtiani, B., D. A. Lockington, and R. E. Volker. 1996. Numerical correction for finite difference solution of the advection–dispersion equation with reaction. *J. Contam. Hydrol.* 23:149–156.

Baker, R. S., and D. Hillel. 1990. Laboratory tests of a theory of fingering during infiltration into layered soils. *Soil Sci. Soc. Am. J.* 54:20–30.

Baver, L. D., W. H. Gardner, and W. R. Gardner. 1972. *Soil Physics*. New York: John Wiley & Sons.

Beven, K. J. 2001. *Rainfall-Runoff Modeling: The Primer*. Chichester: John Wiley & Sons, Ltd.

Beven, K. J., and P. F. Germann. 1982. Macropores and water flow in soils. *Water Resour. Res.* 5:1311–1325.

Booltink, H. W. G., and J. Bouma. 2002a. Bypass flow. In *Methods of Soil Analysis: Part 4, Physical Methods*, ed. J. H. Dane and G. C. Topp. SSSA, Madison, WI. 930–933.

Booltink, H. W. G., and J. Bouma. 2002b. Suction crust infiltrometer. In *Methods of Soil Analysis: Part 4, Physical Methods*, ed. J. H. Dane and G. C. Topp. SSSA, Madison, WI. 926–930.

Bosch, D. D., and L. T. West. 1998. Hydraulic conductivity variability for two sandy soils. *Soil Sci. Soc. Am. J.* 62:90–98.

Bouma, J., and J. H. Denning. 1972. Field measurements of unsaturated hydraulic conductivity by infiltration through gypsum crusts. *Soil Sci. Soc. Proc.* 36:846–847.

Bouwer, H. 1969. Infiltration of water into nonuniform soil. *J. of Irrig. and Drainage Div.*, Proceedings of the ASCE IR4:451–462.

Bouwer, H. 1986. Intake rate: cylinder infiltrometer. In *Methods of Soil Analysis: Part 1, Physical and Mineralogical Methods*. Second Edition, ed. A. Klute. Madison: Am. Soc. of Agron. 825–844.

Bradford, S. A., J. Šimůnek, M. Bettahar, M. Th. van Genuchten, and S. R. Yates. 2003. Modeling colloid attachment, straining, and exclusion in saturated porous media. *Environ. Sci. Technol.* 37:2242–2250.

Brady, N. C., and R. R. Weil. 2008. *The Nature and Property of Soils.* 14th edition. Upper Saddle River, NJ: Pearson Prentice Hall.

Brantley, S. L., M. B. Goldhaber, and K. V. Ragnarsdottir. 2007. Crossing disciplines and scales to understand the critical zone. *Elements.* 3:307–314.

Bresler, E. 1973. Simultaneous transport of solutes and water under transient unsaturated flow conditions. *Water Resour. Res.* 9:975–986.

Brooks, R. H., and A. T. Corey. 1964. Hydraulic properties of porous media. *Colorado State University Hydrology Paper No. 3.* Fort Collins, CO.

Bruce, R. R., J. H. Dane, V. L. Quisenberry, N. L. Powell, and A. W. Thomas. 1983. Physical characteristics of soils in the Southern Region: Cecil. Athens, GA: Georgia Agricultural Experiment Stations.

Brunauer, S., P. H. Emmett, and E. Teller. 1938. Adsorption of gases in multimolecular layers. *J. Am. Chem. Soc.* 60:309–319.

Brunt D. 1952. *Physical and Dynamical Meteorology,* 2nd Edition. Cambridge: Fort Collins. 428.

Buckingham, E. 1907. Studies on the movement of soil moisture. U.S. Dept. of Agr. Bur. of Soils. Bull. 38.

Burdine, N. T. 1953. Relative permeability calculations from pore-size distribution data, *Petr. Trans. Am. Inst. Mining Metall. Eng.,* 198, 71–77.

Campbell, G. S. 1974. A simple method for determining unsaturated conductivity from moisture retention data. *Soil Science* 117:311–314.

Campbell, G. S. 1985. *Soil Physics with BASIC: Transport models for soil-plant systems.* Elsevier. Amsterdam.

Carsel, R. F., and R. S. Parrish. 1988. Developing joint probability distributions of soil water retention characteristics. *Water Resour. Res.* 24:755–769.

Celia, M. A., E. T. Bouloutas, and R. L. Zarba. 1990. A general mass-conservative numerical colution for the unsaturated flow equation. *Water Resour. Res.* 26:1483–1496.

Chang, K. 2008. Test of Mars soil sample confirms presence of ice. New York, NY: New York Times.

Chen, C., and R. J. Wagenet. 1992. Simulation of water and chemicals in macropore soils. Part 1. Representation of the equivalent macropore influence and its effect on soil water flow. *J. Hydrol.* 130:105–126.

Chiang, S. C., D. E. Radcliffe, W. P. Miller, and K. D. Newman. 1987. Hydraulic conductivities of three southeastern soils as affected by sodium, electrolyte concentration, and pH. *Soil Sci. Soc. Am. J.* 51:1293–1299.

Chung S. -O., and R. Horton, 1987. Soil heat and water flow with a partial surface mulch. *Water Resour. Res.,* 23(12): 2175–2186.

Clothier, B., and D. Scotter. 2002. Unsaturated water transmission parameters obtained through infiltration. In *Methods of Soil Analysis: Part 4, Physical Methods.* Ed. J. H. Dane and G. C. Topp. SSSA, Madison, WI. 879–898.

Dane, J. H., and J. W. Hopmans. 2002. Water retention and storage: Laboratory. In *Methods of Soil Analysis: Part 4, Physical Methods,* ed. J. H. Dane and G. C. Topp. SSSA, Madison, WI. 675–720.

Dane, J. H., and G. C. Topp. 2002. In *Methods of Soil Analysis: Part 4, Physical Methods,* Third Edition. SSSA, Madison, WI.

Daniels, R. B., and E. E. Gamble. 1967. The edge effect in some Ultisols in the North Carolina Coastal Plain. *Geoderma* 1:117–124.

Darcy, H. 1856. Les fontaines publiques de la ville de Dijon. Dalmont, Paris.

Dasberg, S., and J. W. Hopmans.1992. Time domain reflectometry calibration for uniformly wetted sandy and clayey loam soils. *Soil Sci. Soc. Am. J.* 56:1341–1345.

de Marsily, G. 1986. *Quantitative Hydrogeology.* London: Academic Press.

de Vries, D. A., 1963. The thermal properties of soils. In *Physics of Plant Environment,* ed. R. W. van Wijk. North Holland, Amsterdam. 210–235.

Donatelli, M., J. H. M. Wösten, and G. Belocchi. 2004. Methods to evaluate pedotransfer functions. In *Development of Pedotransfer Functions in Soil Hydrology*, ed. Y. Pachepsky and W. J. Rawls. Amsterdam: Elsevier. 357–411.

Dupuit, J. 1857. Mouvement de l'eau a travers le terrains permeables, C. R. Hebd. *Seances Acad. Sci.*, 45:92–96.

Dupuit, J. A. 1863. Etudes theoriques et pratiques sur le movement des eaux dans les canaux decouverts e a travers les terrains permeables, 2nd Edition. Dunod. Paris (Cited by Kirkham and Powers, 1972).

Durner, W. 1994. Hydraulic conductivity estimation for soils with heterogeneous por structure. *Water Resour. Res.* 32:211–223.

Elrick, D. E., and W. D. Reynolds. 1992. Infiltration from constant-head well permeameters and infiltrometers. In *Advances in Measurement of Soil Physical Properties: Bringing Theory into Practice*, ed. G. C. Topp, W. D. Reynolds and R. E. Green. SSSA, Madison, WI. 1–24.

Ernst, L. F. 1962. Groundwaterstromingen in de verzadigde zone en hun berekening bij aanwezigheid van horizontale evenwijdige open leidengen (Groundwater flow in the saturated zone and its calculation when horizontal parallel open conduits are present), Versl. Landbouwk. Onderz. 67.1, Pudoc, Wageningen, the Netherlands (in Dutch). 189.

Evett, S. R. 2002. Water and energy balances at soil-plant-atmosphere interfaces. In *Soil Physics Companion*, ed. A. W. Warrick. Boca Raton, FL: CRC Press. 127–188.

Fayer, M. J. 2000. UNSAT-H Version 3.0: Unsaturated Soil Water and Heat Flow Model. Theory, User Manual, and Examples. Pacific Northwest National Laboratory 13249.

Feddes, R. A., E. Bresler, and S. P. Neuman. 1974. Field test of a modified numerical model for water uptake by root systems. *Water Resour. Res.* 10:1199-1206.

Feddes, R. A., P. J. Kowalik, and H. Zaradny. 1978. *Simulation of Field Water Use and Crop Yield*. New York, NY: John Wiley & Sons.

Ferré, P. A., and G. C. Topp. 2002. Time domain reflectometry. In *Methods of Soil Analysis: Part 4, Physical Methods*, ed. J. H. Dane and G. C. Topp. SSSA, Madison, WI. 434–446.

Fetter, C. W. 1988. *Applied Hydrogeology*. 2nd Edition. Columbus, OH: Merrill Publishing Co.

Fiori, A., and D. Russo. 2007. Numerical analysis of subsurface flow and a steep hill slope under rainfall: The role of the spatial heterogeneity of the formation and hydraulic properties. *Water Resour. Res.* 43:W07445.

Flerchinger, G. N., C. L. Hanson and J. R. Wight. 1996. Modeling evapotranspiration and surface energy budgets across a watershed. *Water Resour. Res.* 32:2539–2548.

Flint, A. L., and L. E. Flint. 2002. Particle density. In *Methods of Soil Analysis: Part 4, Physical Methods*, ed. J. H. Dane and G. C. Topp. SSSA, Madison, WI. 229–240.

Flury, M. 1996. Experimental evidence of transport of pesticides through field soils - A review. *J. Environ. Qual.* 25:25–45.

Flury, M., H. Flühler, W. Jury, and J. Leuenberger. 1994. Susceptibility of soils to preferential flow of water: a field study. *Water Resour. Res.* 30:1945–1954.

Food and Agriculture Organization of the United Nations. 1990. Expert consultation on revision of FAO methodologies for crop water requirements, *ANNEX V*, FAO Penman-Monteith Formula, Rome, Italy.

Forchheimer, P. 1930. Hydraulik. B. B. Teubner. Leipzig. (Cited by Kirkham and Powers, 1972).

Freundlich, H. 1909. *Kapillarchemie; eine Darstellung der Chemie der Kolloide und verwandter Gebiete*, Akademische Verlagsgellschaft, Leipzig, Germany.

Gardner, W. D. 1997. The flux of particles to the deep sea: Methods, measurements, and mechanisms. *Oceanography*. 10:116–121.

Gardner, W. R. 1958. Some steady state solutions of the unsaturated moisture flow equation with application to evaporation from a water table. *Soil Science* 85:228–232.

Gardner, W. R., and M. Fireman. 1958. Laboratory studies of evaporation from soil columns in the presence of a water table. *Soil Science* 85:244–249.

Gee, G. W., and D. Or. 2002. Particle-size analysis. In *Methods of Soil Analysis: Part 4, Physical Methods*, ed. J. H. Dane and G. C. Topp. SSSA, Madison, WI. 255–293.

Gerke, H. H., and M. Th.van Genuchten. 1993. A dual-porosity model for simulating the preferential movement of water and solutes in structured porous media. *Water Resour. Res.* 29:305–319.

Glotfelty, D. E., and C. J. Schomburg. 1989. Volatilization of pesticides from soil. In *Reactions and Movement of Organic Chemicals in Soils*, ed. B. L. Sawhney and K. Brown, Spec. Publ. Number 22. SSSA, Madison, WI. 181–208.

Germann, P. F. 1985. Kinematic wave approach to infiltration and drainage into and from soil macropores. *Trans. ASAE* 28:745–749.

Germann, P. F. 1988. Approaches to rapid and far-reaching hydrologic processes in the vadose zone. *J. Contam. Hydrol.* 3:115–127.

Gotfried, B. S. 2007. *Spreadsheet Tools for Engineers Using Excel*. 3rd Edition. Boston, MA: McGraw Hill.

Green, W. H., and G. A. Ampt. 1911. Studies in soil physics: I. The flow of air and water through soils. *J. Agr. Sci.* 4:1–24.

Grossman, R. B., and T. G. Reinsch. 2002. Bulk density and linear extensibility. In *Methods of Soil Analysis: Part 4, Physical Methods*, ed. J. H. Dane and G. C. Topp. SSSA, Madison, WI. 201–228.

Grossman, R. B., D. S. Harms, C. A. Seybold, and J. E. Herrick. 2001. Use-invariant data for soil quality evaluation in the United States. *Journal of Soil and Water Conservation* 56:63–68.

Gupta, H. V., S. Sorooshian, T. S. Hogue, and D. P. Boyle. 2003. Advances in automatic calibration of watershed models. In *Calibration of Watershed Models*, ed. Q. Duan. American Geophysical Union. Washington, DC.

Guy, H. P. 1969. Laboratory Theory and Methods for Sediment Analysis: *U.S. Geological Survey Techniques of Water-Resources Investigations*, Book 5, Chapter C1, 58.

Hanks, R. J., and S. A. Bowers. 1962. Numerical solution of the moisture flow equation for infiltration into layered soils. *Soil Sci. Soc. Proc.* 26:530–534.

Hansen S., H. E. Jensen, N. E. Nielsen and H. Svendsen. 1990. *DAISY: Soil Plant Atmosphere System Model*, NPO Report No. A 10, The National Agency for Environmental Protection, Copenhagen.

Haverkamp, R., M. Kutilek, J. -Y. Parlange, L. Rendon, and M. Krejca. 1985. Infiltration under ponded conditions: 2. Infiltration equations tested for parameter time-dependence and predictive use. *Soil Science* 145:305–311.

Haverkamp, R., M. Vauclin, J. Tovina, P. J. Wierenga, and G. Vachaud. 1977. A comparison of numerical simulation models for one-dimensional infiltration. *Soil Sci. Soc. Am. Proc.* 41:285–294.

Healy, R. W. 1990. Simulation of Solute Transport in Variably Saturated Porous Media with supplemental Information on Modifications to the U.S. Geological Survey's Computer Program VS2DI, Water-Resources Investigation Report 90–4025, U.S. Geological Survey, 125.

Heiberger, T. S. 1996. Simulating the effects of a capillary barrier using the two-dimensional variably saturated flow model SWMS-2D/HYDRUS-2D, PhD Thesis. Oregon State University. Corvallis, OR.

Hendrickx, J. M. H., L. W. Dekker, and O. H. Boersma. 1993. Unstable wetting fronts in water repellent field soils. *J. Environ. Qual.* 22:109–118.

Hendrickx, J. M. H., and M. Flury. 2001. Uniform and preferential flow, Mechanisms in the vadose zone. In Conceptual Models of Flow and Transport in the Fractured Vadose zone. National Research Council. Washington: National Academy Press. 149–187.

Hewlett, J. D., and C. A. Troendle. 1975. Non-point and diffused water sources: A variable source area problem. Utah State University Symposium. American Society of Civil Engineers, New York, NY.

Hignett, C., and S. R. Evett. 2002. Neutron thermalization. In *Methods of Soil Analysis: Part 4, Physical Methods*, ed. J. H. Dane and G. C. Topp. SSSA, Madison, WI. 501–521.

Hillel, D. 1980 *Fundamentals of Soil Physics*. New York, NY: Academic Press.

Hillel, D. 2004. *Introduction to Environmental Soil Physics*. Elsevier Academic Press. Amsterdam, The Netherlands.

Hooghoudt, S. B. 1940. Bijdrage tot de kennis van enige natuutkundige grootheden van de grond (Contribution to the knowledge of several physical soil parameters), Versl. Landbouwk. Onderz. 46 (14) B, 515–707, Wageningen, the Netherlands (in Dutch).

Hopmans, J. W., J. Šimůnek, N. Romano, and W. Durner. 2002. Inverse Modeling of Transient Water Flow. In *Methods of Soil Analysis: Part 4, Physical Methods*, ed. J. H. Dane and G. C. Topp. SSSA, Madison, WI. 936–1008.

Horton, R. E., 1933. The role of infiltration in the hydrologic cycle. *Trans. Am. Geophysical Union* 14:446–460.

Horton, R. E. 1940. An approach towards a physical interpretation of infiltration capacity. *Soil Sci. Soc. Am. Proc.* 5:399–417.

Hubbert, M. K. 1956. Darcy's law and the field equations of the flow of underground fluids. *Am. Inst. Min. Met. Petl. Eng. Trans.* 207:222–239.

Huyakorn, P. S., and G. F. Pinder. 1983. *Computational Methods in Subsurface Flow* London: Academic Press.

Jacques, D., and J. Šimůnek. 2005. User Manual of the Multicomponent Variably-Saturated Flow and Transport Model HP1, Description, Verification and Examples, Version 1.0, SCK•CEN-BLG-998, Waste and Disposal, SCK•CEN, Mol, Belgium, 79.

Jacques, D., J. Šimůnek, D. Mallants, and M. Th. van Genuchten. 2008. Modeling coupled hydrological and chemical processes: Long-term uranium transport following mineral phosphorus fertilization, *Vadose Zone Journal*, doi:10.2136/VZJ2007.0084, Special Issue "Vadose Zone Modeling", 7:698–711.

Jansson P. -E., and L. Karlberg. 2001. Coupled Heat and Mass Transfer Model for Soil-Plant-Atmosphere Systems, Royal Institute of Technology, Department of Civil and Environmental Engineering: Stockholm.

Jarvis, N. J. 1989. A simple empirical model of root water uptake. *J. Hydrol.* 107:57–72.

Jarvis, N.J., 1994. The MACRO model (Version 3.1), Technical description and sample simulations. Reports and Dissertations 19. Dept. Soil Sci., Swedish Univ. Agric. Sci., Uppsala, Sweden, 51 pp.

Jarvis, N. J. 2007. A review of non-equilibrium water flow and solute transport in soil macropores: principles, controlling factors and consequences for water quality. *European Journal of Soil Science*. 58:523–546.

Jaynes, D. B., and E. J. Tyler. 1984. Using soil physical properties to estimate hydraulic conductivity. *Soil Science* 138:298–305.

Jury, W. A., and H. Flühler. 1992. Transport of chemicals through soil: mechanisms, models, and field applications. *Adv. Agron.* 47:141–201.

Jury, W. A., and R. Horton. 2004. *Soil Physics*. Hoboken, NJ: John Wiley & Sons, Inc.

Jury, W. A., and K. Roth. 1990. *Transfer Functions and Solute Movement through Soil*. Birkhäuser Verlag, Basel.

Jury, W. A., W. F. Spencer, and W. J. Farmer, 1983. Behavior assessment model for trace organics in soil: I. Description of model, *J. Environ. Qual.*, 12:558–564.

Jury, W. A., W. F. Spencer, and W. J. Farmer. 1984a. Behavior assessment model for trace organics in soil: III. Chemical classification and parameter sensitivity. *J. Environ. Qual.* 13:567–572.

Jury, W. A., W. F. Spencer, and W. J. Farmer. 1984b. Behavior assessment model for trace organics in soil: IV. Application of screening model. *J. Environ. Qual.* 13:573–579.

Kay, B. D., and D. A. Angers. 2001. In Soil Structure. *Handbook of Soil Science*, ed. M. E. Summner. Boca Raton, FL: CRC Press. A229-A276.

Kirkham, D., and W. L. Powers. 1972. *Advanced Soil Physics*. New York: Wiley.

Knighton, D. 1984. *Fluvial Forms and Processes*. London, England: Edward Arnold. 218.

Kool, J. B., and J. C. Parker. 1987. Development and evaluation of closed-form expressions for hysteretic soil hydraulic properties. *Water Resour. Res.* 23:105–114.

Kostiakov, A. N. 1932. On the dynamics of the coefficient of water percolation in soils and on the necessity of studying it from a dynamic point of view for purposes of amelioration. *Trans. Com. Int. Soc. Soil Sci.* 6th Moscow A:17–21.

Kosugi, K., J. W. Hopmans, and J. H. Dane. 2002. Parametric models. In *Methods of Soil Analysis: Part 4, Physical Methods*, ed. J. H. Dane and G. C. Topp. SSSA, Madison, WI. 739–757.

Kung, K. -J. S. 1990a. Preferential flow in a sandy vadose zone: 2. Mechanism and implications. *Geoderma* 46:59–71.

Kung, K. -J. S. 1990b. Preferential flow in a sandy vadose zone: 1. Field observation. *Geoderma* 46:51–58.

Langmuir, I. 1918. The adsorption of gases on plane surfaces of glass, mica, and platinum. *J. Am. Chem. Soc.* 40:1361–1403.

Leij, F. J., and S. A. Bradford. 1994. 3DADE: A computer program for evaluating three-dimensional equilibrium solute transport in porous media. *Research Report No. 134*, U.S. Salinity Laboratory, USDA, ARS, Riverside, CA.

Leij, F. J., W. B. Russell, and S. M. Lesch. 1997. Closed-form expressions for water retention and conductivity data. *Groundwater*. 35:848–858.

Leij, F. J., M. G. Schaap, and L. M. Arya. 2002. Indirect methods. In *Methods of Soil Analysis: Part 4, Physical Methods*, Chapter 3.6.3, ed. J. H. Dane and G. C. Topp, Third Edition. SSSA, Madison, WI. 1009–1045.

Leij, F. J., and N. Toride. 1997. N3DADE: A computer program for evaluating nonequilibrium three-dimensional equilibrium solute transport in porous media. *Research Report No. 143*, U.S. Salinity Laboratory, USDA, ARS, Riverside, CA.

Leij, F. J., and M. Th. van Genuchten. 2002. Solute transport. In *Soil Physics Companion*, ed. A. W. Warrick. Boca Raton, FL: CRC Press. 189–248.

Lenhard, R. J., J. C. Parker, and J. J. Kaluarachchi. 1991. Comparing simulated and experimental hysteretic two-phase transient fluid flow phenomena. *Water Resour. Res.* 27:2113–2124.

Lenhard, R. J., and J. C. Parker. 1992. Modeling multiphase fluid hysteresis and comparing results to laboratory investigations. In Proc. International Workshop on Indirect Methods for Estimating Hydraulic Properties of Unsaturated Soils, ed. M. Th. van Genuchten, F. J. Leij, and L. J. Lund. University of California, Riverside, CA. 233–248.

Lide, D. R. (ed.) 2002. *Handbook of Chemistry and Physics*. 83rd ed. CRC Press. Boca Raton, FL. 2664 pp.

Lilly, A., A. Nemes, W. J. Rawls, and Ya. A. Pachepsky. 2008. Probabilistic approach to the identification of input variable to estimate hydraulic conductivity. *Soil Sci. Soc. Am. J.* 72:16–24.

Liu, H. H., and J. H. Dane. 1996. Two approaches to modeling unstable flow and mixing of variable density fluids in porous media. Transport in Porous Media 23:219–236.

Luxmoore, R. J. 1981. Micro-, meso-, and macroporosity of soil. *Soil Sci. Soc. Am. J.* 45:671.

Mallants, D., G. Volckaert, and J. Marivoet. 1999. Sensitivity of protective barrier performance to changes in rainfall rate. *Waste Management* 19:467–475.

Marquardt, D. W. 1963. An algorithm for least-squares estimation of nonlinear parameters. *SIAM J. Appl. Math.* 11:431–441.

McCarthy, E. L. 1934. Mariotte's bottle. *Science* 80:100.

McCord, J. T. 1991. Application of second-type boundaries in unsaturated flow modeling. *Water Resour. Res.* 27:3257–3260.

McCord, J. T., and M. T. Goodrich. 1994. Benchmark testing and independent verification of the VS2DT computer code. Sandia National Laboratories. Albuquerque, NM.

McCuen, R. H. 1982. *A Guide to Hydrologic Analysis Using SCS Methods*. Prentice-Hall, Inc., Englewood Cliffs, NJ.

McInnes, K. J. 2002. Soil heat. In *Methods of Soil Analysis: Part 4, Physical Methods*, ed. J. H. Dane and G. C. Topp. SSSA, Madison, WI. 1183–1199.

Miller, E. E., and R. D. Miller. 1956. Physical theory for capillary flow phenomena. *J. Appl. Phys.* 27:324–332.

Miller, W. P., and D. E. Radcliffe. 1992. Soil Crusting in the Southeastern U.S. In *Soil Crusting: Chemical and Physical Processes*, ed. M. E. Sumner and B. A. Stewart. Boca Raton, FL: S. Lewis Publ. 233–266.

Millington, R. J., and J. P. Quirk. 1961. Permeability of porus media. *Trans. Faraday Soc.* 57:1200–1207.

Moldrup, P., T. Yamaguchi, D. E. Rolston, K. Vestergaard, and J. Hansen. 1994. Removing numerically induced dispersion from finite difference models for solute and water transport in unsaturated soils. *Soil Science* 157, no. 2:153–161.

Molz, F. J. 1981. Models of water transport in the soil-plant system: A review. *Water Resour. Res.* 17:1245–1260.

Monteith, J. L. 1981. Evaporation and surface temperature, *Quarterly J. Royal Meteor. Soc.*, 107, 1–27.

Monteith, J. L., and M. H. Unsworth. 1990. *Principles of Environmental Physics*. London: Edward Arnold.

Mualem, Y. 1976. A new model for predicting the hydraulic conductivity of unsaturated porous media. *Water Resour. Res.* 12:593–622.

Murray F. W. 1967. On the computation of saturation vapor pressure, *J. Appl. Meteor.*, 6, 203–204.

Nemes, A., M. G. Schaap, F. J. Leij, and J. H. M. Wösten. 2001. Description of the unsaturated soil hydraulic database UNSODA version 2.0. *J. Hydrol.* 251:151–162.

Neuman, S. P., R. A. Feddes, and E. Bresler. 1974. Finite element simulation of flow in saturated-unsaturated soils considering water uptake by plants, Third Annual Report, Project No. A10-SWC-77, Hydraulic Engineering Lab., Technion, Haifa, Israel.

Nielsen, D. R., J. W. Biggar, and K. T. Erh. 1973. Spatial variability of field measured soil water properties. *Hilgardia* 42:215–259.

Nkedi-Kizza, P., J. W. Biggar, H. W. Selim, M. Th. van Genuchten, P. J. Wierenga, J. M. Davidson, and D. R. Nielsen. 1984. On the equivalence of two conceptual models for describing ion exchange during transport through an aggregated Oxisol. *Water Resour. Res.* 20:1123–1130.

National Research Council (NRC). 2001. Basic research opportunities in earth science. Washington, DC: National Academies Press.

National Resource Conservation Service (NRCS). 2004. *National Engineering Handbook Part 630*. Natural Resource Conservation Service. Available online at http://policy.nrcs. usda.gov/viewerFS.aspx?hid=21422 (verified 1/2/2009).

National Resource Conservation Service NRCS. 2009. National Soil Information Service. Natural Resource Conservation Service. Available online at http://soils.usda.gov/technical/nasis/documents/metadata/nmdoview.html (verified 1/2/2009).

Pachepsky, Y. A., W. J. Rawls, and H. S. Lin. 2006. Hydropedology and pedotransfer functions. *Geoderma* 131:308–316.

Paniconi, C., A. A. Aldama, and E. F. Wood. 1991. Numerical evaluation of iterative and noniterative methods for the solution of the nonlinear Richards equation. *Water Resour. Res.* 27:1147–1163.

Papiernik, S. K., S. R. Yates, and J. Gan. 2002. Processes governing transport or organic solutes. In *Methods of Soil Analysis: Part 4, Physical Methods*, ed. J. H. Dane and G. C. Topp. SSSA, Madison, WI. 1451–1479.

Parker, J. C., and M. Th. van Genuchten. 1984. Flux-averaged and volume-averaged concentrations in continuum approaches to solute transport. *Water Resour. Res.* 20:866–872.

Parkhurst, D. L., and C. A. J. Appelo. 1999. User's guide to PHREEQC (Version 2) – A computer program for speciation, batch-reaction, one-dimensional transport and inverse geochemical calculations. Water-Resources Investigations, Report 99–4259, Denver, CO, USA, 312.

Parlange, J. -Y., R. Haverkamp, and J. Touma. 1985. Infiltration under ponded conditions: 1. Optimal analytical solution and comparison with experimental observations. *Soil Science* 139:305–311.

Parsons, L. R., and W. M. Bandaranayake. 2009. Performance of a new capacitance soil moisture probe in a sandy soil. *Soil Sci. Soc. Am. J.* 73:1378–1385.

Pennell, K. D. 2002. Specific surface area. In *Methods of Soil Analysis: Part 4, Physical Methods*, ed. J. H. Dane and G. C. Topp. SSSA, Madison, WI. 295–315.

Perrochet, P., and D. Berod. 1993. Stability of the standard Crank-Nicolson-Galerkin scheme applied to the diffusion-convection equation: some new insights. *Water Resour. Res.* 29(9), 3291–3297.

Perroux, K. M., and I. White. 1988. Designs for disc permeameters. *Soil Sci. Soc. Am. J.* 52:1205–1215.

Perianez, R. 2005. Modeling the dispersion of radionuclides by a river plume: Application to the Rhone River. *Continental Shelf Research* 25:1583–1603.

Philip, J. R. 1957. The theory of infiltration. 1. The infiltration equation and its solution. *Soil Science* 83:345–357.

Philip, J. R. 1966. Plant water relations: Some physical aspects. *Ann. Rev. Plant Physiol.* 17:245–268.

Pruess K. 1991. Tough2 – A General-Purpose Numerical Simulator for Multiphase Fluid and Heat Flow, Report LBL-29400, Lawrence Berkeley Laboratory, Berkeley, CA.

Pruess, K., and J. S. Y. Wang. 1987. Numerical modeling of isothermal and non-isothermal flow in unsaturated fractured rock - a review. In Flow and Transport Through Unsaturated Fractured Rock. Geophysics Monograph 42, ed. D. D. Evans, and T. J. Nicholson. Washington: American Geophysical Union. 11–22.

Quisenberry, V. L., and R. E. Phillips. 1976. Percolation of surface-applied water in the field. *Soil Sci. Soc. Am. J.* 40:484–489.

Rawls, W. J., and D. L. Brakensiek. 1985. Prediction of soil water properties for hydrologic modeling. In Watershed management in the eighties. Proc. Irrig. Drain. Div., Denver, CO 30 Apr.-1 May, 1985, ed. E. B. Jones and T. J. Ward. Reston: ASCE. 293–299.

Rawls, W. J., D. L. Brakensiek, and K. E. Saxton. 1982. Estimation of soil water properties. *Trans. ASAE.* 1316–1320, 1328.

Rawls, W. J., T. J. Gish, and D. L. Brakensiek. 1991. Estimating soil water retention from soil physical properties and characteristics. *Adv. Soil Sci.*, 9, 213–234.

Reynolds, W. D., and D. E. Elrick. 2002a. Constant head soil core (tank) method. In *Methods of Soil Analysis: Part 4, Physical Methods*, ed. J. H. Dane and G. C. Topp. SSSA, Madison, WI. 804–808.

Reynolds, W. D., and D. E. Elrick. 2002b. Falling head soil core (tank) method. In *Methods of Soil Analysis:. Part 4, Physical Methods*, ed. J. H. Dane and G. C. Topp. SSSA, Madison, WI. 809–812.

Reynolds, W. D., D. E. Elrick, and E. G. Youngs. 2002c. Ring or cylinder infiltrometers (vadose zone). In *Methods of Soil Analysis: Part 4, Physical Methods*, ed. J. H. Dane and G. C. Topp. SSSA, Madison, WI. 818–843.

Reynolds, W. D., and D. E. Elrick. 2002d. Constant head well permeameter (vadose zone). In *Methods of Soil Analysis: Part 4, Physical Methods*, ed. J. H. Dane and G. C. Topp. SSSA, Madison, WI. 844–858.

Reynolds, W. D., S. R. Vieira, and G. C. Topp. 1992. An assessment of the single-head analysis for the constant head well permeameter. *Can J. Soil Sci.* 72:489–501.

Richards, L. A. 1931. Capillary conduction of fluid through porous mediums. *Physics* 1:318–333.

Ritzi, R. W., and P. Bobeck. 2008. Comprehensive principles of quantitative hydrogeology established by Darcy (1856) and Dupuit (1857). *Water Resour. Res.* 44:W10402.

Robinson, D. A., S. B. Jones, J. M. Wraith, D. Or, and S. P. Friedman. 2003. A review of advances in dielectric and electrical conductivity measurement in soils using time domain reflectometry. *Vadose Zone Journal* 2:444–475.

Romano, N., and A. Santini. 2002. Water retention and storage: Field. In *Methods of Soil Analysis: Part 4, Physical Methods*, ed. J. H. Dane and G. C. Topp. SSSA, Madison, WI. 721–738.

Russo, D., and E. Bresler. 1980. Field determinations of soil hydraulic properties for statistical analyses. *Soil Sci. Soc. Am. J.* 44:697–702.

Russo, D., and M. Bouton. 1992. Statistical analysis of spatial variability in unsaturated flow parameters. *Water Resour. Res.* 28:1911–1925.

Scanlon, B. R., B. J. Andraski, and J. Bilskie. 2002. Miscellaneous methods for measuring matric or water potential. In *Methods of Soil Analysis: Part 4, Physical Methods*, ed. J. H. Dane and G. C. Topp. SSSA, Madison, WI. 643–670.

Schaap, M. G., F. J. Leij, and M. Th. van Genuchten. 1998. Neural network analysis for hierarchical prediction of soil water retention and saturated hydraulic conductivity. *Soil Sci. Soc. Am. J.* 62:847–855.

Schaap, M. G., and F. J. Leij. 1998. Database-related accuracy and uncertainty of pedotransfer functions, *Soil Science* 163:765–779.

Schaap, M. G., and F. J. Leij. 2000. Improved prediction of unsaturated hydraulic conductivity with the Mualem-van Genuchten model. *Soil Sci. Soc. Am. J.* 64:843–851.

Schaap, M. G., F. J. Leij, and M. Th. van Genuchten. 2001. Rosetta: a computer program for estimating soil hydraulic parameters with hierarchical pedotransfer functions. *Journal of Hydrology* 251:163–176.

Schijven, J., and J. Šimůnek. 2002. Kinetic modeling of virus transport at field scale, *J. of Contam. Hydrology*, 55(1-2), 113-135.

Schoeneberger, P., and A. Amoozegar. 1990. Directional saturated hydraulic conductivity and macropore morphology of a soil-saprolite sequence. *Geoderma* 46:31–49.

Scott, P. S., G. J. Farquhar, and N. Kouwen. 1983. Hysteresis effects on net infiltration, Advances in Infiltration. Publ. 11–83. Am. Soc. Agri. Eng., St. Joseph, MI. 163–170.

Selim, H. M., R. Schulin, and H. Flühler. 1987. Transport and ion exchange of calcium and magnesium in an aggregated soil. *Soil Sci. Soc. Am. J.* 51(4), 876–884.

Simmons, C. S., D. R. Nielsen, and J. W. Biggar. 1980. Scaling of field-measured soil water properties. *Hilgardia* 47:101–122.

Šimůnek, J., and J. W. Hopmans. 2002. Parameter Optimization and Nonlinear Fitting. In *Methods of Soil Analysis: Part 4, Physical Methods*, Chapter 1.7, ed. J. H. Dane and G. C. Topp, Third edition. SSSA, Madison, WI. 139–157.

Šimůnek, J., N. J. Jarvis, M. Th. van Genuchten, and A. Gärdenäs. 2003. Nonequilibrium and preferential flow and transport in the vadose zone: review and case study. *Journal of Hydrology* 272:14–35.

Šimůnek, J., M. Šejna, H. Saito, M. Sakai, and M. Th. van Genuchten. 2008. The HYDRUS-1D Software Package for Simulating the Movement of Water, Heat, and Multiple Solutes in Variably Saturated Media, Version 4.0, HYDRUS Software Series 3, Department of Environmental Sciences, University of California Riverside, Riverside, CA, USA.

Šimůnek, J., M. Šejna, and M. Th. van Genuchten. 1999a. The HYDRUS-2D software package for simulating two-dimensional movement of water, heat, and multiple solutes in variably saturated media. Version 2.0. IGWMC - TPS-53. International Ground Water Modeling Center, Colorado School of Mines, Golden, CO.

Šimůnek, J., M. Šejna, and M. Th. van Genuchten. 1998. The HYDRUS-1D software package for simulating the one-dimensional movement of water, heat, and multiple solutes in variably-saturated media. Version 2.0, *IGWMC - TPS - 70,* International Ground Water Modeling Center, Colorado School of Mines, Golden, CO. 202.

Šimůnek, J., and D. L. Suarez. 1994. Major ion chemistry model for variably saturated porous media, *Water Resour. Res.* 30(4), 1115–1133.

Šimůnek, J., D. L. Suarez, and M. Šejna, 1996. The UNSATCHEM software package for simulating one-dimensional variably saturated water flow, heat transport, carbon dioxide production and transport, and multicomponent solute transport with major ion equilibrium and kinetic chemistry, Version 2.0, *Research Report No. 141*, U.S. Salinity Laboratory, USDA, ARS, Riverside, CA, 186.

Šimůnek, J., and M. Th. van Genuchten. 2006. Contaminant Transport in the Unsaturated Zone: Theory and Modeling, Chapter 22 . In *The Handbook of Groundwater Engineering*, Ed. Jacques Delleur, Second Edition, CRC Press. 22.1–22.46.

Šimůnek, J., M. Th. van Genuchten, M. Šejna, N. Toride, and F. J. Leij. 1999. The STANMOD computer software for evaluating solute transport in porous media using analytical solutions of convection-dispersion equation. Versions 1.0 and 2.0, *IGWMC - TPS - 71*, International Ground Water Modeling Center, Colorado School of Mines, Golden, Colorado, 32 pp.

Šimůnek, J., M. Th. van Genuchten, and M. Šejna. 2006. The HYDRUS Software Package for Simulating Two- and Three-Dimensional Movement of Water, Heat, and Multiple Solutes in Variably-Saturated Media, Technical Manual, Version 1.0, PC Progress, Prague, Czech Republic.

Šimůnek, J., M. Th van Genuchten, M. Šejna, N. Toride, and F. J. Leij. 1999b. The STANMOD computer software for evaluating solute transport in porous media using analytical solutions of convection-dispersion equation. Versions 1.0 and 2.0, *IGWMC-TPS-71,* International Ground Water Modeling Center, Colorado School of Mines, Golden, CO, 32.

Sisson, J. B. 1987. Drainage from layered field soils: Fixed gradient models, *Water Resour. Res.* 23:2071–2075.

Skaggs, T. H., and F. J. Leij. 2002. Solute transport: Theoretical background. In *Methods of Soil Analysis: Part 4, Physical Methods*. Chapter 6.3, ed. J. H. Dane and G. C. Topp, Third edition. SSSA, Madison, WI. 1353–1380.

Skaggs, T. H., T. J. Trout, and J. Šimůnek. 2004. Comparison of HYDRUS-2D simulations of drip irrigation with experimental observations. *J. Irrig. Drain. Eng.* 130:304–310.

Sklash, M. G., M. K. Stewart, and A. J. Pearce. 1986. Storm runoff generation in humid headwater catchments2. A case study of hillslope and low-order stream response. *Water Resour. Res.* 22:1273–1282.

Skopp, J. M. 2000. Physical properties of primary particles. In *Handbook of Soil Science*, ed. M. E. Sumner. Boca Raton, FL: CRC Press. A3-A18

Smith, G. D. 1985. *Numerical solution of partial differential equations: Finite difference methods.* Oxford: Clarendon Press.

Soil Survey Division Staff. 1993. *Soil Survey Manual.* Soil Conservation Service. U.S. Department of Agriculture Handbook 18.

Starr, J. L., and I. C. Paltineanu. 2002. Capacitance devices. In *Methods of Soil Analysis: Part 4, Physical Methods*, ed. J. H. Dane and G. C. Topp. SSSA, Madison, WI. 463–474.

Steenhuis, T. S., M. Winchell, J. Rossing, J. A. Zollweg, and M. F. Walter. 1995. SCS runoff equation revisited for variable-source runoff areas. *J. Irrig. and Drainage Eng.* 234–238.

Tetens, O. 1930. Uber einige meteorologische Begriffe, *Z. Geophys.*, 6, 297–309.

Tiedje, J. M. 1988. Ecology of denitrification and dissimilatory nitrate reduction to ammonium. In *Biology of Anaerobic Microorganisms*, ed. A. J. B. Zehnder. New York, NY: John Wiley & Sons 174–244.

Tietje, O., and V. Hennings. 1996. Accuracy of the saturated hydraulic conductivity prediction by pedo-transfer functions compared to the variability with FAO textural classes. *Geoderma* 69:71–84.

Tindall, J. A., and J. R. Kunkel. 1999. *Unsaturated Zone Hydrology*. Upper Saddle River, NJ: Prentice Hall.

Topp, G. C., J. L. Davis, and A. P. Annan. 1980. Electromagnetic determination of soil water content: Measurements in coaxial transmission lines. *Water Resour. Res.* 16:574–582.

Topp, G. C., and P. A. Ferré. 2002a. Thermogravimetric using convective oven-drying. In *Methods of Soil Analysis: Part 4, Physical Methods*, ed. J. H. Dane and G. C. Topp. SSSA, Madison, WI. 422–424.

Topp, G. C., and P. A. Ferré. 2002b. Gravimetric use in microwave oven-drying. In *Methods of Soil Analysis: Part 4, Physical Methods*, ed. J. H. Dane and G. C. Topp. SSSA, Madison, WI. 425–428.

Topp, G. C., and P. A. Ferré. 2002c. Water content. In *Methods of Soil Analysis: Part 4, Physical Methods*, ed J. H. Dane and G. C. Topp. SSSA, Madison, WI. 417–422.

Topp, G. C., and W. D. Reynolds. 1998. Time domain reflectometry: A seminal technique for measuring mass and energy in soil. *Soil Tillage Res.* 47:125–132.

Toride, N., and F. J. Leij. 1996. Convective-dispersive stream tube model for field-scale solute transport: II. Examples and calibration. *SSSJA* 60:352–361.

Toride, N., F. J. Leij, and M. Th. van Genuchten. 1999. The CXTFIT code for estimating transport parameters from laboratory or field tracer experiments. Version 2.0, *Research Report No. 137*, U.S. Salinity Laboratory, USDA, ARS, Riverside, CA.

Tugel, A. J., J. E. Herrick, J. R. Brown, M. J. Mausbach, W. Puckett, and K. Hipple. 2005. Soil change, soil survey, and natural resources decision making: A blueprint for action. *Soil Sci. Soc. Am. J.* 69:738–747.

US Environmental Protection Agency (USEPA). 2009. Ozone layer depletion - Regulatory programs. http://www.epa.gov/Ozone/mbr/ (verified 7/19/2009).

van Dam, J. C., J. Huygen, J. G. Wesseling, R. A. Feddes, P. Kabat, P. E. V. van Valsum, P. Groenendijk, and C. A. van Diepen. 1997. Theory of SWAP, version 2.0. Simulation of water flow, solute transport and plant growth in the Soil- Water- Atmosphere- Plant environment. Dept. Water Resources, WAU, *Report 71*, DLO Winand Staring Centre, Wageningen, Technical Document 45.

van Genuchten, M. Th. 1974. Mass transfer studies in sorbing porous media. Ph.D. thesis. New Mexico State University, Las Cruces, NM.

van Genuchten, M. Th. 1980a. A closed form equation for predicting the hydraulic conductivity of unsaturated soils. *Soil Sci. Soc. Am. J.* 44:892–898.

van Genuchten, M. Th. 1980b. Determining transport parameters from solute displacement experiments. *Research Report No. 118*, U.S. Salinity Laboratory, USDA, ARS, Riverside, CA.

van Genuchten, M. Th. 1985. Convective-dispersive transport of solutes involved in sequential first-order decay reactions. *Computers and Geosciences.* 11:129–147.

van Genuchten, M. Th. 1987. A numerical model for water and solute movement in and below the rootzone. *Research Report No. 121*, U.S. Salinity Laboratory, USDA, ARS, Riverside, CA.

van Genuchten, M. Th., and W. A. Jury. 1987. Progress in unsaturated flow and transport modeling. *Reviews of GeoPhysics* 25:135–140.

van Genuchten, M. Th., F. J. Leij, and S. R. Yates. 1991. The RETC code for quantifying the hydraulic functions of unsaturated soils. EPA/600/2–91/065. Available online at www.ars.usda.gov/Main/docs.htm?docid=15992 (verified 21 December, 2007).

van Genuchten, M. Th., and P. J. Wierenga. 1976. Mass transfer studies in sorbing porous media: I. Analytical solutions. *Soil Sci. Soc. Am. J.* 40:473–479.

van Hoorn, J. W. 1997. Drainage for salinity control, Dept. of Water Resour., Wageningen Agricultural University, Wageningen, the Netherlands.

Vepraskas, M. J., and J. P. Williams. 1995. Hydraulic conductivity of saprolite as a function of sample dimensions and measurement technique. *Soil Sci. Soc. Am. J.* 59:975–981.

Verburg K., P. J. Ross, and K. L. Bristow. 1996. SWIM v2.1 User Manual, Divisional Report 130, CSIRO.

Vereecken, H., J. Maes, J. Feyen, and P. Darius. 1989. Estimating the soil moisture retention characteristic from texture, bulk density, and carbon content. *Soil Science* 148:389–403.

Vereecken, H., J. Maes, and J. Feyen. 1990. Estimating unsaturated hydraulic conductivity from easily measured soil properties. *Soil Science* 149:1–12.

Vervoort, R. W., D. E. Radcliffe, and L. T. West. 1999. Soil structure and preferential solute flow. *Water Resour. Res.* 35:913–928.

Vogel, T., M. Císlerová, and J. W. Hopmans. 1991. Porous media with linearly variable hydraulic properties. *Water Resour. Res.* 27:2735–2741.

Vrugt, J. A., M. T. van Wijk, J. W. Hopmans, and J. Šimůnek. 2001. One-, two-, and three-dimensional root water uptake functions for transient modeling. *Water Resour. Res.* 37:2457–2470.

Walter, M. T., J. -S. Kim, T. S. Steenhuis, J. -Y. Parlange, A. Heilig, R. D. Braddock, J. S. Selker, and J. Boll. 2000. Funneled flow mechanisms in a sloping layered soil: Laboratory investigations. *Water Resour. Res.* 36:841–849.

Warrick, A. W. 1992. Models for disc infiltrometers. *Water Resour. Res.* 28:1319–1327.

Warrick, A. W., and P. Broadbridge. 1992. Sorptivity and macroscopic capillary length relationships. *Water Resour. Res.* 28:427–431.

Warrick, A. B. 2003. *Soil Water Dynamics*. Oxford University Press.

White, I., and M. J. Sully. 1987. Macroscopic and microscopic capillary length and time scales from field infiltration. *Water Resour. Res.* 23:1514–1522.

White, R. E. 1985. The influence of macropores on the transport of dissolved and suspended matter through soil. In *Advances in Soil Science*. New York, NY: Springer-Verlag.

Wooding, R. A. 1968. Steady infiltration from a shallow circular pond. *Water Resour. Res.* 4:1259–1273.

Wraith, J. M., and D. Or. 1998. Nonlinear parameter estimation using spreadsheet software. *J. Nat. Resour. Life Sci. Educ.* 27:13–19.

Yeh, G. T., and V. S. Tripathi. 1990. HYDROGEOCHEM: A coupled model of hydrological and geochemical equilibrium of multicomponent systems. Rep. ORNL-6371, Oak Ridge Natl. Lab., Oak Ridge, TN.

Young, M. H. 2002. Piezometry. In Methods of Soil Analysis: Part 4, Physical Methods, ed. J. H. Dane and G. C. Topp. SSSA, Madison, WI. 547–573.

Young, M. H., and J. B. Sisson. 2002. Tensiometry. In Methods of Soil Analysis: Part 4, Physical Methods, ed. J. H. Dane and G. C. Topp. SSSA, Madison, WI. 575–608.

Zangar, C. N. 1953. Flow from a test hole located above groundwater level. In Theory and problems of water percolation. Engineering Monograph No. 8. Bureau of Reclamation. U.S. Dept. Of Interior. 69–71.

Index